Automatic Speech Analysis and Recognition

NATO ADVANCED STUDY INSTITUTES SERIES

Proceedings of the Advanced Study Institute Programme, which aims at the dissemination of advanced knowledge and the formation of contacts among scientists from different countries

The series is published by an international board of publishers in conjunction with NATO Scientific Affairs Division

A Life Sciences B Physics	Plenum Publishing Corporation London and New York
C Mathematical and Physical Sciences	D. Reidel Publishing Company Dordrecht, Boston and London
D Behavioural and Social Sciences E Engineering and Materials Sciences	Martinus Nijhoff Publishers The Hague, London and Boston
F Computer and Systems Sciences G Ecological Sciences	Springer Verlag Heidelberg

Series C – Mathematical and Physical Sciences

Volume 88 – Automatic Speech Analysis and Recognition

Automatic Speech Analysis and Recognition

Proceedings of the NATO Advanced Study Institute held at Bonas, France, June 29 - July 10, 1981

edited by

JEAN-PAUL HATON
CRIN, Université de Nancy I, France

D. Reidel Publishing Company

Dordrecht : Holland / Boston : U.S.A. / London : England

Published in cooperation with NATO Scientific Affairs Division

Library of Congress Cataloging in Publication Data

CIP

NATO Advanced Study Institute (1981 : Bonas, France)
Automatic speech analysis and recognition.

(NATO advanced study institutes series. Series C, Mathematical and physical sciences ; v. 88)
"Published in cooperation with NATO Scientific Affairs Division."
Includes bibliographical references and index.
1. Automatic speech recognition–Congresses.
I. Haton, Jean-Paul, 1943- II. North Atlantic Treaty Organization. Division of Scientific Affairs. III. Title. IV. Series.
TK7895.S65N37 1981 001.53'4 82-9135
ISBN-13: 978-94-009-7881-2 e-ISBN-13: 978-94-009-7879-9
DOI: 10.1007/978-94-009-7879-9

Published by D. Reidel Publishing Company
P.O. Box 17, 3300 AA Dordrecht, Holland

Sold and distributed in the U.S.A. and Canada
by Kluwer Boston Inc.,
190 Old Derby Street, Hingham, MA 02043, U.S.A.

In all other countries, sold and distributed
by Kluwer Academic Publishers Group,
P.O. Box 322, 3300 AH Dordrecht, Holland

D. Reidel Publishing Company is a member of the Kluwer Group

TABLE OF CONTENTS

PREFACE vii

ACKNOWLEDGEMENTS viii

LIST OF PARTICIPANTS ix

§ 1. SPEECH ANALYSIS AND TRANSMISSION

*H. WAKITA / Linear Prediction of Speech and its Application to Speech Processing 1

*V.R. VISWANATHAN, J. MAKHOUL and R. SCHWARTZ / Medium and Low Bit Rate Speech Transmission 21

*W.J. HESS / Algorithms and Devices for Pitch Determination of Speech Signals 49

§ 2. ACOUSTICS AND PHONETICS

*G. MERCIER / Acoustic-phonetic Decoding and Adaptation in Continuous Speech Recognition 69

*V.W. ZUE / Acoustic-phonetic Knowledge Representation : Implications from Spectrogram Reading Experiments 101

L. MICLET, Y. GRENIER and J. LE ROUX / Phonetic Recognition by Linear Prediction : Experiences at ENST 121

B. LOCHSCHMIDT / Acoustic-phonetic Analysis Based on an Articulatory Model 139

G. RUSKE and T. SCHOTOLA / The Efficiency of Demisyllable Segmentation in the Recognition of Spoken Words 153

P. DEMICHELIS, R. DE MORI and P. LAFACE / Interaction between Auditory, Syllabic and Lexical Knowledge in a Speech Understanding System 165

§ 3. APPLICATIONS

*S.E. LEVINSON / Speaker Independent Connected Word Recognition 179

J.S. BRIDLE, M.D. BROWN and R.M. CHAMBERLAIN / An Algorithm for Connected Word Recognition 191

C. CAGNOULET, M. COUVRAT and D. JOUVET / SERAPHINE : A connected Word Recognition System 205

*J. ALLEN / Implementation of Models for Speech Recognition with very Large Scale Integrated Circuit Technology 217

§ 4. SYSTEMS

*F. JELINEK / Self-organized continuous Speech Recognition 231

H.W. HEIN / The Erlangen Speech Understanding Project 239

J.P. HATON, J.M. PIERREL and S. SABBAGH / Semantic and Pragmatic Processing in Continuous Speech Understanding 253

G. PERENNOU / The ARIAL II Speech Recognition System 269

§ 5. MISCELLANEOUS

*P. CORSI / Speaker Recognition : A Survey 277

*J.C. SIMON / Complexity of Algorithms and Pattern Recognition 309

*W. LEA / Selecting, Designing, and Using Practical Speech Recognizers 331

INDEX 369

---

* Are Review Papers

PREFACE

This book is the result of the second NATO Advanced Study Institute on speech processing held at the Château de Bonas, France, from June 29th to July 10th, 1981.

This Institute provided a high-level coverage of the fields of speech transmission, recognition and understanding, which constitute important areas where research activity has recently been associated with actual industrial developments. This book will therefore include both fundamental and applied topics.

Ten survey papers by some of the best specialists in the field are included. They give an up-to-date presentation of several important problems in automatic speech processing. As a consequence the book can be considered as a reference manual on some important areas of automatic speech processing. The surveys are indicated by a * in the table of contents.

This book also contains research papers corresponding to original works, which were presented during the panel sessions of the Institute.

For the sake of clarity the book has been divided into five sections :

1. Speech Analysis and Transmission :

An emphasis has been laid on the techniques of linear prediction (LPC), and the problems involved in the transmission of speech at various bit rates are addressed in details.

2. Acoustics and Phonetics :

One of the major bottleneck in the development of speech recognition systems remains the transcription of the continuous speech wave into some discrete strings or lattices of phonetic symbols. Two survey papers discuss this problem from different points of view and several practical systems are also described.

3. Applications :

This section deals with application systems, especially in the framework of connected word recognition. Practical problems rela-

J.-P. Haton (ed.), Automatic Speech Analysis and Recognition, vii–viii.

ted to the use, the evaluation and also to the design of such systems using VLSI custom circuits are also presented.

4. Continuous Speech Recognition Systems :

Most of the research effort is presently devoted to continuous speech recognition and understanding. In this section several systems are described ; they correspond to various approaches of the problem.

5. Miscellaneous :

This section presents a survey of three important, unrelated problems :
- speaker recognition and verification, which already corresponds to actual applications,
- complexity of algorithms in the framework of pattern recognition, and especially speech recognition.
- the practical problems associated with the use of speech recognition systems.

ACKNOWLEDGEMENTS

I would like to thank the editing committee for helping with the publication of this book.

The 1981 Bonas meeting has been made possible through the financial support and the framework provided by the NATO Scientific Affairs Division.

The additional support of the US Army European Office and of the CRIN (University of Nancy I) are also gratefully acknowledged.

As the Director of the NATO Institute, I would like to thank all the participants for their valuable participation in the discussions and panel presentations which finally made a great success of this Institute, and also J.C. SIMON, animateur of the Centre Culturel de Bonas, for the excellent conditions we found in Bonas.

Director of the NATO ASI : Jean-Paul HATON, Professor, C.R.I.N. University of Nancy I

Editing Committee : J. ALLEN (MIT, USA), J.P. HATON (University of Nancy I, France), W. HESS (T.U. Munich, West-Germany), S. LEVINSON (Bell Laboratories, USA), V. VISVANATHAN (BBN, USA).

LIST OF PARTICIPANTS

ALLEN, J.	MIT, CAMBRIDGE, MASS. 02139, U.S.A.
ASTIER, A.	R.N.U.R., B.P. 103, 92109 Boulogne Billancourt, France.
BAUDRY, M.	Institut de Programmation, Univers. Paris VI, 4, place Jussieu, 75005 Paris, France.
BELLISSANT, C.	Laboratoire IMAG, Univers. Grenoble, B.P. 53 X, Grenoble 38047, France.
BENNETT, D.C.	University of London, Malet Street, LONDON WC/E 7HP, U.K.
BRASSARD, J.P.	Université de Montréal, CP 6128, Succursale "A", Montreal P.Q. H3C-3J7, Canada.
CALIKOGLU, D.	Middle East Technical University (O.D.T.U.) Electrical Engineering Department, Ankara, Turkey.
CAPPELLI, A.	Instituto di Linguistica Computazionale CNR, via della Faggiola, 32 56100 Pisa, Italy.
CARBONELL, N.	CRIN, Université de Nancy I, B.P. 239, 54506 Vandoeuvre Cédex, France.
CARRION PEREZ M.C.	Universidad de Granada, Granada, Spain.
CHAINTREUIL, S.	Univers. Paris Sud, laboratoire Cristallographie et Physique des Matériaux, Bât. 490, 91405 Orsay, France.
CHAMBERLAIN, R.M.	Joint speech Research Unit, Princess Elizabeth Way, Cheltenham, Gloucester,England GL52 5AJ, U.K.
CHEVAL, E.	MATRA, D.T.T.V., Laboratoire traitement du signal, B.P. n°1, 78146 Velizy-Villacoublay, France.
CHOLLET, G.	Inst. de Phonétique, Univers. de Provence, 29 Av. Robert Schumann, 13621 Aix en Provence, France.
CLERMONT, F.	Speech Technology Laboratory, 3888 State Street, Santa Barbara, California 93105,USA.
CORSI, P.	IBM Research Center, 5600 Cottle road, San José, California 95193, USA.
DE MORI, R.	Université de Turin, C.M. d'Azeglio 42, 10125 Torino, Italy.
DI BENEDETTO, M.D.	IBM France, Centre scientifique, 36 Av. Raymond Poincaré, 75116 Paris, France.
DUPEYRAT, B.	CEA, DEIN/SIR CEN Saclay, 91191 Gif sur Yvette France.
ESKENAZI, M.S.	LIMSI, Bâtiment 508, B.P. 30, 91406 Orsay, France.
FIRE, P.	US Navy Office of Naval Research Edison House 223/231 Old Marylebone Road, London NWI 5th, U.K.

HAMMOND, J.	School of Oriental and African Studies, Malet street, London WCIS 7HP, U.K.
HATON, J.P.	CRIN, Université de NANCY I, B.P. 239, 54506 Vandoeuvre Cédex, France.
HEIN, H.W.	Universität Erlangen-Nuernberg Informatik 5 (Mustererkennung), Martenstrasse 3, D-8520 Erlangen,R.F.A.
HESS, W.	Technische Universität Munchen, Postfach 202420, 8000 München 2, W. Germany.
D'HUMIERES, D.	Groupe de Physique des solides de l'Ecole Normale Supérieure, 24 rue Lhomond, 75231 Paris Cédex 05, France.
JELINEK, F.	IBM Watson Research center, Yorktown Heights, N.Y. 10598, USA.
JUNQUA, J.C.	CRIN, Université de NANCY I, B.P. 239, 54506 Vandoeuvre Cédex, France.
KIELCZEWSKI, GM.	University of Warsaw,00-901 Warsaw,PKIN pok 850-00-901, Warsaw, Poland.
KNAPPE, U.	Oelschlägern 21 D-3300 Braunschweig, W. Germany.
KOSMIDOU, O.	University of Thrace, Xanthi, Greece.
LEA, W.	Speech Communication Research Laboratory, 806 W Adams Boulevard, Los Angeles, California 90007, USA.
LE GUENNEC, L.	C.N.E.T. T.S.S/RCP, 22300 Lannion, France.
LEVINSON, S.	Bell Laboratories, 600 Mountain Avenue, Murray Hill, NJ 07974, USA.
LINGGARD, R.	Department of Electrical and Electronic Engineering, The Queen's University of Belfast, Stranmillis road, Belfast BT9 5AH, N. Ireland
LITTEL, B.	Siemens AG ZT ZFE FL KMS 13, Otto-Hahn-Ring 6, 8000 München 83, W. Germany.
LOCHSCHMIDT, B.F.	Forschunginstitut des DBP beim Fernmeldetechnischen Zentralamt, am Kavalleriesand 3 D-6100, Darmstadt, W. Germany.
MARCHAL, A.	Université de Montreal CP 6128 Succursale "A", Québec H3C 3J7, Canada.
MARI, J.F.	CRIN, Université de Nancy I, B.P. 239, 54506 Vandoeuvre Cédex, France.
MARQUET P.	CIT-ALCATEL, 10 rue Louis Lormand, 78320 Le Mesnil-St-Denis, France.
MARSLEN-WILSON W.	Max-Planck Institut für Psycholinguitik, 6522 BC Nijmegen, Pays-Bas.
MATHESON, J.M.R.	Cambridge University Engineering Department, Trumpington Street, Cambridge CB2 IPZ, U.K.
MERCIER, G.	CNET A, 22301 Lannion, France.
MERTZIOS, B.G.	Democritus University of Thrace, Polytechnic School, Greece.
MICLET, L.	ENST, 46 rue Barrault, 75634 Paris cédex 13, France.

MOURADI, A.	Faculté des sciences de Rabat, B.P. 1014, Rabat, Marocco.
MUELLER, J.V.	Speech communication Research Laboratories 2513 Anabas ave, San Pedro, California, USA.
NEEL, F.D.	LIMSI CNRS, B.P. 30 bâtiment 508, 91406 Orsay, France.
NOETH, E.V.	Friedrich-Alexander-Universität Informatik 5 (Mustererkennung), Martenstrasse 3 D-8520, Erlangen, W. Germany.
PERENNOU, G.	CERFIA, Univers. Paul Sabatier, 118 route de Narbonne, 31062 Toulouse Cédex, France.
PIERACCINI,R.	CSELT, via G.R. Romali 274, 10100 Torino, Italy.
PIERREL, J.M.	CRIN, Université de Nancy I, B.P. 239, 54506 Vandoeuvre Cédex, France.
PISTER, C.	CRIN, Université de Nancy I, B.P. 239, 54506 Vandoeuvre Cédex, France.
REGRAGUI, F.	Faculté des sciences de Rabat, 214 ave John Kennedy, Rabat, Maraocco.
RODET, X.J.	IRCAM, 31 rue Saint Merri, 75004 Paris, France.
RONNEBRINCK, W.	Institut für Nachtrichtentechnik, TU Braunschweig Postfach 3329, 3300 Braunschweig, RFA.
RUBIO, A.	Universidad de Granada, GRANADA, Spain.
RUSSELL, M.J.	Royal Signals and Radar Establishement, B 109 R.S.R.E. North. Leigh Sinton road, Malvern, Worcs, U.K.
SABBAGH, S.	132 boulevard du 11 janvier, Casablanca, Marocco
SAUTER, L.	C.G.E., 91 Marcoussis, France.
SCHULZE, E.	Heinrich-Hertz-Institut für Nachrichtentechnik, Einstenufer 37 D-1000, Berlin 10, R.F.A.
SENEFF, S.	MIT. Cambridge 02239, USA.
SIMON, J.C.	Univers. Paris VI, Institut de Programmation, 4 place Jussieu, 75230 Paris Cédex 05, France.
SIROUX, J.	CNET, 22301 Lannion, France.
SOKOL, G.M.	European Research Office, 223 Old Marylebone road, London NWI 5th, U.K.
TRAN CAO, T.	Université de Liège, Institut Montefire, B-28 Sart Tilman B-4000 Liège, Belgium.
TUFFELLI, D.P.	E.N.S.E.R., 23 rue des Martyrs, 38031 Grenoble, France.
TYLER, P.	Max-Planck Institut für Psycholinguistik, 6522 BC Nijmegen, Pays-Bas.
VALLE ARROYO, F.	University of Oviedo, Depto de Psicologia, Edifico Valdès Salas, Oviedo, Spain.
VAROUFAKIS, S.	Nuclear Research Center Democritus, Aghia Paraskevi-Attikis, Greece.
VICENZI, C.	Elettronica san Giorgo Elsag S.P.A., via Hermada, 6 16151 Genova sestri, Italy.

VIDAL RUIZ, E.	Centre de Informatica Universidad de Valencia, Ada docter Moliner S/N, Burjasot Valencia, Spain
VISVANATHAN, V.	BBN Moulton street, Cambridge, Mass. 02238, USA.
WAKITA, H.	Speech technology Lab. , 3888 state street, Santa Barbara, California 93105, USA.
WEDGWOOD, R.E.	Cambridge University Engineering Dept., Trumpington street, Cambridge CB2 IPZ, U.K.
WITTENBURG, P.M.	Max-Planck-Institut für Psycholinguistics Bergen Dalse weg 79 NL 6522 BC, Nijmegen, Pays-Bas.
WU, F.	Univers. Paris XI, laboratoire de cristallographie et physique des solides, bât. 490, 91405 Orsay Cédex, France.
ZUE, V.	MIT, Cambridge 02139, USA.

LINEAR PREDICTION OF SPEECH AND ITS APPLICATION TO SPEECH PROCESSING

Hisashi Wakita

Speech Technology Laboratory

Abstract. This paper gives a tutorial review of the linear prediction method, especially focusing on its application to acoustical analysis of sampled speech. The paper discusses the estimation of the formant frequencies and the fundamental frequencies from sampled speech waves by the use of linear prediction. Advantages and disadvantages of the method are discussed together with the problems in estimating the above parameters. Application examples to acoustic analysis, segmentation, and vowel normalization are also given.

1. INTRODUCTION

The recent development in digital techniques has brought substantial innovations to methods and techniques for acoustical analysis of speech sounds. Particularly, the linear prediction method has been widely accepted as an efficient and effective technique for speech analysis and synthesis. Since the concept of linear prediction was first applied to speech analysis (1),(5), it has been fairly thoroughly studied theoretically and experimentally (8),(9),(15). This paper will present a brief review of the linear prediction method, an evaluation of the method in estimating the formant frequencies and the fundamental frequencies from sampled speech waves, and then present several applications to actual speech processing. This paper is intended for the readers with various backgrounds, thus those who are interested in more detailed aspects of the linear prediction method are encouraged to read some of the references listed at the end of the paper.

J.-P. Haton (ed.), Automatic Speech Analysis and Recognition, 1–19.

2. LINEAR PREDICTION MODEL AND METHOD

In the linear prediction of speech, a very simplistic model of speech production is assumed as given in Fig. 1. The transfer

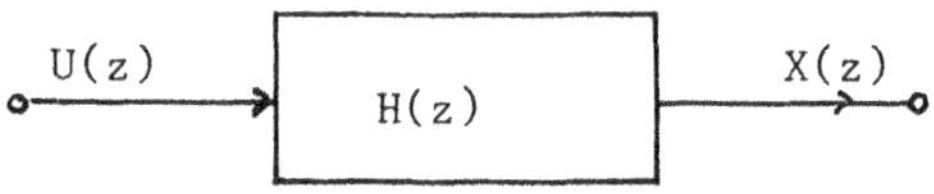

Fig. 1. Speech production model for linear prediction.

function H(z) is given by

$$H(z) = \frac{G}{1 + \sum_{i=1}^{M} a_i z^{-i}} \qquad (a_0=1) \qquad (1)$$

where a_i's are called the filter coefficients and G is a gain constant. Since the transfer function has only poles, the system is an all-pole model. Thus, the signal x_n is given as a linear combination of past values and the excitation u_n,

$$x_n = -\sum_{i=1}^{M} a_i x_{n-i} + Gu_n . \qquad (2)$$

Letting

$$F(z) = \sum_{i=1}^{M} a_i z^{-i}, \qquad (3)$$

then,

$$X(z) = \frac{GU(z)}{1 - F(z)} . \qquad (4)$$

The diagram in Fig. 1 can be redrawn as in Fig. 2 to represent a speech production model.

In the linear prediction analysis, the inverse process for the part indicated by the dotted line in Fig. 2 is optimally estimated on a certain criterion. Thus an analysis model is represented as in Fig. 3. Since the excitation source in Fig. 2 is

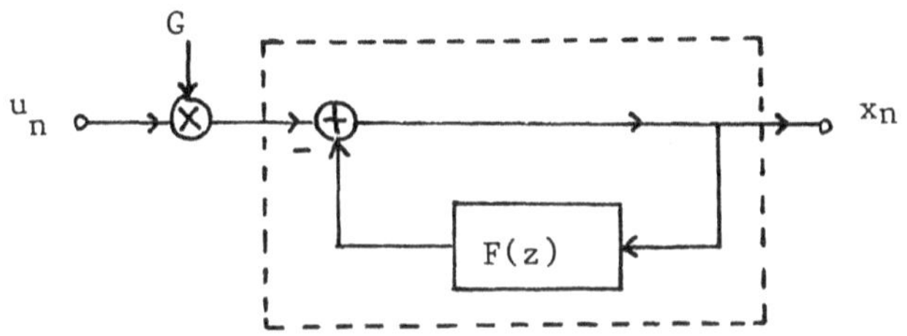

Fig. 2. A time domain representation of the speech production model.

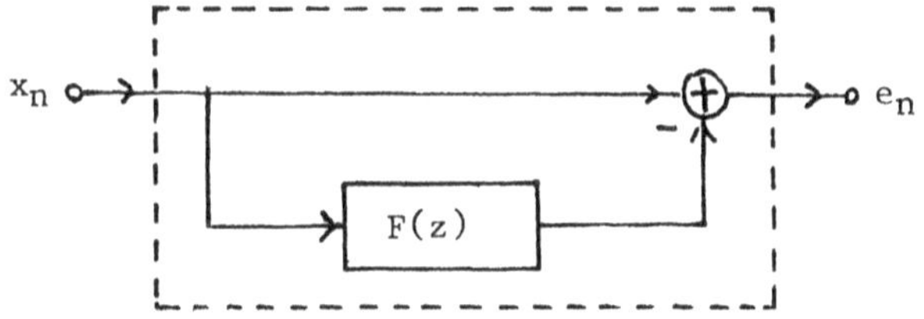

Fig. 3. A linear prediction analysis model.

normally unknown for the speech signals, the signal x_n is predicted from a linearly weighted summation of the past samples. Let the estimated sample of x_n be

$$\hat{x}_n = -\sum_{i=1}^{M} a_i x_{n-i} , \qquad (5)$$

then the error between the actual sample x_n and the predicted one $\hat{x}_n$ is given by

$$e_n = x_n - \hat{x}_n = \sum_{i=0}^{M} a_i x_{n-i} . \; (a_0=1) \qquad (6)$$

The parameters a_i are determined in such a way so that the total squared error summed over a certain interval is minimized with respect to each of the parameters. If x_n is assumed to be deterministic, the total squared error E is given by

$$E = \sum_{n=n_0}^{n_1} e_n^2 = \sum_{n=n_0}^{n_1} \left[\sum_{i=0}^{M} a_i x_{n-i} \right]^2 \qquad (7)$$

$$= \sum_{n=n_0}^{n_1} \sum_{i=0}^{M} \sum_{j=0}^{M} a_i x_{n-i} x_{n-j} a_j \qquad (8)$$

where n_0 and n_1 are the index limits to evaluate the error minimization. Letting

$$c_{ij} = \sum_{n=n_0}^{n_1} x_{n-i} x_{n-j} , \qquad (9)$$

then

$$E = \sum_{i=0}^{M} \sum_{j=0}^{M} a_i c_{ij} a_j . \qquad (10)$$

Now E is minimized by setting the partial derivative of E with respect to a_j, j=1,2,...,M, to zero, that is,

$$\frac{\delta E}{\delta a_j} = 0 = 2 \sum_{i=0}^{M} a_i c_{ij} . \qquad (11)$$

Since $a_0=1$, equation (11) is given by

$$\sum_{i=1}^{M} a_i c_{ij} = -c_{ij} \qquad (12)$$

There are two specific cases of interest to estimate the parameters; those are referred to as autocorrelation method and covariance method, respectively. The difference between the two methods is in the choice of the index limits to compute c_{ij}.

In the autocorrelation method, the index limits n_0 and n_1 are chosen to be infinity. However, in practice, a finite period is needed for analysis, and thus the signal is normally windowed so that the signal becomes zero outside the window. Letting the windowed signal x'_n be

$$x'_n = \begin{cases} x_n w_n & 0 \leq n \leq N-1 \\ 0 & \text{otherwise} \end{cases} \qquad (13)$$

where w_n is a window function, then the autocorrelation function R(i) is given by

$$R(i) = \sum_{n=0}^{N-i-1} x'_n x'_{n+i} \tag{14}$$

Then c_{ij} in equation (12) is replaced by $R(|i-j|)$, since $R(i)$ is an even function.

In the covariance method, the error E is minimized over a finite period, typically $n_0=0$ and $n_1=N-1$. Thus,

$$c_{ij} = \sum_{n=0}^{N-1} x_{n-i} x_{n-j} \; . \tag{15}$$

The correlation matrix in the autocorrelation method is a symmetric Toeplitz matrix and it satisfies the positive definiteness. Thus, the stability of the resulting digital filter is guaranteed, whereas the stability is not guaranteed in the covariance method. The covariance method, however, gives better estimation of the parameters for a short time segment (less than or equal to a pitch period). The covariance method approaches the autocorrelation method as the time interval becomes larger.

The solution for the parameters in the simultaneous equations given by equation (12) reduces to find an efficient method for inverting the the correlation matrix. There are several approaches to it (8),(9),(5). According to Robinson's recursive procedure (14), the relations between a_i's obtained in the passage between recursive steps m and m+1 are given by

$$\left.\begin{aligned} a_0^{(m+1)} &= a_0^{(m)} = 1 \\ a_i^{(m+1)} &= a_i^{(m)} + k_m a_{m-i+1}^{(m)} \\ a_{m+1}^{(m+1)} &= k_m \end{aligned}\right\} \tag{16}$$

where the parenthesized superscripts of the filter coefficients denote the recursion step and k_m is given by

$$k_m = \frac{\sum_{i=0}^{m} a_i^{(m)} R(m+1-i)}{\sum_{i=0}^{m} a_i^{(m)} R(i)} \tag{17}$$

Thus, the parameters a_i's are recursively obtained as $a_i^{(M)}$'s. The parameters k_m are called the reflection coefficients, or the PARCOR coefficients since the k_m is also given by the partial correlation coefficient of the forward and backward prediction errors at the m-th recursive stage (9). The reflection coefficients also have correspondence to the reflection coefficients μ_m of a discrete acoustic tube, in which μ_m is given by

$$\mu_m = \frac{S_m - S_{m+1}}{S_m + S_{m+1}} \tag{18}$$

where S_m is the cross-sectional area of the m-th acoustic tube section (15). Since the reflection coefficients are limited to less than ± 1, they are used more frequently than the filter coefficients. The filter coefficients are used when the spectral envelopes are computed. Since they are the dicrete impulse response of the inverse filter, their Fourier transform (normally with the sufficient number of zeros added to maintain a good resolution) gives an optimal spectral estimation for an input spectrum.

3. ANALYSIS CONDITIONS

Proper analysis conditions for the linear prediction method are important to ensure satisfactory results. The analysis conditions to be noted are (1) sampling frequency, (2) the number of coefficients, (3) time window and length (frame size), (4) window shift, and (5) preemphasis. The sampling frequency determines the frequency range of interest. The frequency range must be less than or equal to half the sampling frequency(normally the latter is chosen). The number of coefficients is dependent on the frequency range to be chosen. When the frequency range is exactly half the sampling frequency (F_s kHz), a good rule of thumb for the number of coefficients is from F_s+2 to F_s+4. The reason for this appears to be that there will be about $F_s/2$ resonances in the frequency band limited by $F_s/2$, provided that F_s is given in units of 1 kHz. Each resonance requires two coefficients for its representation, and so about F_s coefficients will be needed to account for the expected resonances in the analysis band. In addition, two to four coefficients are normally used for approximating the spectral slope due to the excitation source.

The analysis conditions (3) and (4) vary depending upon which of two different methods of linear prediction is used, the autocorrelation method or covariance method. The autocorrelation method requires a window length of at least 1.5 pitch periods and a Hamming window is recommended to suppress the spectral disturbances in the high frequency region due to the edge effect of the time window. The covariance method, on the other hand, does not require any particular time window, and the window length can be less than a pitch period. Thus this method can be used for pitch-synchronous analysis of speech sounds. When a window length of less than a pitch period is chosen, care must be taken since the analysis results vary depending upon what portion of the pitch period is chosen for analysis. This method is particularly useful for extracting the true vocal tract characteristics by choosing the glottis-closed portion of the speech waves. The major disadvantage of the covariance method is that there is theoretically no guarantee for obtaining a stable transfer function for the inverse filter, and thus a more sophisticated algorithm is required to automatically process the cases of instability. Also a more sophisticated algorithm is needed for automatically segmenting the speech wave into pitch-synchronous intervals.

The window shift in the covariance method, thus, involves a more complicated procedure than it does in the autocorrelation method. In the latter method, the window shift is rather arbitrary, depending upon the speech samples to be analyzed. The shift can be greater than the window length for steady-state sounds, whereas, for speech sounds in which the formant frequencies are rapidly changing, a smaller window shift will be better for obtaining smooth contour of the formant frequencies.

A 6 dB/octave preemphsis is recommended for formant analysis. This is accomplished by taking the backward differencing of the sampled speech. The purpose of the preemphasis is to enhance the spectral peaks in the high frequency region. The 6 dB/octave preemphasis also roughly compensates the -12 dB/octave glottal source characteristics and the +6 dB/octave lip radiation characteristics.

4. ESTIMATION OF FORMANT FREQUENCIES

As mentioned before, the Fourier transform of the filter coefficients gives the frequency characteristics of the inverse filter the inverse of which are the frequency characteristics of the vocal tract filter. Thus the procedure for obtaining the smooth spectral envelope by use of the linear prediction method is given by the block diagram shown in Fig. 4. The speech signal is first digitized at some sampling frequency after being passed through a lowpass filter to limit the frequency band according to the sampling frequency. Linear prediction analysis is then performed using predetermined analysis conditions, and resulting in a set of filter coefficients for each speech segment analyzed. Smooth spectral envelopes are computed from the output of the Fourier transform of the filter coefficients with added zeros. As a result of linear prediction analysis, the residual signal, which

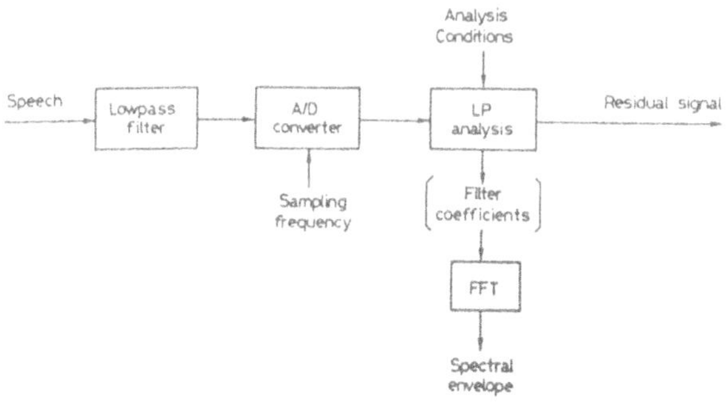

Fig. 4. A block diagram to compute the smooth spectral envelopes of speech sounds by the linear prediction method.

is an error signal given by equation (6), is saved for detecting pitch periods as will be described later.

An example of analysis results is shown in Fig. 5. This

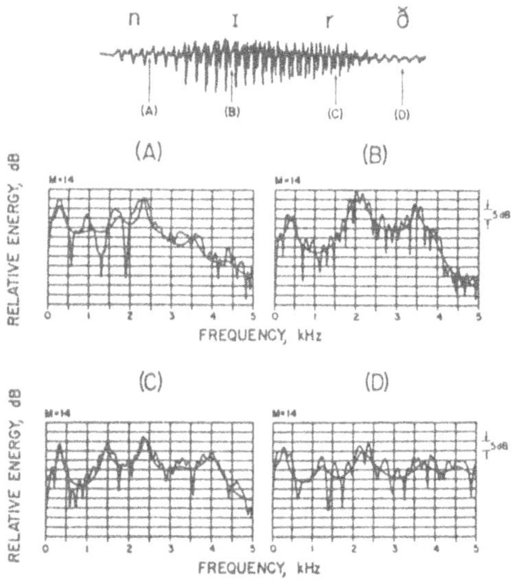

Fig. 5. An example of linear prediction analysis (sampling frequency 10 kHz; number of coefficients 14; frame size 20 msec with a Hamming window and +6 dB/octave preemphasis).

example is a part of a sentence "Near the boat..." and the spectral envelope estimation for /n/, /I/, /r/, and /ð/ are shown in the figure together with the direct Fourier transform of the corresponding speech waves. It is seen that spectral peaks are well approximated by the extracted spectral envelope. However, the spectral dips due to antiresonances as in the sound /n/ are ignored in the linear prediction method, in which the nasal tract is not considered. It should be noted that the linear prediction method was developed as a method for efficient speech analysis-synthesis telephony on the basis of the fact that the human ear is insensitive to spectral dips. Thus ignorance of spectral dips is not a major problem as far as analysis-synthesis telephony is concerned. However, if one is interested in a more accurate estimation of spectral dips as well as peaks, a new model has to be developed, which is currently investigated by some researchers.

The formant frequencies are estimated from the smooth spectral envelope by finding the location of the spectral peaks by a peak-picking method. Although this method is simple and worthwhile, it presents problems when two peaks are close together or merged into a broad peak. Another method is to compute the exact locations of the peaks by solving for the roots of the transfer function of the inverse filter. In both methods, the spectral peaks do not always correspond to the formant frequencies, and thus a certain algorithm to automatically select formant peaks has to be designed (10). For both methods, a careful inspection of the analysis results is recommended before further processing of the formant frequencies is initiated.

5. ACCURACY OF FORMANT ESTIMATION

It is rather difficult to determine the accuracy of formant estimation for natural utterances, since there is no way of accurately measuring the vocal tract configuration to compute its resonances while a sound is being produced. Chandra and Lin (3) made an evaluation of the autocorrelation and covariance methods by using synthetic vowels. In their study, vowels in the 'h-d' context were synthesized by a simulated formant synthesizer, and the two linear prediction methods were applied to analyze those synthetic vowels. As analysis conditions in this case, the sampling frequency was 10 kHz and the number of coefficients was 12. The results of their study are shown in Figs. 6 and 7. Fig 5 shows the estimation error (in Hz) of the first three formant frequencies for both methods applied pitch-synchronously and pitch-asynchronously. For the pitch-synchronous case, the window length coincided with the segment position between the two pitch pulses. For the pitch-asynchronous case, the window length of 24 msec was arbitrarily chosen on the speech waves. The results indicate that the pitch-synchronous covariance method gives better accuracy

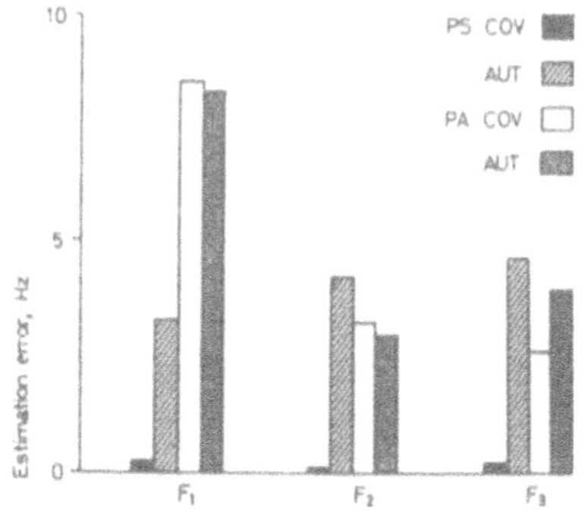

Fig. 6. Evaluation of formant frequency estimation by autocorrelation (AUT) and covariance (COV) methods for pitch-synchronous(PS) and pitch-asynchronous(PA) cases.

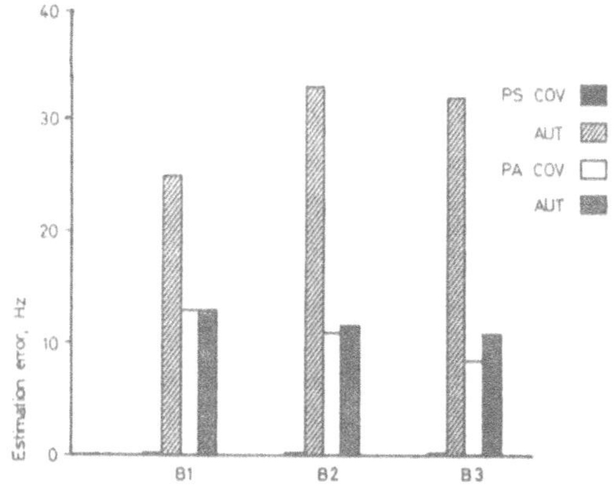

Fig. 7. Evaluation of formant bandwidth estimation by autocorrelation (AUT) and covariance (COV) methods for pitch-synchronous(PS) and pitch-asynchronous(PA) cases.

than the others. In the pitch-asynchronous case, when the window length becomes greater than one and a half pitch period, the two methods give similar accuracy. The pitch-synchronous autocorrelation method resulted in the worst accuracy. This is more so in estimating formant bandwidths as shown in Fig. 7.

For natural utterances, it is anticipated that the accuracy of estimating formant frequencies and bandwidths becomes worse than for the synthetic sounds. Especially, it is anticipated that the result of the pitch-synchronous case will become worse, because the condition at the glottis varies during one pitch period for natural utterances, whereas the glottal condition for this particular synthesizer. When the glottal condition varies during a chosen analysis segment,the resulting formant frequencies will probably be the average of the instantaneous formant frequencies. The result obtained by Chandra and Lin indicates that the pitch-synchronous covariance method gives more accurate estimates of formant frequencies and their bandwidths than the pitch-asynchronous autocorrelation method. Although the estimation accuracy of the formant bandwidths is not well known, it is known that the bandwidth estimates are sometimes too narrow and too broad. If the bandwidth information is needed, it has to be carefully checked against the direct Fourier transform of the corresponding sampled speech.

6. PROBLEMS IN FORMANT ESTIMATION

Since the estimation of formant frequencies is made from the envelope estimation of speech spectra, the accuracy of estimation is highly dependent on harmonic density. The more sparse the harmonic density becomes as pitch goes up, the more difficult the estimation of formant frequencies becomes. This is a rather inherent problem in the estimation of vocal tract resonances from given speech waves, irrespective of method. In many cases, the linear prediction method works well for speech sounds with fundamental frequencies of up to approximately 250 Hz. For female speakers and children with fundamental frequencies higher than 250 Hz, difficult cases of formant estimation are frequently observed. Formant estimation becomes impossible as the pitch becomes extremely high, in which case harmonics are picked up as spectral peaks. In case the exact vocal tract resonances need to be known, some other methods may have to be used. One approach to it is to use an external excitation with a low fundamental frequency, which often leads to a successful estimation of fundamental frequencies.

7. FUNDAMENTAL FREQUENCY ESTIMATION

In inverse filtering in the linear prediction method, most of the vocal tract characteristics are filtered out into the filter coefficients. The residual signal, the output of the inverse filter, still contains the information on the excitation source. A typical residual signal for the vowels is shown in Fig. 8. It is seen that large errors synchronous with pitch periods occur. A typical approach to computing the periodicity from this kind of waveform is to compute the autocorrelation function as shown in Fig. 9. Two conspicuous spikes are found in the autocorrelation, one at the origin and one at a distance of one pitch period from the origin. The fundamental frequency is then given by the reciprocal of the pitch period.

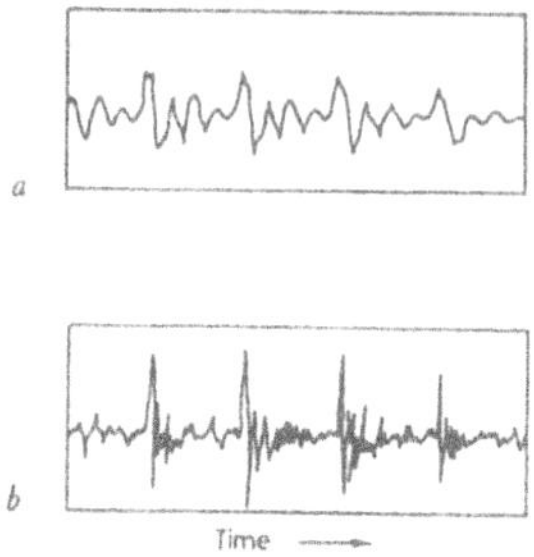

Fig. 8. a Speech waves. b The residual signal after linear prediction analysis.

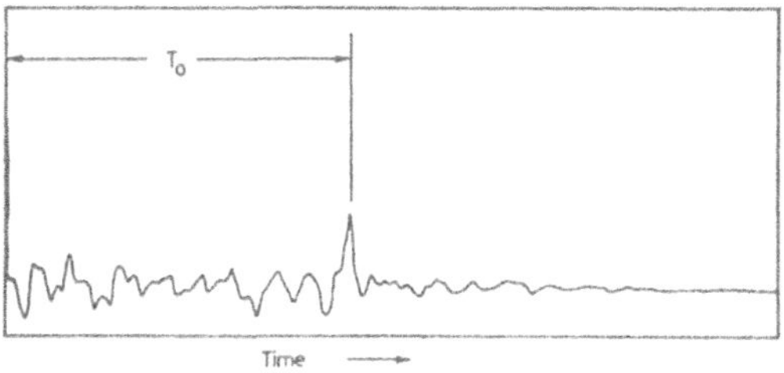

Fig. 9. Autocorrelation function of the residual signal of Fig. 8.

It has been shown that the linear prediction method is quite efficient and effective for estimating the formant frequencies. However, how accurate and reliable the extraction of fundamental frequency is an intriguing question, since there are many other techniques for estimating the fundamental frequency. Rabiner et al. (13), in their study of the comparative performance of several pitch detection algorithms, point out the following major problems in detecting the fundamental frequency: (1) glottal excitation is not perfectly periodic; (2) defining the exact beginning and end of each period is difficult; (3) the distinction between unvoiced portions and low level voiced portions is difficult; (4) there is an interaction between the vocal tract and the glottal excitation. The above problems are intrinsic in any of the pitch detection methods. However, evaluation of several pitch detection methods indicates some differences in their performance.

8. ACCURACY IN FUNDAMENTAL FREQUENCY ESTIMATION

Let us take the following pitch detection methods from the study by Rabiner et al.: (1) autocorrelation method with clipping (time domain method); (2) cepstrum method (frequency domain method); (3) linear prediction 'SIFT'(Simplified Inverse Filter Tracking) method (time-frequency method). The types of errors can be categorized into (a)voiced-to-unvoiced error,(b) unvoiced-to-voiced error, (c) gross error in which the error in detecting the pitch period is greater than a certain threshold, and (d) fine error in which the error in detected pitch period is less than the threshold.

The above three methods were tested against 6 speakers (3 males, 2 females, and 1 child) by using four monosyllabic nonsense words and four sentences. The analysis results were compared with the standard pitch contours which were carefully measured by using semi-automatic pitch detector. The results for the first three types of errors are shown in Figs. 10-12. The results are shown

10a

10b

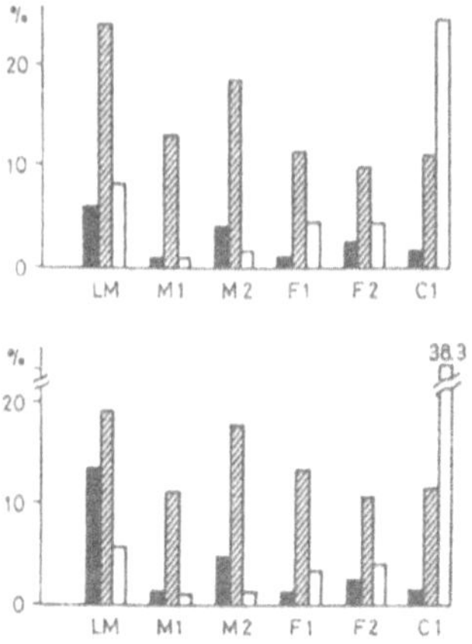

Fig. 10. Voiced-to-unvoiced errors for three pitch detection methods: autocorrelation(■), cepstrum(▨), and SIFT(□).
a Unsmoothed. b Smoothed. LM=Low pitch male; M1,M2=males; F1,F2=females; C1=child. The ordinate shows the percentage error rate against the total number of voiced intervals.

both for unsmoothed (raw data) and smoothed cases. In the smoothed case a nonlinear smoothing technique was applied to the raw data (11). It is seen that the nonlinear smoothing generally improves the accuracy: particularly, the gross errors are substantially improved. It is also seen that all three methods are somewhat speaker dependent. For the voiced-to-unvoiced errors, the error rate of the cepstrum method is much higher than the others except for the child speaker. For the unvoiced-to-voiced errors, on the other hand, the error rate of the cepstrum method is better than

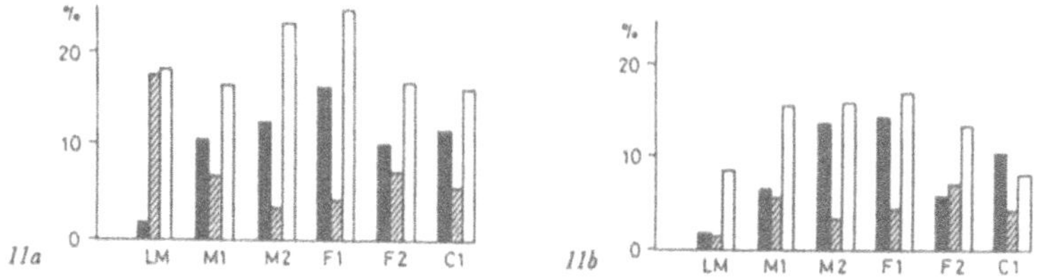

Fig.11. Unvoiced-to-voiced error for three pitch detection methods: autocorrelation(■), cepstrum(▨), and SIFT(□).
a Unsmoothed. b Smoothed. LM=Low-pitched male; M1,M2=males; F1,F2=females; C1=child. The ordinate shows the percentage error rate against the total number of unvoiced intervals.

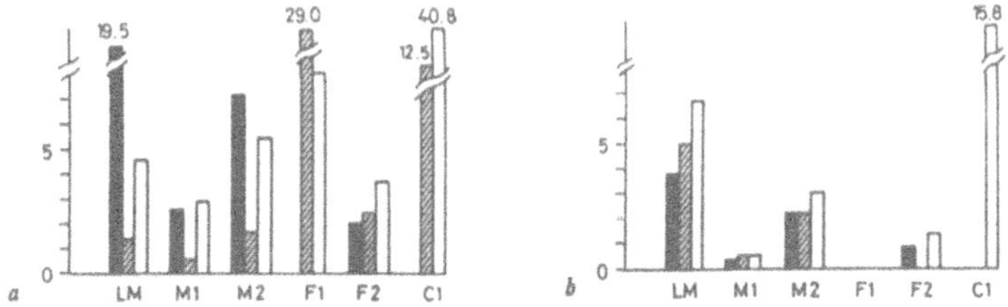

Fig. 12. Gross errors for three pitch detection methods: autocorrelation(■), cepstrum(▨), SIFT(□). a Unsmoothed.
b Smoothed. LM=Low-pitched male; M1,M2=males; F1,F2=females; C1=child. The ordinate shows the average number of samples.

the others except for one of the female speakers for the smoothed case. In overall performance evaluation, there seems to be not much difference between the performance of the autocorrelation and linear prediction methods, except that the linear prediction method resulted in an exceedingly poor performance for the child speaker for the unvoiced-to-voiced and gross errors.

8. OTHER RELATED TOPICS

The filter in the linear prediction model in Fig. 1 contains the contribution from the glottal characteristics and the radiation effect at the lips as well as the vocal tract characteristics. Since the model assumes a linear system, those factors can be separated and changed in order as shown in Fig. 13. If the glottal and radiation characteristics can be eliminated by a proper preprocessing of the speech, the true vocal tract characteristics can be obtained by the linear prediction method. One of the im-

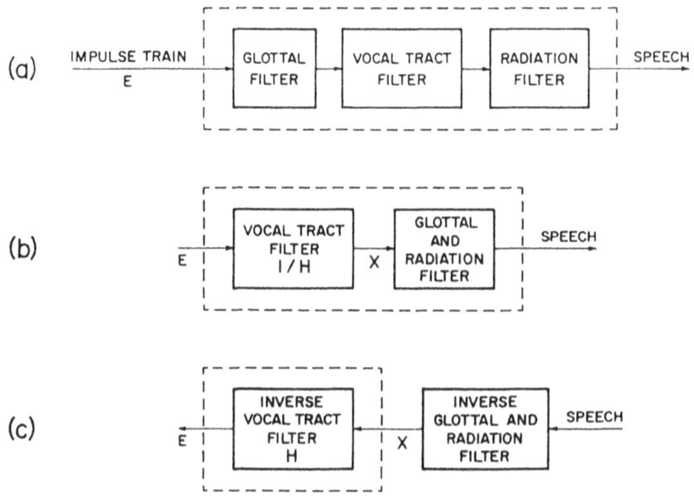

Fig. 13. Block diagrams to obtain the vocal tract characteristics by eliminating the glottal and radiation effects.

portant features of the linear prediction method is that a set of reflection coefficients obtained for a given speech segment gives an acoustic tube shape which has a frequency characteristic identical to the vocal tract characteristics extracted from this speech segment. In this case, the acoustic tube is represented by a concatenation of cylindrical sections of different cross-sectional areas. Consequently, if the analysis conditions are properly chosen after preprocessing sampled speech to eliminate the glottal and radiation characteristics, the acoustic representation thus obtained is expected to be a good approximation to the vocal tract area function which denotes the cross-sectional areas along the vocal tract from the glottis to the lips (15),(17).

Another interesting topic is the use of the linear prediction parameters for speech synthesis. The synthesizer can be the synthesis part of the linear prediction analysis-synthesis telephony. This type of synthesizer is now commercially available in the form of an LSI chip. Low bit-rate, high quality synthesis would assure the usefulness of the synthesizer. Since the formant frequencies and bandwidths constitute the roots of the inverse filter transfer function, they can be related to the filter coefficients. The reflection coefficients are also related to the filter coefficients in the mathematical formulation of linear prediction. Thus, those parameters mentioned above are interchangeable for each other, and any of these parameters can be used for the linear prediction synthesizer.

It should also be noted that the linear prediction parameters and the residual signal are very useful for automatic speech recognition (4),(12). They will particularly be effective for acoustic feature detection, segmentation, labeling, and speaker normalization.

9. APPLICATION EXAMPLES

In concluding this tutorial paper, several examples taken from the author's past studies will be given below to demonstrate the application of the linear prediction method to acoustical analysis of speech.

9.1 Formant variability within a speaker(2)

Fig. 14 shows the F_1-F_2 distribution for 17 unrounded vowel types produced by a female phonetician in order to study the variability of formant frequencies. In this study, 30 repetitions of 30 dif-

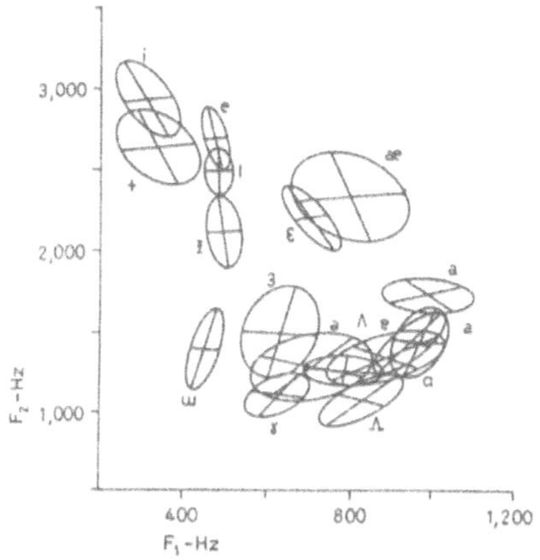

Fig. 14. F_1-F_2 distribution of 17 unrounded vowel types produced by a female speaker. Ellipses represent two standard deviations.

ferent isolated vowel utterances (900 in total) were analyzed by the autocorrelation method (sampling frequency 10 kHz; number of coefficients 12; window size 25 msec with Hamming window and +6 dB/octave preemphasis) and formant frequencies were carefully inspected by displaying the results vowel by vowel on the display terminal. Approximately 5 % of apparently wild data were excluded for further processing.

9.2 Formant variability among speakers and vowel normalization(16)

The example in Fig. 15a shows the F_1-F_2 distribution of nine American English vowels spoken by 26 speakers (14 males and 12 females) in order to study the variability of formant frequencies among male and female speakers. Vowels were produced in the con-

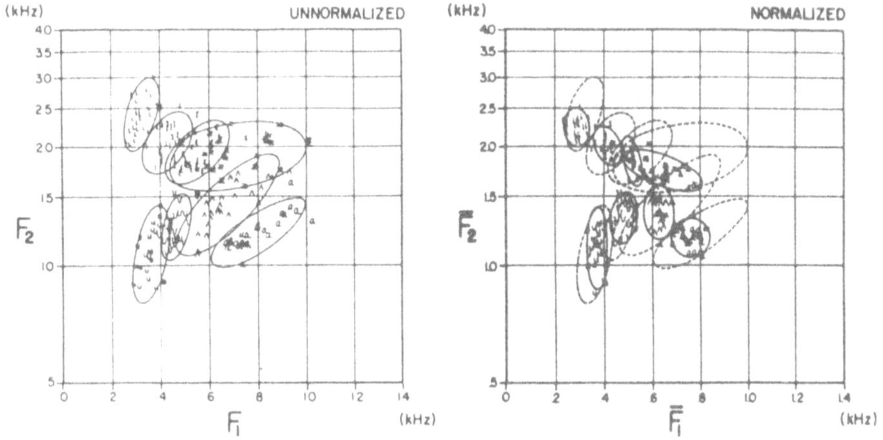

Fig. 15. F_1-F_2 distribution of nine American English vowels produced by 26 speakers(14 males and 12 females). Ellipses represent two standard deviations. a Unnormalized. b Normalized.

text of 'h-d' and the autocorrelation method was applied to analyze the vowel portions with the same analysis conditions as in the previous example. The formant frequencies which were averaged over the most steady-state five frames were used to represent each vowel. Fig. 15b shows the distribution of F_1 and F_2 after they were normalized to an acoustic tube of 17 cm long in order to eliminate a part of speaker variation. In this case, the instantaneous vocal tract length was computed from the estimated first four formant frequencies and bandwidths.

9.3 Segmentation and feature detection(6)

Fig. 16 is an example in which the linear prediction area functions were used to automatically segment speech into vowel-like and non-vowel-like intervals. The area functions, combined with the speech energy function, give sufficient cues for the first stage of segmentation without obtaining spectral information which is more time consuming. In this case, the autocorrelation method

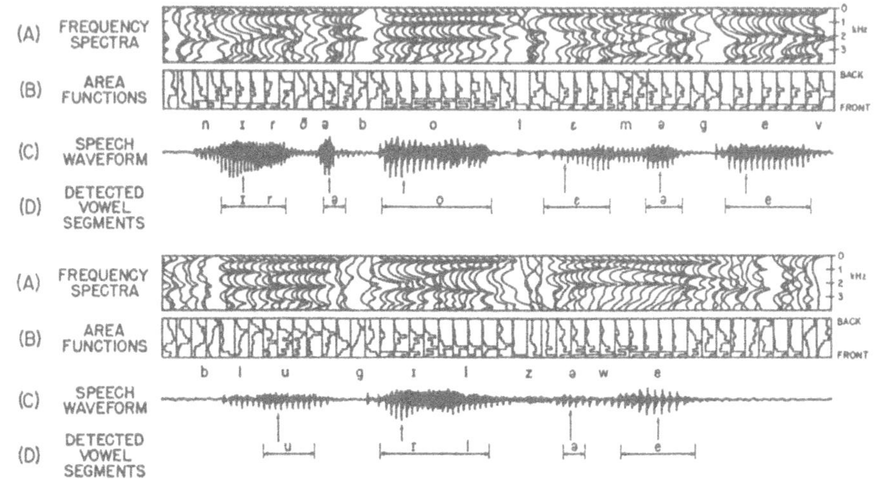

Fig. 16. An example of segmenting the vowel-like intervals for the sentence utterance 'Near the boat, Emma gave blue-gills away'.

was also used (sampling frequency 10 kHz; number of coefficients 14; window size 15 msec with a Hamming window and +6 dB/octave preemphasis). The relatively short analysis window length was used in this example for detecting the bursts of plosives, and the window shift was 12.8 msec. In this line of study, the area functions were also used to effectively detect the nasal segments (7).

10. CONCLUSION

The concept and evaluation of the linear prediction method were described in this paper. Because of its tutorial nature, the descriptions in some cases may be inadequate from the theoretical point of view. This paper is the revised version of the paper appeared in Phonetica (Vol. 37, No. 1-2, 1980).

REFERENCES

(1) Atal, B., and Schroeder, M.R.: 1967, 1967 Conf. Commun. and Process., pp. 360-361
(2) Broad, D.J., and Wakita, H.: 1978, Speech communication with computers, Macmillan, London, pp. 52-92.
(3) Chandra, S., and Lin, W.: 1974, IEEE Trans.ASSP22, pp. 403-415.

(4) Itakura, F.: 1975, IEEE Trans. ASSP23, pp. 67-72
(5) Itakura, F., and Saito, S.: 1966, Technical Report 3107, Electrical Commun. Res. Lab., NTT.
(6) Kasuya, H., and Wakita, H.: 1979, IEEE Trans. ASSP27, pp. 319-327.
(7) Kasuya, H., and Wakita, H.: 1977, Contributed papers Vol. 1, 9th ICA, p. 486.
(8) Makhoul, J.: 1975, Proc. IEEE 63, pp. 561-580.
(9) Markel, J.D., and Gray, A.H.: 1976, Linear prediction of speech, Springer.
(10) McCandless, S.S.: 1974, IEEE Trans. ASSP22, pp. 135-141.
(11) Rabiner, L.R., Sambur, M.R., and Schmidt, C.E.: 1975, IEEE Trans. ASSP23, pp. 552-557.
(12) Rabiner, L.R., and Schafer, R.W.:1978, Digital processing of speech signals, Prentice-Hall.
(13) Rabiner, L.R., Cheng, M.G., Rosenberg, A.E., and McGonegal, C.A.: 1976, IEEE Trans. ASSP24, pp. 399-418.
(14) Robinson, E.A.: 1967, Statistical communication and detection, Hafner, pp. 274-279.
(15) Wakita, H.: 1973, IEEE Trans. AU 21, pp. 417-427.
(16) Wakita, H.: 1977, IEEE Trans. ASSP 25, pp. 183-192.
(17) Wakita, H.: 1979, IEEE Trans. ASSP 27, pp. 281-285.

MEDIUM AND LOW BIT RATE SPEECH TRANSMISSION

Vishu R. Viswanathan, John Makhoul and
Richard Schwartz

Bolt Beranek and Newman Inc.
10 Moulton St.
Cambridge, MA 02238

ABSTRACT

We describe five types of speech coding systems with transmission bit rates spanning the range from 16,000 bits/sec (b/s) down to 100 b/s: adaptive predictive coders at 16 kb/s, baseband coders at 9.6 kb/s, linear predictive coders at 1.5-2.4 kb/s, clustering vocoders at 600-800 b/s, and diphone-based phonetic vocoders at 100 b/s. For each type of coders, we describe the coder configuration, discuss the important coding issues, and present the results available to date.

1. INTRODUCTION

Digital speech transmission is an important area of application in the speech processing field. In this paper, we describe several types of speech coding (or data compression) systems for use in medium and low bit rate speech transmission. For medium bit rate transmission, we consider 16 kb/s (kilobits/sec) adaptive predictive coders (APC) and 9.6 kb/s baseband coders (BBC). For low bit rate transmission, we consider three types of coders: (1) linear predictive coders (LPC) at 1.5-2.4 kb/s, (2) clustering vocoders at 600-800 b/s, and (3) diphone vocoders at about 100 b/s.

The choice of a speech coder for a specific application is governed by factors such as available channel bandwidth, desired quality of audio output, computational complexity of the coder, and power requirement of the transmission system. The 16 kb/s APC coders are chosen for use in applications that require faithful reproduction of the quality of the input speech.

J.-P. Haton (ed.), Automatic Speech Analysis and Recognition, 21–48.

Speech coders at 9.6 kb/s may be chosen for transmission over regular telephone wirelines, since this bit rate represents a reasonable upper limit for achieving reliable speech communication. LPC coders at about 2.4 kb/s are used in speech communication over HF channels. Speech coders at 600-800 b/s have possible applications in Navy and mining systems, since the communication channels available in these systems cannot support higher data rates. Finally, the 100 b/s coders may be used in applications characterized by one of the following three conditions: low-power communication to avoid detection, sufficient power per bit to reliably "burn through" jamming networks, or use under extreme atmospheric conditions and in highly shielded environments.

Although speech data compression techniques are presented in this chapter from the viewpoint of speech transmission applications, these techniques can be used to accomplish (1) efficient storage and playback of speech, which may be required, for example, in voice response systems such as database enquiry, dial-up information, and computer-aided instruction systems and (2) efficient storage of speech templates or model parameters, which may be required in speech or speaker recognition systems and in speech synthesis systems.

For each of the five types of coders, we discuss important coding issues and summarize the results available to data. Before we proceed to describe these coders in order, as background material we discuss two topics that are common to these coders: synthesis model and linear prediction analysis.

2. SYNTHESIS MODEL

2.1 A General Synthesis Model

In each of the five types of speech compression systems that we are considering, the receiver employs the general synthesis model shown in Fig.1(a). In this model, the output synthetic speech r(t) is generated as the result of applying a time-varying excitation signal u(t), which has a flat spectral envelope, as input to a time-varying spectral shaping filter H(z). The corresponding analysis explicitly or implicitly required with the use of this synthesis model is shown in Fig.1(b). The input speech signal s(t) is inverse-filtered using A(z) (see Section 3 below), which is proportional to the reciprocal of H(z), to produce the residual or error signal e(t).

The five types of coders, described in Sections 4-8, differ in the way they obtain a model of the excitation signal u(t)

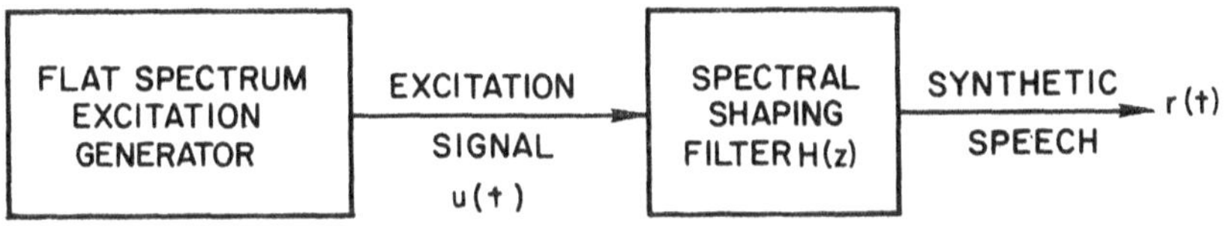

(a) Implementation at the receiver

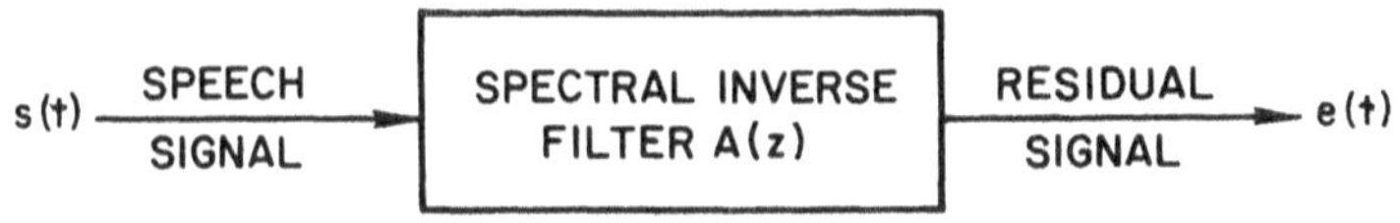

(b) Analysis at the transmitter

Fig. 1. General synthesis model employed by the five types of speech coders.

from the residual signal e(t). At one extreme, the APC coders represent the excitation signal in terms of the samples of the entire residual waveform. Data compression is achieved in such waveform coders through quantization and encoding of the residual samples.

At another extreme, three low-bit-rate coders derive the excitation signal from a pitch-excited source, which traditionally has been the following binary pulse/noise model: periodic sequence of pulses (separated by the pitch period) for voiced sounds and random noise sequence for unvoiced sounds. Thus, the binary excitation model for a given speech sound is specified by its voiced/unvoiced (V/U) status and by the pitch frequency if it is voiced. Recently, an improved pitch-excited source, called mixed-source model, has been proposed. This model is described below in Section 2.2.

As a compromise between the above two extremes, the BBC coder extracts and encodes only a low-frequency portion (or baseband) of the residual; for speech resynthesis, it regenerates the highband through some nonlinear operation on the baseband.

2.2 A Mixed-Source Model

Using the binary pulse/noise excitation model leads to several problems including unnatural and "buzzy" speech quality and severe speech-quality degradation due to errors made in the V/U decision. The recently proposed mixed-source model overcomes the problems of the binary model (1).

The new model allows for simultaneous pulse and noise excitations. The model divides the spectrum into a low-frequency region and a high-frequency region, with the pulse source exciting the low region and the noise source exciting the high region. The cutoff frequency FC that separates the two regions is adaptively varied in accordance with the changing speech signal.

At the receiver, the excitation signal is realized by passing the pulse sequence through a lowpass filter with cutoff FC and the noise sequence through a highpass filter with the same cutoff frequency FC, then adding the outputs of the two filters. The resulting mixed-source excitation signal is applied to the spectral shaping filter. For unvoiced sounds (FC=0), the model uses a pure noise excitation. The model has only two parameters: the cutoff frequency FC and the pitch period when FC>0. Since small changes in FC do not seem to be perceptible, it is sufficient to quantize FC into 2-3 bits for transmission purposes.

In both informal and formal listening tests conducted on speech synthesized using the mixed-source model, the model was found to largely eliminate the "buzzy" quality of the synthesized speech, perform better for female speech, and result in a certain "fullness" in perceived speech quality that was absent with the binary voiced/unvoiced synthesis (1, 2).

3. LINEAR PREDICTION ANALYSIS

All five types of coders described in the paper use the well-known linear prediction method for the all-pole modeling of the speech signal (3). In this method, it is assumed that a speech sample s(n) can be approximately predicted as a linear combination of a number of immediately preceding samples:

$$\tilde{s}(n) = - \sum_{k=1}^{p} a(k)\ s(n-k), \tag{1}$$

where $\tilde{s}(n)$ is the predicted value of s(n), a(k), $1 \leq k \leq p$, are real constants known as predictor coefficients, and p is the order of the predictor. The residual e(n) is the error between the actual value and the predicted value, i.e., $e(n) = s(n) - \hat{s}(n)$. As shown in Fig.1(b), e(n) is obtained by passing the speech signal through the inverse filter A(z); for the linear prediction method, A(z) is given by

$$A(z) = 1 + \sum_{k=1}^{p} a(k)\, z^{-k}. \qquad (2)$$

The predictor coefficients are obtained by minimizing the mean-squared value of the residual signal with respect to a(k), over a frame of speech that is 10-25 ms long in duration. The linear prediction analysis is usually carried out at the rate of once every 10-30 ms.

Thus, with the linear prediction model, the spectral shaping filter H(z) in Fig.1(a) is the all-pole filter given by

$$H(z) = G/A(z), \qquad (3)$$

where G is the filter gain chosen such that the energy of the synthesized speech is approximately equal to the energy of the input speech.

4. ADAPTIVE PREDICTIVE CODERS

4.1 Coder Description

Figure 2 depicts the APC coder. The feedback structure, which constitutes the transmitter, encodes the sampled input speech S(z) in terms of the quantized residual $\hat{W}(z)$, the spectral-inverse filter A(z) given by (2), and a 3-tap pitch-inverse filter C(z) given by

$$C(z) = 1 + c1\, z^{-(M-1)} + c2\, z^{-M} + c3\, z^{-(M+1)}, \qquad (4)$$

where M is the pitch period in number of samples. The spectral predictor A(z)-1 is designed to remove the short-term redundancy due to non-flat speech spectrum, while the pitch predictor C(z)-1 is designed to remove the long-term redundancy due to pitch

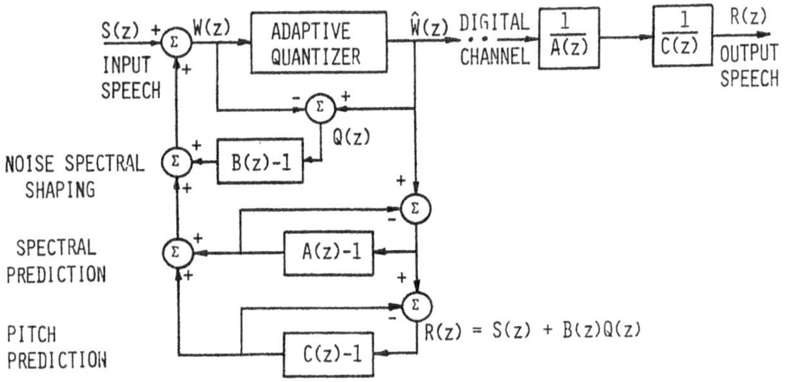

Fig. 2. Adaptive predictive coder with spectral prediction, pitch prediction, and noise spectral shaping. S(z), R(z), etc. are z-transforms of the corresponding discrete-time signals s(n), r(n), etc.

periodicity during voiced sounds. The filter B(z) in the feedback structure in Fig. 2 is designed to shape the spectrum of the quantization noise $Q(z) = \hat{W}(z) - W(z)$ in a way that yields a perceptually more pleasing output speech (4, 5) The APC coder encodes input speech one frame at a time. For each frame, the samples of the quantized APC residual $\hat{W}(z)$ are transmitted to the receiver along with the parameters of the adaptive quantizer, the spectral predictor coefficients $\{a(k), 1 \leq k \leq p\}$, the pitch M, and the pitch predictor coefficients {c1,c2,c3}. Notice that the parameters of the filter B(z) are not transmitted.

At the receiver, the decoded signal is filtered by the all-pole spectral filter 1/A(z) and then by the pitch-synthesis filter 1/C(z), to produce the output speech R(z). In the absence of channel bit-errors, it can be shown that

$$R(z) = S(z) + B(z)\,Q(z) \; . \tag{5}$$

This expression demonstrates the noise-shaping role of the filter B(z).

Below, we discuss two issues that are important in the design of 16 kb/s APC coders: coding of the APC residual and adaptive noise shaping. Following that, we present the results of a recent speech-quality optimization study of APC coders.

4.2 Coding of the APC Residual

For the discussion given in this section, unless stated otherwise, we assume that pitch prediction is not used. Also, the APC residual W(z) is quantized using Adaptive PCM (APCM). This quantizer has an adaptive gain multiplier G followed by an optimum (minimum mean-square error) uniform or nonuniform unit-variance quantizer. G is transmitted to the receiver for decoding the APC residual.

We describe below five methods for the quantization and encoding of the APC residual. For 16 kb/s speech transmission, the average number of available bits per residual sample is usually about 2 bits. Each of the five residual coding methods described below attempts to limit the extent of clipping errors in a different manner. At the end of this subsection, we present additional coding schemes obtained by combining two or more of these five coding schemes.

Entropy Coding (EC.) This scheme uses a uniform quantizer and a large number of levels to avoid clipping completely. The scheme uses a variable-length Huffman or entropy code to yield an average codelength of about 2 bits/sample (4). Variable-to-fixed rate conversion, which is required to interface the entropy-coded residual data to a fixed-rate channel, may be accomplished by forcing the total bits per frame (about 25 ms) to be equal to a predetermined fixed value (6). The speech obtained using the entropy coding scheme has very good quality but has some perceivable granular noise.

Pitch Prediction (PP). This scheme uses pitch prediction within the APC feedback loop as discussed above, a nonuniform quantizer, and a fixed-length code for residual samples. With the autocorrelation method (3) of computing the pitch predictor parameters, the 1-tap filter is guaranteed to be stable; however, this is not the case for the 3-tap filter. In one study, the 3-tap filter was unstable for about 8% of the frames (6, 7). The instabilities cause pops and beeps in the output speech, when the quantizer uses less than 4 levels per sample. A successful remedy to this problem is to check the stability of the filter each frame and to use the 1-tap filter, whenever the 3-tap filter is unstable. The 3-tap case equipped with this remedy produces a 1-2 dB higher signal-to-quantization-noise (S/Q) ratio and more clarity of speech than the 1-tap case (6, 7).

Segmented Quantization (SQ). This scheme employs a nonuniform quantizer and a fixed-length code. The analysis frame is divided into several equal-length segments, and one value of quantizer gain G is computed for each segment. This makes the quantizer step size adapt rapidly to local variations in the residual samples. The SQ scheme is most effective when used along with pitch prediction (6, 7).

Segmented Quantization with Bit Allocation (SQ-BA). This scheme is the same as the SQ scheme, with the difference that it uses a variable-length code (6). A segment with a larger residual energy is coded with more bits.

Pitch-Adaptive (PA) Scheme. This scheme is a pitch-synchronous version of the SQ-BA scheme. Voiced frames are divided into individual pitch periods, and each pitch period is subdivided into a small fixed number (e.g., 4) of approximately equal-length segments (6, 7). As in the SQ-BA scheme, larger-energy segments are coded with more bits.

Composite Schemes. Composite schemes obtained by combining the individual ones can lead to significantly better speech quality and higher S/Q ratios. The composite schemes that we have found to be the best are conveniently denoted, in terms of the previously used abbreviations, as: EC-PP, PP-SQ, PP-SQ-BA, and PA-PP. For example, the EC-PP scheme uses pitch prediction and entropy coding. Except for the EC-PP scheme, all other schemes produced varying amounts of low-level pops and clicks. These discrete distortions were generally reduced when we preprocessed the input speech with a preemphasis filter $(1-0.4z^{-1})$ and postprocessed the output speech with the corresponding 1-pole deemphasis filter (6, 8).

4.3 Adaptive Noise Shaping

Of several methods of adaptive shaping of the quantization-noise spectrum, the pole-zero noise shaping method produces generally the best results (6, 7, 8). For this method, the noise-shaping filter B(z) is chosen as the special pole-zero filter $A(z/\alpha)/A(z)$, where $0<\alpha<1$ is a constant (5); $A(z/\alpha)$ is obtained from the expression (2) with z substituted by z/α. Use of this noise-shaping method produces about 2-3 dB decrease in S/Q ratio, but results in perceivable improvement in speech quality (6, 7). The speech quality improvement was the biggest for the entropy-coded APC systems.

4.4 Speech Quality Optimization

In one study of 16 kb/s APC coders with input speech corrupted by broadband noise and distortions typical in long-distance telephone circuits and with error-free transmission, it was found that using the entropy coding scheme (without pitch prediction) produced output speech perceptually indistinguishable from input speech, for signal-to-noise ratios of corrupted input speech in the range 10-30 dB (4).

In another investigation, an extensive parameter optimization and tradeoff study of 16 kb/s APC coders was conducted (6, 7, 8). Summarizing the results of this study, for high-quality input speech and error-free synchronous channels, the APC coder with entropy coding and 3-tap pitch prediction produced output speech with perceived quality essentially indistinguishable from that of the input speech (7). For noisy channels with a 1% bit-error rate, the APC coder with 3-tap pitch prediction and 3-segment segmented quantization produced the best speech quality (8). This PP-SQ system devotes about 1000 b/s for error-protection (via Hamming code) of important transmission parameters. Quite impressively, the perceived speech quality of this coder degrades only slightly as the channel bit-error rate is varied from 0% to 1%. This robust PP-SQ APC coder has been implemented on the CSPI MAP-300 array processor for full-duplex real-time speech communication (9).

5. BASEBAND CODERS

5.1 Coder Description

Figure 3 shows the overall block diagram of a BBC system based on the linear prediction method of spectral analysis. In the figure, we assume that the bandwidth of the input speech signal is W Hz and that the width of the baseband is B Hz (B<W). In the work reported here, the ratio W/B is assumed to be an integer, denoted by L and called the "number of bands". The terms in parentheses below the signal path in Fig. 3 represent the Nyquist frequency (half the sampling rate) at each of those points. Along the lower branch of the transmitter in Fig. 3(a), LPC parameters are extracted, quantized, and encoded. Along the upper branch, the linear prediction residual is computed by inverse-filtering the speech signal with the filter A(z) given by (2). A baseband is then extracted by lowpass filtering the residual. The extracted baseband signal is then decimated (down-sampled) to its Nyquist rate, quantized, and encoded.

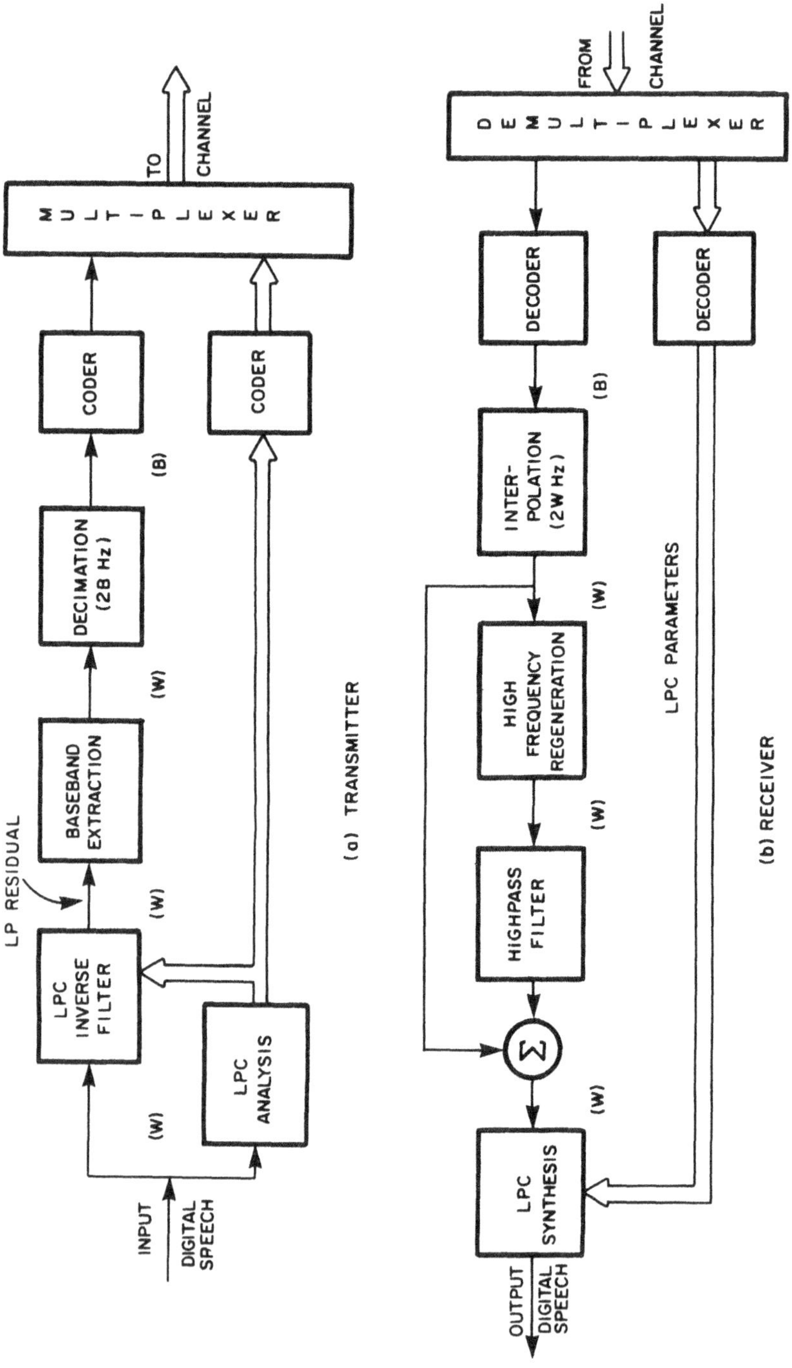

Fig. 3. Block diagram of a baseband coder.

The receiver of the BBC coder is depicted in Fig. 3(b). The received data are decoded into the baseband signal (top branch) and LPC parameters (bottom branch). The process of high-frequency regeneration (HFR) operates on the baseband to produce a signal with a fullband width W Hz. The HFR process generally involves some nonlinear operation on the baseband (such as waveform rectification), followed by spectral flattening of the distorted signal for the purpose of LPC synthesis. The baseband part of the nonlinearly regenerated fullband signal is. in general, distorted relative to the received baseband. This problem is corrected by reintroducing the received baseband, as shown in Fig. 3(b). The output of the HFR process is highpass filtered at a cutoff frequency of B Hz and added to the received baseband signal. The resulting signal is then applied to the all-pole LPC synthesizer 1/A(z), to produce the reconstructed output speech.

The data rate and the output speech quality of a BBC system are determined by the following items: (a) estimation and coding of LPC parameters; (b) baseband width or equivalently the number of bands, L=W/B; (c) coding of the baseband residual; and (d) high-frequency regeneration. In one study, an extensive investigation of these and other issues was conducted. With the goal of maximizing the output speech quality for a synchronous data rate of 9.6 kb/s (10,11,12) . Below, we discuss the issues of baseband coding and high-frequency regeneration and present the results of the speech-quality optimization study.

5.2 Coding of the Baseband Residual

The baseband residual can be coded using any of the available waveform coding techniques. Below, we consider the two techniques: Adaptive PCM (APCM) and Adaptive Predictive Coder (APC).

<u>Adaptive PCM Coder</u>. The APCM coder has a gain multiplier G followed by an optimum nonuniform unit-variance quantizer. G is varied adaptively by setting its value equal to the reciprocal of the rms value of the baseband residual over the current frame. A Laplacian quantizer has been found to produce slightly better speech quality than Gaussian and gamma quantizers (10, 12).

<u>APC Coder</u>. The APC coder shown in Fig. 4 has a feedback structure with an APCM quantizer placed in the forward path and an adaptive pitch predictor placed in the feedback path around the quantizer. The quantizer gain G in Fig. 4 is computed from the "second residual" that is obtained by filtering the baseband residual with the pitch-inverse filter C(z). Both 1-tap and 3-tap pitch predictors were considered in the study mentioned

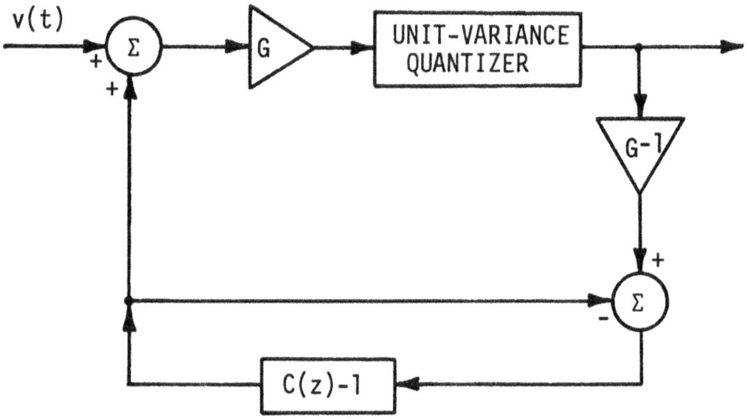

Fig. 4. Adaptive predictive coding of the baseband residual v(t). C(z) is the pitch-inverse filter.

above (10, 12). The 3-tap filter C(z) is given by (4). For the 1-tap filter, c1 and c3 in (4) are both set to zero. Notice that M in (4) is now the pitch period in number of baseband samples.

Experimental investigations have shown that the APC coder produces a significant improvement in speech quality relative to the APCM coder (10, 12). Under the constraint of a fixed data rate of 9.6 kb/s, the 1-tap APC coder yields slightly better speech quality than the 3-tap coder (10, 12).

5.3 High-Frequency Regeneration

The HFR problem is stated as follows: Given the baseband residual, generate high frequencies such that the fullband excitation signal has a flat spectrum and proper harmonic structure for voiced sounds. The well-known waveform rectification method of accomplishing HFR produces audible roughness or background noises in the output speech and is computationally rather expensive (11, 13). Below, we describe two other HFR methods.

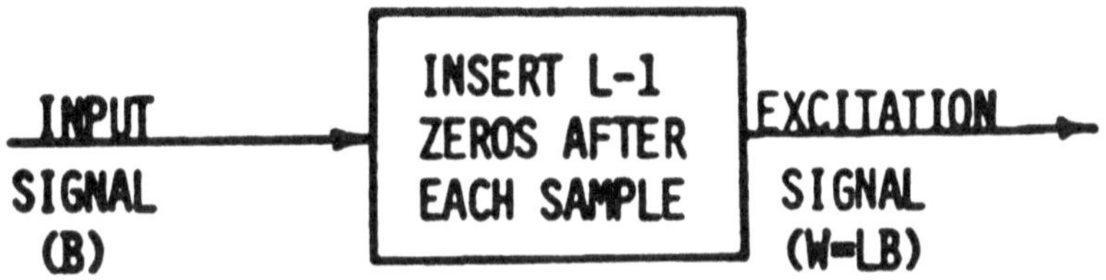

(a) Method

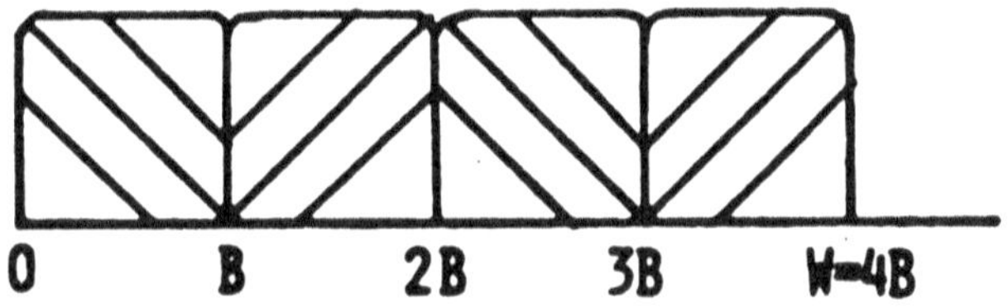

(b) Frequency-domain interpretation for the example L=4

Fig. 5. High-frequency regeneration by spectral folding.

<u>Spectral Folding</u>. In this method, which is depicted in Fig. 5(a), the HFR process reduces to inserting L-1 zeros after every baseband sample (13, 14). This upsampling operation aliases the baseband into the highband, as shown in Fig. 5(b). Clearly, the spectral folding method is quite simple to perform. It produces generally good speech quality. However, the output speech contains "tonal noises" that increase with L and pitch of the speaker (11, 13).

Recently, several HFR methods were proposed to overcome the problems of the simple spectral folding method (10, 15). The best of these methods is described below.

<u>Perturbed Spectral Folding</u>. Simple spectral folding exhibits a spectral regularity (as may be seen in Fig. 5(b)), which may be responsible for some of the tonal noises. The perturbed spectral folding method breaks up this spectral regularity. This method is illustrated in Fig. 6. In the lower

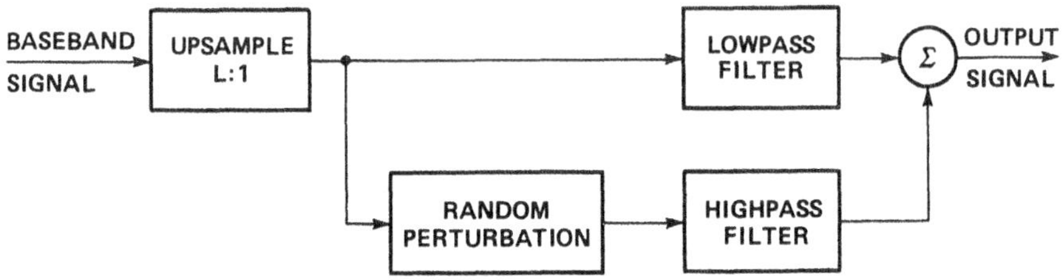

Fig. 6. Perturbed spectral folding method of high-frequency regeneration.

branch in Fig. 6, a small number of selected nonzero samples of the upsampled baseband residual are randomly perturbed by simply interchanging them with adjacent (zero) samples. The perturbation procedure is such that smaller samples have a higher probability of being perturbed. The perturbed signal is then highpass filtered and added to the received baseband residual at the full sampling rate of 2W Hz (top branch in Fig. 6), to produce the fullband excitation signal. The perturbed spectral folding method essentially adds a signal-correlated noise to the spectrally-folded residual. Perceptually, this additive noise has the effect of masking the tonal noises. The additive noise also causes slight roughness.

In a formal subjective speech quality test of several HFR methods including waveform rectification and spectral folding, the perturbed spectral folding method was found to produce the best speech quality (10, 12).

5.4 Speech-Quality Optimization

An extensive investigation of the BBC systems was carried out to maximize the output speech quality (10, 11, 12). This investigation included optimization of parameters of individual methods for dealing with each of the various aspects of the coder, such as a given HFR scheme, a given baseband coding method, etc. Under the constraint of a fixed data rate of 9.6 kb/s, tradeoffs were considered between various coder parameters such as number of quantizer levels/sample, number of bands L, frame size, etc. In this investigation, informal listening tests were used to judge the relative speech quality of competing coders.

Below, we briefly describe the optimized coder for each of the two cases considered in the investigation mentioned above: (1) W = 4 kHz, error-free transmission channel (11); and (2) W = 3.33 kHz, channel bit-error rate = 1% (10, 12). The optimized coder in either case employs 1-tap APC for baseband coding and perturbed spectral folding for HFR. For Case 1, the optimized coder has the following additional specifications: Frame size = 25 ms; LPC order = 10; L = 4; and number of quantization levels per baseband sample = 12. Similarly, the optimized coder for Case 2 has the following additional specifications: Frame size = 27 ms; LPC order = 8; L = 3; and number of bits per baseband sample = 3. The second coder devotes about 11.5% of the total data rate for error protection of important transmission parameter data (12), and it has been recently implemented on the CSPI MAP-300 array processor for full-duplex real-time speech communication (9).

6. LINEAR PREDICTIVE CODERS

6.1 Coder Description

In an LPC coder, the input speech is analyzed one frame at a time to extract the excitation and spectral parameters: the V/U status (or the mixed-source cutoff frequency), the pitch frequency if the frame is voiced, the gain G in Equation (3), and the p predictor coefficients a(k) in (1). The value of the LPC order p depends upon the bandwidth of the input speech signal; p is typically 8-14. These coder parameters are extracted at a fixed rate in the range 33-100 fps (frames per second). The extracted parameters are quantized, encoded, and transmitted. At the receiver, the output speech is produced using the synthesis model discussed in Section 2; the parameters of the excitation signal and the spectral shaping filter are updated using the received parameter data. Below, we discuss quantization of LPC parameters and describe several techniques to lower the transmission data rate of LPC coders without causing any additional speech-quality degradation at the output, by exploiting the redundancies that exist in the speech.

6.2 Quantization of LPC Parameters

Logarithmic quantization of pitch and gain yields satisfactory results. There are, however, a number of ways the coefficients of the linear prediction all-pole filter can be quantized. We investigated the relative quantization properties of several alternate but equivalent sets of parameters and showed that the reflection coefficients have the best quantization properties (16). Under a minimax spectral error criterion, we also derived an optimal quantization procedure for the reflection coefficients. In this derivation, we employed a fundamental and useful concept called spectral sensitivity of a linear prediction parameter (reflection coefficient or any other equivalent parameter), which is a measure of the extent by which the spectrum of the linear predictor is altered due to an incremental change in the value of that parameter. The optimal quantization of the reflection coefficients requires first transforming them nonlinearly to log area ratios (LARs)

$$g(i) = 10 \log_{10}[1 + K(i)]/[1-K(i)], \tag{6}$$

where K(i) are the reflection coefficients and then quantizing the LARs uniformly. Assuming that each LAR has a flat spectral sensitivity behavior over its range of values, we developed a method for optimally allocating a given number of bits among the p LARs; this method calls for the use of the same quantization step size for the different LARs (16). Subsequent work, in which we employed experimentally obtained spectral sensitivity characteristics and probability histograms of LARs, has shown that better LAR quantization is achieved using unequal step sizes, with smaller step sizes for LARs with lower indices (17).

6.3 Variable-Order Linear Prediction

It is known that different speech sounds can be represented adequately by different order linear predictors. Thus, a method that is better than sending a fixed number of parameters for every frame is to transmit for every frame the minimum number of predictor parameters that adequately represent the speech spectrum in that frame. This method, when implemented with an information theoretic criterion to determine the "optimal" order, produces average savings of about 10% in the transmission rate (3, 18). A simplified version of this method is to employ a fixed higher order for voiced frames and a fixed lower order for unvoiced frames. For example, the U.S. Government standard 2.4 kb/s LPC-10 coder uses 10 coefficients for voiced frames and 4 coefficients for unvoiced frames (19).

6.4 Variable Frame Rate Transmission

In our work on developing minimally redundant narrowband speech transmission systems, we have used quite successfully the concept of variable frame rate (VFR) transmission (17, 18, 20). In a VFR scheme, the LPC model parameters (LARs, pitch, and gain) are transmitted only when the properties of the speech signal have changed sufficiently over the transmission interval since the preceding transmission; the parameters for the untransmitted frames are regenerated at the receiver through linear interpolation between the parameters of the two adjacent transmitter frames. For example, speech parameters may be transmitted less often during steady-state portions of speech and more often during rapid speech transitions.

For the input speech signal lowpass filtered at 5 kHz and sampled at 10 kHz, we extracted LPC parameter data at a rate of 100 fps. The first version of our VFR scheme reduced the transmission frame rate from 100 fps to an average of about 37 fps, with only a slight reduction in the quality of the resynthesized speech relative to the case when all the available 100 fps data were transmitted (18).

In an effort to reduce the average frame rate further, without any speech quality degradation, we subsequently developed a more efficient VFR scheme, based on the following functional perceptual model of speech (17, 21):

1. Speech can be represented in terms of LPC parameters extracted at a minimal set of perceptually significant time points (or frames) not necessarily equally spaced.

2. Between any two such time points, the parameters vary linearly.

3. The location of these points is obtained independently for pitch, gain, and spectral (or LAR) parameters.

Our requirement was that the quality of the resynthesized speech based on this model should be no worse than that of the unreduced or the full 100 fps case. We experimentally demonstrated the validity of the above model by using a manual, trial-and-error scheme, and we achieved a lower limit for the frame rate of about 2 transmissions per phoneme, or about 24 fps (21). We then developed a fully automatic scheme which meets the model requirements and which uses only the information about the transmission parameters: LARs, pitch, and gain (17). To transmit to the receiver the timing information as to when the parameters are transmitted, the VFR scheme transmits a 3-bit header for every analysis frame. The first header bit is 1 if pitch is transmitted for that frame; similarly, the second and third header bits are used to indicate if gain and LARs, respectively, are transmitted for that frame. For an analysis rate of 100 fps, the header transmission requires 300 b/s.

Using the new VFR scheme, we obtained an average transmission rate of about 2100 b/s for continuous speech (i.e., not including silences). In a formal subjective speech-quality test, the speech quality of the above VFR system was found to be equal to or better than that of the 5700 b/s fixed-rate (100 fps) system (17, 20).

6.5 Variable Wordlength Encoding

After the LARs, pitch, and gain are quantized, the quantized levels are encoded into bits and transmitted. Encoding is the last step to redundancy removal before transmission. We have used Huffman coding to remove the redundancy due to the nonflat probability distribution of the quantized levels for each of the parameters (18). This technique encodes the higher-probability levels with fewer bits, thus producing a variable wordlength code with a minimum average wordlength. Using Huffman coding, we achieved savings of about 20% in the average transmission rate. It should be pointed out that no compromise whatsoever in speech quality is made when employing Huffman coding, because the coding does not result in any information loss; it merely transmits the information more efficiently.

6.6 Discussions

When the compression techniques described in Sections 6.3 to 6.5 are used together, the average transmission rate reduces to about 1500 b/s for continuous speech (17, 18). With explicit silence detection and elimination from transmission, this rate should reduce to less than 1000 b/s for conversational speech.

The variable-rate data resulting from the use of the above-described compression techniques is naturally suited to packet-switched asynchronous channels (e.g., ARPANET computer network), since they can handle variable-length bit streams. For applications involving synchronous or fixed-rate channels, the variable-rate data must be interfaced to the channel via a variable-to-fixed rate conversion mechanism involving a tandem of transmit and receive buffers (22, 23).

In one study involving the design of a robust LPC coder for operating over 2.4 kb/s noisy channels with bit-error rates of up to 5%, the techniques of VFR transmission and variable-order linear prediction were employed to achieve bit savings to be used for error-protection (via Hamming code) of important transmission parameters (22, 23).

7. CLUSTERING VOCODER

In a clustering vocoder, the spectral parameters are not quantized individually as in the LPC coder; rather the spectral parameters for a single frame are treated as one entity, which is compared against a stored set of reference spectral parameter patterns or templates. The template that, in some predetermined sense, is "closest" to or best matches the computed parameter set is identified, and its code is transmitted to the receiver. The receiver reconstructs the speech signal by using this surrogate spectral pattern. If the ensemble of stored templates is reasonably small in number, the transmitted code contains only a small number of bits, which produces a low data rate transmission of speech.

The cluster-coding approach has been used in the development of a 900 b/s channel vocoder (24) and, more recently, in the development of 800-1200 b/s LPC coders (25). In an on-going project at BBN, the techniques of cluster coding and variable frame rate transmission (see Section 6.4) have been combined to produce satisfactory speech quality and intelligibility at about 700 b/s; in this system, the spectral parameters (LARs) are coded via 10-bit cluster coding.

8. DIPHONE VOCODER

The block diagram for this vocoder is shown in Fig. 7. The vocoder extracts and transmits a sequence of phonemes. It also transmits the phoneme durations and a single pitch value for each voiced phoneme to preserve the intonation in the input speech. For an average speaking rate of 12 phonemes per second, the identity, duration, and pitch of the phonemes can be transmitted at an average rate of about 100 b/s (26).

The basic speech unit chosen for use in both the analyzer and synthesizer is the diphone. A diphone is defined as the region from the middle of one phoneme to the middle of the next phoneme. The diphone is a natural unit for synthesis, because the coarticulatory influence of one phoneme does not usually extend much further than half way into the next phoneme. Since diphone junctures are often at articulatory steady states, minimal smoothing is required between adjacent diphones. Also, the difficult task of representing phoneme transitions by acoustic-phonetic rules is avoided.

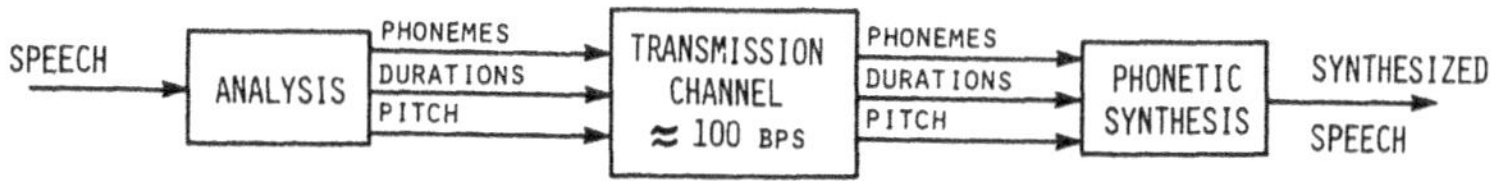

Fig. 7. Block diagram of a diphone vocoder.

Below, we describe the results of the work, performed as part of an on-going project at BBN, on the diphone synthesis and diphone analysis components of the phonetic vocoder (27, 28). At the present time, both components are being designed for a single speaker.

8.1 Diphone Synthesis

Since the design of the analyzer depends critically on the synthesizer model, we discuss the synthesizer first. Fig. 8 shows a block diagram of the synthesis program. The phonetic synthesizer (27) uses the transmitted phonemes, duration, and pitch values, to produce a sequence of control parameters (LPC parameters, voicing, pitch, gain, mixed-source cutoff frequency) for an LPC synthesizer. These parameters are stored in a large inventory of templates - one for each diphone. A template contains, for each 10 ms frame, 14 LAR coefficients and a value of gain. The diphone inventory has been selected to account for many of the stronger coarticulation effects in English (27). The diphone templates are extracted from a carefully constructed set of short utterances. The current total diphone synthesis inventory is approximately 2850 diphones.

The first step in the synthesizer is to translate the input phoneme sequence into a diphone sequence. In some cases, there are multiple diphone templates for a single phoneme pair. The template used is determined by the surrounding phonetic context. Each of the diphone templates is warped in time so that the input phoneme duration requirements are satisfied. The time-warping takes into account the relative inelasticity of the phoneme transition region. Next, the program smooths between

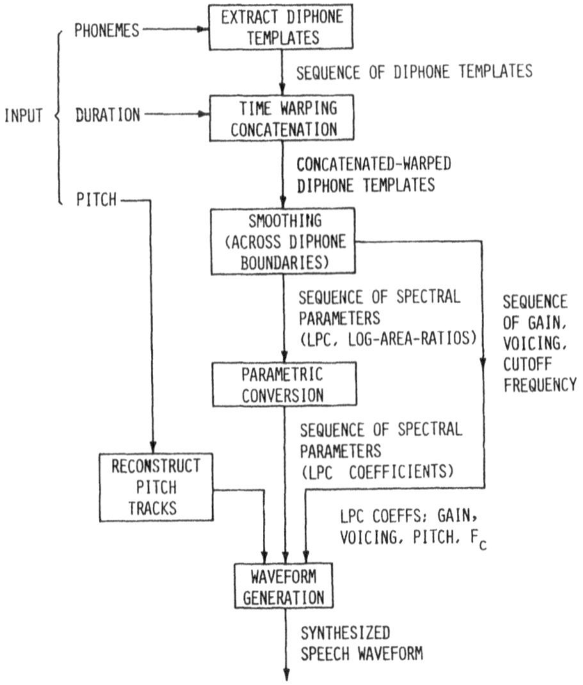

Fig. 8. Diphone synthesis method.

consecutive diphone templates to minimize gain and spectral discontinuities. The smoothing algorithm is designed to preserve the original parameter tracks intact where possible. Continuous pitch tracks are reconstructed by linear interpolation of the sequence of single pitch values, one for each phoneme. The cutoff frequency and voicing flag for each frame are determined by rule from the phonemes being synthesized. The resulting continuous parameter tracks (specified every 10 ms) are used to control an LPC speech synthesizer.

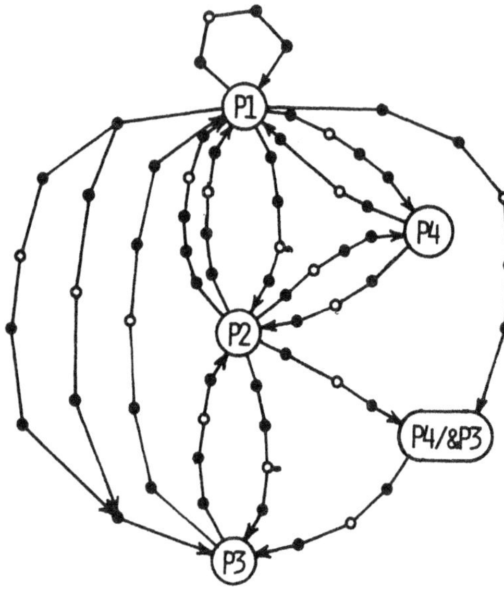

Fig. 9. Example of diphone network for four phonemes.

8.2 Diphone Analysis (Recognition)

The analyzer takes into account the synthesis model by using a network of diphone templates to recognize the sequence of phonemes in the input speech. The diphone network consists of nodes and directed arcs. An example of a simple network is shown in Fig. 9. There are two types of nodes: phoneme nodes and spectrum nodes. The phoneme nodes (shown as labelled circles) correspond to the midpoints of the phonemes; there is one such node for each phoneme. These phoneme nodes are connected by diphone templates. Each diphone template is represented in the network as a sequence of spectrum nodes (shown as dots). When two or more consecutive spectra in the original diphone template are very similar, they are represented by a single spectrum node in the network. The open dots indicate the first spectrum node in the original diphone template that is at or past the labelled phoneme boundary. Note that, in Fig. 9, the diphone template P1-P2 is distinct from template P2-P1. Also note the possibility of diphones of the type P1-P1. The network allows for two or more templates going from one phoneme to another (e.g., P2-P1). Branching and merging of paths within a template is also allowed (e.g., P1-P3). Finally, the network allows the specification of diphones in context. The node P4/&P3 represents the phoneme P4 followed only by P3. Thus the template P2-P4/&P3 can be different from unconditioned template P2-P4. The generation and training of the network is discussed further in the next section. The analyzer chooses the sequence of templates that best matches the input speech according to a weighted Euclidean LAR distance measure. When phonemes are incorrectly recognized, the above procedure will still produce output speech that is spectrally close to the input. Therefore, the "incorrect" phonemes are expected to differ from the "true" phonemes in a minimal number of features.

The network matcher uses a dynamic programming algorithm which attempts to find the sequence of templates in the network that best matches the input. The basic operation of the program begins by updating each "theory" by the addition of the newest input frame. A theory consists of a detailed account of how a sequence of input frames is aligned with the network, along with a total score for that correspondence. The program allows more than one input frame to be aligned with a single spectrum node, or conversely, up to two network nodes to be aligned with a single input frame. Therefore, each theory is replaced by at least three new theories: one that corresponds to the new input frame being aligned with the same node as the previous frame, and one for each permutation of one or two successive network nodes immediately following that most recent node. The program then discards all theories whose score is less than the best theory by more than a threshold.

To decide on the phonemes to transmit, the program examines all the remaining theories after updating each theory with the newest input frame. If all the theories have a common beginning, the phonemes corresponding to this common beginning are transmitted. Experiments have shown that preserving a few hundred theories for each frame assures that the program will find a path through the network that is close to optimum. As the template statistics are updated through training, it is hoped that the number of theories required for this "beam search" will decrease somewhat. Thus, the analyzer requires a variable lag between the input and the transmitted output. The program can also be given a maximum delay parameter which forces a decision if the transmission delay exceeds a threshold. These two features allow the analyzer to operate in a continuous mode, rather than on one sentence at a time.

8.3 Network Generation and Training

Initially, the network was generated by duplicating all the templates used by the diphone synthesizer. At the present time, the network is being augmented by the addition of new paths as a result of extensive training. This process allows the network to represent a variety of pronunciations for each diphone and to account for multiple speakers.

An initial version of the diphone recognition program has been designed and implemented, but will need extensive training before conclusions can be drawn. However, preliminary results are encouraging. The goal of this on-going work is to achieve a phoneme recognition accuracy of over 80%, which is required to produce intelligible speech (26).

9. SUMMARY

We described two types of compression systems for medium bit rate speech transmission: 16 kb/s APC and 9.6 kb/s BBC and three types of systems for low bit rate transmission: 1.5-2.4 kb/s LPC, 600-800 b/s clustering vocoder, and 100 b/s diphone vocoder. Considering the quality of the transmitted speech, APC produces speech that is approximately equivalent to the input speech; BBC produces good speech quality that may be termed "communications quality"; and the three low-bit-rate systems produce varying degrees of synthetic speech quality.

In some applications, the channel may call for the use of a wide range of bit rates as a means of coping with varying data traffic conditions. In such applications, a multirate speech coder may be used; embedded in this coder design are several speech coders at different bit rates (29).

ACKNOWLEDGMENTS

The authors would like to thank their colleagues M. Berouti, E. Blackman, K. Field, A. Higgins, A.W.F. Huggins, S. Roucos, W. Russell, and J. Wolf, for their contributions to various aspects of this research. The work reported in this paper was sponsored by different agencies of the Department of Defense including the Advanced Research Projects Agency and the Defense Communications Agency.

REFERENCES

1. J. Makhoul, R. Viswanathan, R. Schwartz and A.W.F. Huggins, "A Mixed-Source Model for Speech Compression and Synthesis." J. Acoust. Soc. Amer., Vol. 64, Dec. 1978, pp. 1577-1581.

2. A.W.F. Huggins, R. Schwartz, R. Viswanathan and J. Makhoul, "Subjective Quality Testing of a New Source Model of LPC Vocoders," Presented at the 96th Meeting of the Acoust. Soc. Amer., Honolulu, Hawaii, Nov. 27-Dec. 1, 1978.

3. J.Makhoul, "Linear Prediction: A Tutorial Review," Proc. IEEE, Vol. 63, April 1975, pp. 561-580.

4. J. Makhoul and M. Berouti, "Adaptive Noise Spectral Shaping and Entropy Coding in Predictive Coding of Speech," IEEE Trans. Acoustics, Speech and Signal Processing, Vol. ASSP-27, Feb. 1979, pp. 63-73.

5. B.S. Atal and M.R. Schroeder, "Predictive Coding of Speech Signals and Subjective Error Criteria," IEEE Trans. Acoustics, Speech and Signal Processing, Vol. ASSP-27, June 1979, pp. 247-254.

6. R. Viswanathan, W. Russell, and A. Higgins,, "Design and Real-Time Implementation of a Robust APC Coder for Speech Transmission over 16 kb/s Noisy Channels." BBN Report No. 4565, Vol. I: Algorithm Design and Simulation, AD No. A096091, Final Report, Contract DCA100-79-C-0037, Dec. 1980.

7. R. Viswanathan, W. Russell, A. Higgins, M. Berouti and J. Makhoul, "Speech-Quality Optimization of 16 kb/s Adaptive Predictive Coders," IEEE International Conf. Acoustics, Speech and Signal Processing, Denver, CO, April 1980, pp. 520-525.

8. R. Viswanathan, W. Russell, and A. Higgins, "Noisy Channel Performance of 16 kb/s APC Coders," IEEE International Conf. Acoustics, Speech and Signal Processing, Atlanta, GA, April 1981, pp. 615-618.

9. J. Wolf and K. Field, "Real-Time Speech Coder Implementation at 9.6 and 16 kb/s," IEEE International Conf. Acoustics, Speech and Signal Processing, Atlanta, GA, April 1981, pp. 607-610, (A revised version has been accepted for publication in the April 1982 issue of IEEE Trans. Communication.).

10. R. Viswanathan, J. Wolf, L. Cosell, K. Field, A. Higgins, and W. Russell, "Design and Real-Time Implementation of a Baseband LPC Coder for Speech Transmission Over 9600 Bps Noisy Channels," Final Report, Contract No. DCA100-79-0003, Bolt Beranek and Newman Inc., BBN Report No. 4327, Vol I: ADA083079, Vol II: ADA083238, Feb. 1980.

11. R. Viswanathan, W. Russell, and J. Makhoul, "Voice-Excited LPC Coders for 9.6 Kbps Speech Transmission," IEEE International Conf. Acoustics, Speech and Signal Processing, Washington, DC, April 1979, pp. 568-571.

12. R. Viswanathan, A. Higgins, W. Russell and J. Makhoul, "Baseband LPC Coders for Speech Transmission Over 9.6 kb/s Noisy Channels," IEEE International Conf. Acoustics, Speech and Signal Processing, Denver, CO, April 1980, pp. 348-351.

13. J. Makhoul and M. Berouti, "High-Frequency Regeneration in Speech Coding Systems," IEEE International Conf. Acoustics, Speech and Signal Processing, Washington, DC, April 1979, pp. 428-432.

14. J. Makhoul and M. Berouti, "Predictive and Residual Encoding of Speech," J. Acoust. Soc. Amer., Dec. 1979, pp. 1633-1641.

15. A. Higgins, R. Viswanathan, and W. Russell, "New High-Frequency Regeneration (HFR) Techniques for Voice-Excited Speech Coders," J. Acoust. Soc. Amer., Vol. 66, Nov. 1979, pp. S22 (Abstract).

16. R. Viswanathan and J. Makhoul, "Quantization Properties of Transmission Parameters in Linear Predictive Systems," IEEE Trans. Acoustics, Speech and Signal Processing, Vol. ASSP-23, June 1975, pp. 309-321.

17. R. Viswanathan, J. Makhoul and A.W.F. Huggins, "Speech Compression and Evaluation," Final Report, Contract No. MDA903-75-C-0180, Bolt Beranek and Newman Inc., BBN Report No. 3794, ADA055019, April 1978.

18. J. Makhoul, R. Viswanathan, L. Cosell and W. Russell, "Natural Communications with Computers: Speech Compression Research at BBN," Bolt Beranek and Newman Inc., Report No. 2976, Vol. II, Dec. 1974.

19. T.E. Tremain. J.W. Fussell, R.A. Dean, B.M. Abzug, M.D. Cowing and P.W. Boudra, Jr., "Implementation of Two Real-Time Narrowband Speech Algorithms," Proc. EASCON '78, Washington, D.C., September 1978, pp. 698-708.

20. R. Viswanathan, J. Makhoul, R. Schwartz, and A.W.F. Huggins, "Variable-Frame-Rate Transmission: A Review of the Methodology and Application to Narrowband LPC Speech Coding," Accepted for publication in IEEE Trans. Comm., April 1982.

21. R. Viswanathan, J. Makhoul and R. Wicke, "The Application of a Functional Perceptual Model of Speech to Variable-Rate LPC Systems." IEEE International Conf. Acoustics, Speech and Signal Processing, Hartford. CT, May 1977, pp. 219-222.

22. R. Viswanathan, E. Blackman and J. Makhoul, "Variable Frame Rate Narrowband Speech Transmission over Fixed Rate Noisy Channels," EASCON'77 Record, Sept. 1977, p. 23-4.

23. E. Blackman, R. Viswanathan, W. Russell and J. Makhoul, "Narrowband LPC Speech Transmission over Noisy Channels." IEEE International Conf. Acoustics, Speech and Signal Processing, Washington, D.C., April 1979, pp. 60-63.

24. C.P. Smith, "Perception of Vocoder Speech Processed by Pattern Matching," J. Acoust. Soc. Amer., Vol. 46, No. 6, (Part 2) 1969, pp. 1562-1571.

25. A. Buzo, A.H. Gray Jr., R.M. Gray, and J.D. Markel, "Speech Coding Based Upon Vector Quantization," IEEE Trans. Acoustics, Speech, and Signal Processing, Vol. ASSP-28, Oct. 1980, pp. 562-574.

26. J. Makhoul, C. Cook, R. Schwartz, and D. Klatt, "A Feasibility Study of a Very Low Rate Speech Compression System," Report No. 3508 (NTIS NO. AD A044400), Bolt Beranek and Newman Inc., Cambridge, MA, Feb. 1977.

27. R. Schwartz, J. Klovstad, J. Makhoul, D. Klatt and V. Zue, "Diphone Synthesis for Phonetic Vocoding," IEEE International Conf. Acoustics, Speech and Signal Processing, Washington, DC, April 1979, pp. 891-894.

28. R. Schwartz, J. Klovstad, J. Makhoul, and J. Sorensen, "A Preliminary Design of a Phonetic Vocoder Based on a Diphone Model," IEEE International Conf. Acoustics, Speech and Signal Processing, Denver, CO, April 1980, pp. 32-35.

29. M. Berouti and J. Makhoul, "An Embedded-Code Multirate Speech Transform Coder," Proc. 1980 Int. Conf. Acoustics, Speech, and Signal Processing, Denver, CO, April 1980, pp. 356-359.

Algorithms and Devices for Pitch Determination of Speech Signals

Wolfgang J. Hess

Lehrstuhl für Datenverarbeitung, TU München
Munich, Fed. Rep. of Germany

Abstract. *In this paper the various pitch determination methods and algorithms (PDAs) are grouped into two major classes: time-domain PDAs and short-term analysis PDAs. The short-term analysis PDAs leave the signal domain by a short-term transformation. They supply a sequence of average pitch estimates from consecutive frames. The individual algorithm is characterized by the short-term transform it applies. The time-domain methods, on the other hand, track the signal period by period. Extraction and isolation of the fundamental harmonic, and investigation of the temporal signal structure are the two extremes between which most of these PDAs are found. After the review of these principles the paper finally discusses different application-oriented aspects, i.e., the role of the PDA in phonetics, education, phoniatrics, and speech communication systems.*

1. INTRODUCTION

Pitch (i.e., fundamental frequency F_0 and fundamental period duration T_0) takes on a key position in the acoustic speech signal. The prosodic information of an utterance is predominantly determined by this parameter. The ear is by an order of magnitude more sensitive to changes of fundamental frequency than to changes of other speech signal parameters (Flanagan, 1972; Flanagan and Saslow, 1958). The quality of

---

This paper is the revised version of a paper presented at the First European Conference on Signal Processing, Lausanne, September 1980.

J.-P. Haton (ed.), Automatic Speech Analysis and Recognition, 49–67.

vocoded speech is essentially influenced by the quality and faultlessness of the pitch measurement (Gold, 1977). Hence the importance of this parameter claims for good and reliable measurement methods.

At the first glance the task looks simple: one has just to detect the fundamental frequency of a quasi-periodic signal. Dealing with speech signals, however, the assumption of quasi-periodicity is often far from reality. Figure 1 shows an arbitrary (but typical) example of a speech signal. For a number of reasons, the task of pitch determination must be counted among the most difficult problems in speech analysis.

1) In principle, speech is a nonstationary process; the momentary position of the vocal tract may change abruptly at any time. This leads to drastic variations in the temporal structure of the signal, even between subsequent pitch periods.
2) Due to the flexibility of the human vocal tract and the wide variety of voices, there exist a multitude of possible temporal structures. Narrow-band formants at low harmonics (especially at the second or third harmonic) are an additional source of trouble.
3) For an arbitrary speech signal uttered by an unknown speaker, the fundamental frequency can vary over a range of almost four octaves (50 to 800 Hz). Especially for female voices, F_0 thus often coincides with the first formant (the latter ranging from about 200 Hz to 1400 Hz). This causes problems when inverse filtering techniques are applied.
4) The excitation signal itself is not always regular. Even under normal conditions, i.e., when the voice is neither hoarse nor pathologic, the glottal waveform exhibits occasional irregularities (Fujimura, 1968; Lieberman, 1963). In addition, the voice may temporarily fall into vocal fry (Hollien, 1974) which is a nonpathologic mode of voice excitation with irregular intervals between subsequent glottal pulses.
5) Additional problems arise in speech communication systems where the signal is often distorted or band-limited (for instance, in the telephone channel).

Literally hundreds of methods for pitch determination have been developed. This paper will give a unified survey of the prevailing principles. As to the realization, we shall not distinguish between a hardware device (whether analog or digital) and an algorithmic solution: they are all regarded as pitch determination algorithms (PDAs). Nor will we distinguish as to what is actually measured: the fundamental frequency F_0, the period duration T_0, or individual periods. F_0 and T_0 are interconnected by the relation

$$F_0 = 1/T_0$$

In case individual periods are measured, T_0 will be momentarily defined as the elapsed time from the beginning of one period until the beginning of the next one.

Pitch determination is one subproblem of voice source parameter determination. The other big subproblem is "where the periods are".

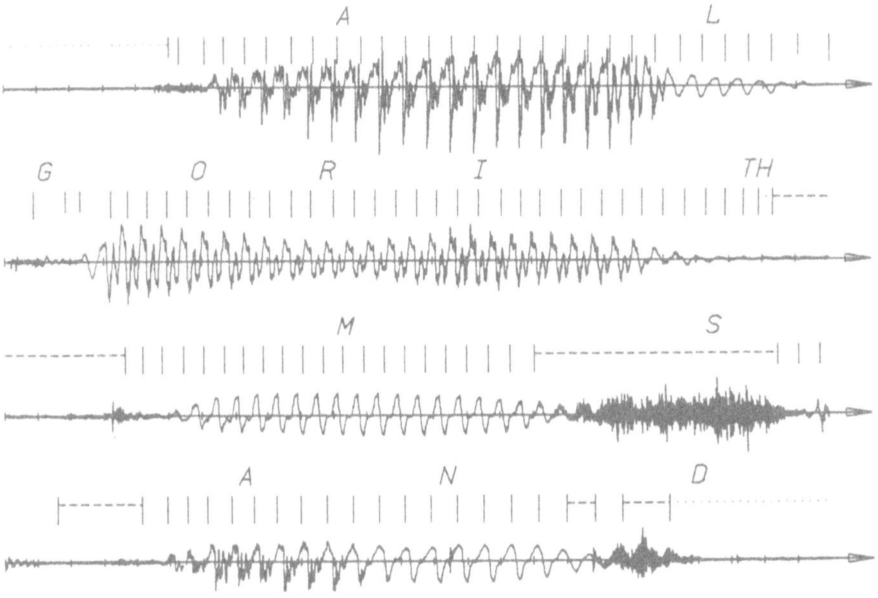

Fig.1. Example of a speech signal (after pitch determination). Beginning of the utterance "Algorithms and devices for pitch determination". Speaker: male; undistorted signal. Scale: 250 ms per line. The analysis was done using the algorithm by Hess (1979). ------ Voiceless, ... Pause, |||||| pitch period boundaries ("markers"). Markers indicated by short lines were found irregular.

This task, also referred to as voiced/unvoiced discrimination, will not be discussed here. In most PDAs it is actually done apart from the real pitch determination. So we will assume in the remainder of this paper that the voiced/unvoiced decision has been done already, and that there are only voiced signals at the input of the PDA to be analyzed.

In the following, we will first categorize the different methods of pitch determination. After that, the various categories will be briefly discussed. Finally we will shortly review possible application areas and the various requirements and presumptions associated therewith.

2. CATEGORIZING THE VARIOUS PRINCIPLES

A PDA can be subdivided into three steps of processing: a) the preprocessor, b) the basic extractor, and c) the postprocessor (McKinney, 1965). The basic extractor performs the very measurement: it converts the input signal into a series of pitch estimates. The task of the pre-

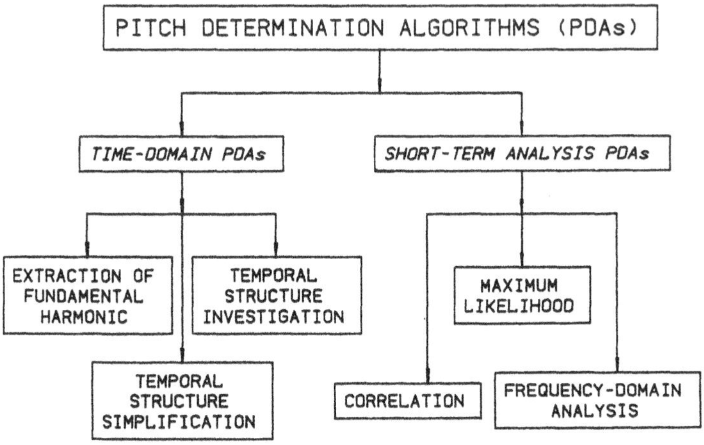

Fig.2. Principles of pitch determination

processor is data reduction in order to facilitate the operation of the basic extractor. The postprocessor operates in a more application-oriented way: error correction, smoothing of the pitch contour, and graphic display are some of its typical tasks.

The existing PDA principles can be split up into two gross categories when the input signal of the basic extractor is taken as a criterion. If this signal has the same time base as the original speech signal, the PDA operates in time domain. In all other cases, the time domain has been left somewhere in the preprocessor. Since the speech signal is time-variant, this cannot be done other than by a short-term transformation. Accordingly, we have the two PDA categories: a) time-domain PDAs, and b) short-term analysis PDAs. The schematic overview of Fig.2 already adds some aspects to this general classification that will be discussed in more detail in the following sections.

3. SHORT-TERM ANALYSIS PDAs

3.1 Principle of short-term analysis. In any short-term analysis (STA) PDA a short-term transformation is performed in the preprocessor step. For this the speech signal is split up into a series of frames (Fig.3). An individual frame is obtained by taking a limited number of consecutive samples from the starting point, K, to the ending point, (K+N-1). The frame length, N, is chosen short enough so that the parameter(s) to be measured can be assumed approximately constant within the frame. On the other hand, N must be large enough to guarantee that the parameter remains measurable. For most STA PDAs a frame thus contains two or three complete periods at least. In extreme cases, when F_0 changes abruptly, or when the signal is irregular, the contra-

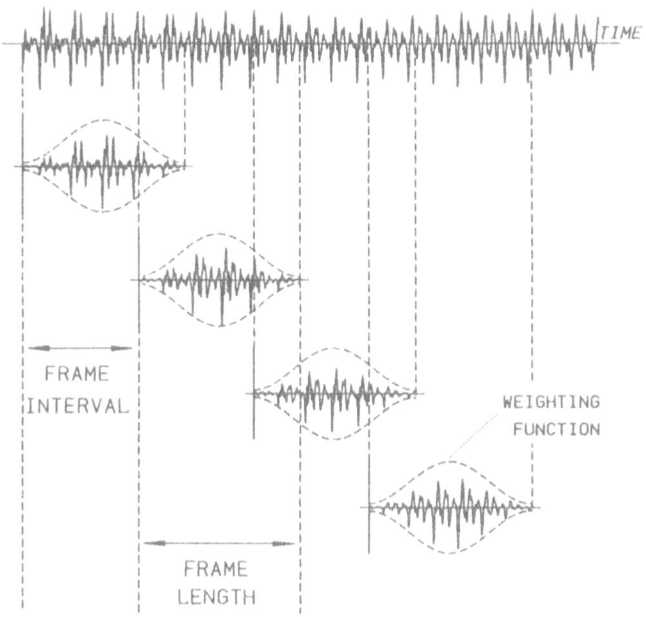

Fig.3. Principle of short-term analysis

diction of these two conditions can be a source of errors. The frame interval M, i.e., the distance between two consecutive frames (or its reciprocal, the frame rate), is determined in such a way that any significant parameter change is documented in the measurements. Like the original signal, the sequence of parameter measurements forms a sampled signal which obeys the sampling theorem, although the sampling rate, compared to that of the original signal, can be much lower. For transmission purposes it might be appropriate not to use a constant frame rate, but to adapt the frame interval to the course of the signal parameters. For the measurement task, however, the frame rate is assumed constant. Hence short-term parameter extraction can be regarded as a special case of downsampling.

The short-term transformation, so to speak, is intended to behave like a concave mirror which focuses all the scattered information on pitch, as far as it is available within the frame, into one single peak in the spectral domain. This peak is then determined by a peak detector (as the usual implementation of the basic extractor in this type of PDAs). Hence the output signal of the basic extractor is a sequence of average pitch estimates, not a period-by-period tracking of the signal. The short-term transform causes the phase relations between spectral domain and original signal to be lost. At the same time, however, the algorithm loses much of its sensitivity to phase distortions and signal degradation; in particular, it gets insensitive to low-

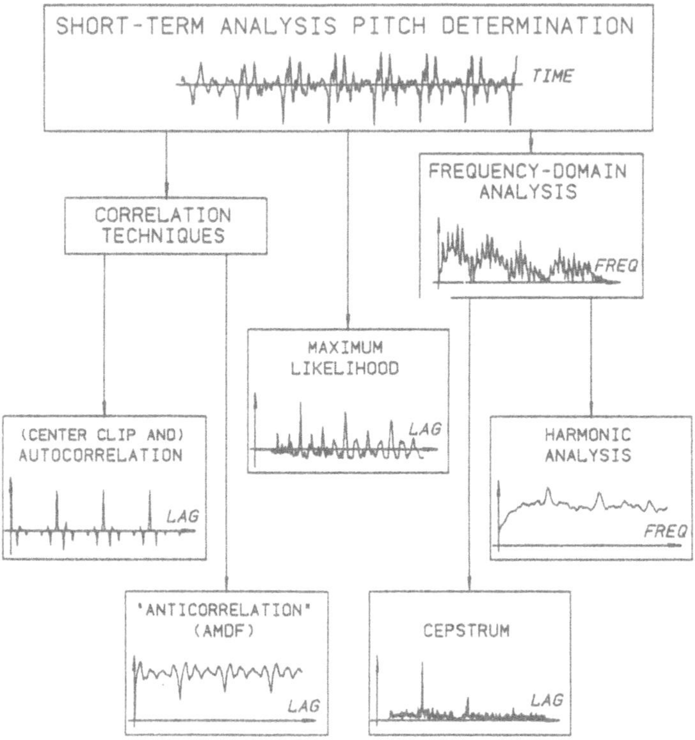

Fig.4. Methods of short-term analysis (STA) pitch determination

frequency band limitation, as it occurs in the telephone channel. Unfortunately the increased reliability of the algorithm is accompanied by an increased computing effort (which is at least one order of magnitude higher than for a time-domain PDA). Most of this effort goes into the numeric calculation of the transform. Besides the search for reliability, the search for a fast implementation has therefore been an important issue in the design of STA PDAs. The recent technological developments towards high speed and low-cost digital hardware solutions, however, decrease the relevance of this aspect.

3.2 The individual PDAs. Not all the known spectral transforms show the desired focusing effect. All the ones which do are in some way related to the power spectrum: correlation techniques, frequency-domain techniques, and the maximum likelihood approach (Fig.4). Among the correlation techniques we find the well known autocorrelation function (ACF) which became successful in pitch determination as soon as it was combined with center clipping of the signal (Sondhi, 1968; Galand et al., 1976; Rabiner, 1977). The computation of the ACF is significantly speeded up when additional peak clipping is performed (Dubnowski et al., 1976).

The counterpart of the ACF is given by the "anticorrelation", i.e., the average magnitude difference function (AMDF):

$$\mathrm{AMDF}\ (k) = \sum_{n} |x(n) - x(n+k)|$$

When the lag (delay time) k equals the period duration T_0, the AMDF will show a strong minimum. Since the AMDF needs no multiplication, it is fairly fast. In addition, the definition of the AMDF does not require (quasi-)stationarity; it can thus cope with very short frames of one pitch period or even less (Moorer, 1974; Ross et al., 1974). The AMDF principle has also been successfully applied to the LPC residual (Un and Yang, 1977).

The frequency-domain methods are also split up into two groups. Direct determination of F_0 as the location of the lowest peak in the power spectrum is unreliable. It is thus preferred to investigate the harmonic structure of the signal. One way to do this is spectral compression, which computes the fundamental frequency as the greatest common divider of all harmonics. The power spectrum is compressed along the frequency axis by a factor of two, three etc. and then

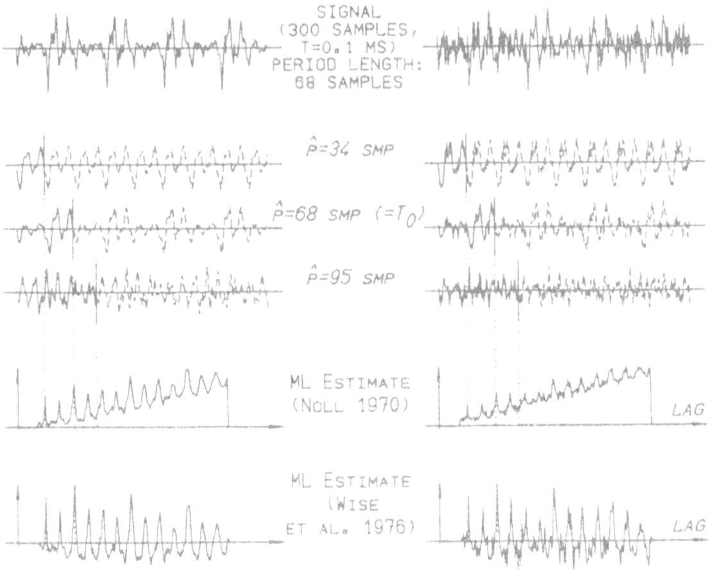

Fig.5. Example of a noise resistive PDA: the maximum likelihood method. Left part of figure: undistorted signal; right part: same signal with noise added, overall SNR: 0 dB. All the signals were normalized before plotting.

added to the original power spectrum. This operation gives a peak at F_0 resulting from the coherent additive contribution of the higher harmonics (Schroeder, 1968; Noll, 1970; Terhardt, 1979). A corresponding (analog!) technique was developed in the time domain using a filter bank (Miller, 1970). An alternative method which applies a majority decision principle for the distance of adjacent spectral peaks was proposed by Seneff (1978). The second frequency domain technique leads back into the time domain. Instead of transforming the power spectrum itself (which would lead to the ACF), however, the inverse transform is performed on the logarithmic power spectrum. This results in the well known "cepstrum" (Noll, 1967), which shows a distinct peak at the "quefrency" (lag) T_0.

Finally we have to mention the least-squares ("maximum likelihood") approach. This is originally a mathematical procedure to separate a periodic signal of unknown period duration T_0 (Noll, 1970) from Gaussian noise within a finite signal. Since neither the speech signal is periodic nor the background noise (plus the aperiodic components of the speech signal itself) can be expected as Gaussian, the approach has to be slightly modified in order to work in a PDA (Wise et al., 1976; Friedman, 1977, 1978). This PDA as well as the spectral compression method are extremely noise-resistive. Even at a signal-to-noise ratio of 0 dB (where the speech gets unintelligible) some of these PDAs can still operate. This has been verified by Noll (1970) as well as by Irwin (1979) for the maximum likelihood PDA (Fig. 5).

In summary, short-term analysis PDAs provide a sequence of average pitch estimates rather than a measurement of individual periods. They are not very sensitive to phase distortions and to absence of the fundamental harmonic. On the other hand, the computing effort is relatively high.

4. TIME-DOMAIN PITCH DETERMINATION METHODS

This group of PDAs is less homogenous than the short-term analysis methods. One possibility to split them up is according to the way how the burden of data reduction is distributed among the preprocessor and the basic extractor. Doing this, we find most time-domain PDAs between two extremes (Fig.6):

1) The whole burden is imposed on the preprocessor. In the extreme case, only the waveform of the first harmonic is offered to the basic extractor.
2) The whole burden is imposed on the basic extractor, which then has to cope with the whole complexity of the temporal signal structure. In the extreme case, the preprocessor is totally omitted.

The time-domain PDA is principally able to track the signal period by period. At the output of the basic extractor we find a sequence of period boundaries (pitch "markers"). Since the local information on pitch has to be taken from each period individually, time-domain PDAs are usually more sensitive to local signal degradations and thus less reliable than the majority of their short-term analysis counter-

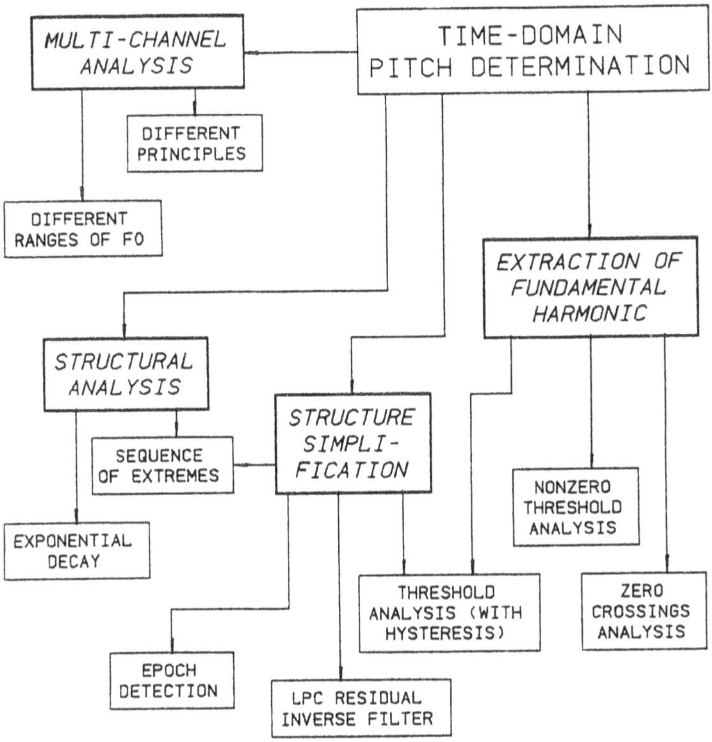

Fig.6. Principles of time-domain PDAs

parts. On the other hand, time-domain PDAs are at least in principle able to operate correctly when the signal itself is irregular due to temporary voice perturbation.

4.1 Structural Analysis. The pitch period is the truncated response of the vocal tract to an individual glottal impulse. Since the vocal tract behaves like a lossy linear system, its impulse response consists of a sum of exponentially damped oscillations. It is therefore to be expected that the magnitude of the significant peaks in the signal is greater at the beginning of the period than versus the end (Fig.7). Appropriate investigation of the signal peaks (maxima and/or minima) will thus lead to an indication of periodicity.

There are problems associated with this approach, however. First, the frequencies of the dominant damped waveforms are determined by the local formant pattern and may change abruptly. Second, the damping of the formants, particularly of a low first formant, is often quite weak and can be overrun by temporary changes of the signal level. Third, if the signal is phase distorted, different formants may

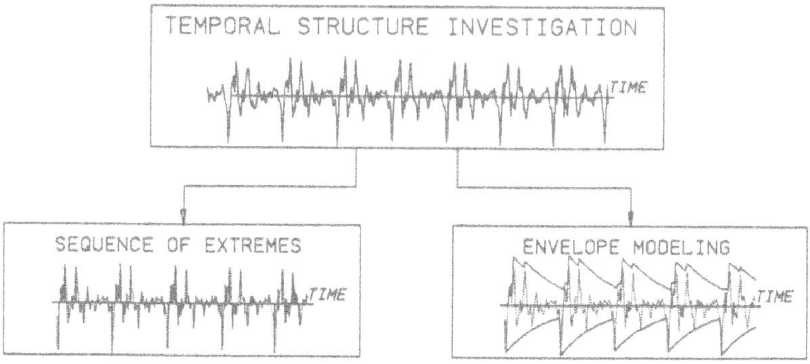

Fig.7. Time-domain PDAs: temporal structure investigation

be excited at different points in time. These problems are surmountable, but they lead to relatively complicated algorithmic solutions which have to regard a great variety of temporal structures. Since most of the program instructions are decisions, however, these PDAs run relatively fast. The usual way to carry out the analysis is the following (Reddy, 1967; Miller, 1975; Baudry and Dupeyrat, 1976; Tucker and Bates, 1978).

1) Do a moderate low-pass filtering to remove the influence of higher formants.
2) Determine all the local maxima and minima.
3) Exclude those extremes which are found insignificant until one significant point per period is left.
4) Reject obviously incorrect points by local correction.

For restricted applications these PDAs can be remarkably simple (Gold and Rabiner, 1969; Tillmann, 1978).

A different solution has been realized in the analog domain (Dolansky, 1955; Filip, 1969; Winckel, 1964; Zurcher, 1977). The envelope of the period is modeled by a diode-resistance-capacitance circuit with a short rise time constant and a comparatively long decay time constant. This circuit emphasizes the principal peaks of the signal and suppresses all the others. Unfortunately its performance strongly depends on the proper adjustment of the decay time constant. For that reason this relatively simple device works well only for a restricted range of F_0 (about 2 octaves). A manual range switch or the preset of an expected value is required if a wider range of F_0 is to be analyzed.

<u>4.2 Fundamental Harmonic Detection.</u> F_0 can be detected in the signal via the waveform of the fundamental harmonic. This harmonic is extracted from the signal by extensive low-pass filtering in the preprocessor. The basic extractor can then be relatively simple. Figure 8 shows the principle of three basic extractors: the zero crossings basic extractor (ZXABE) as the simplest device, the nonzero threshold basic

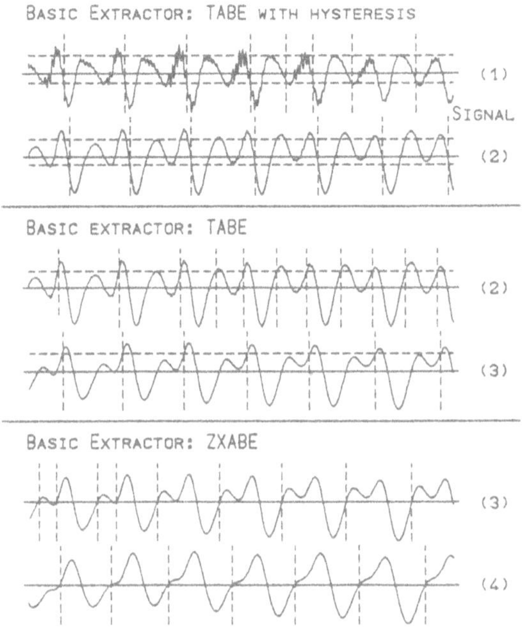

Fig.8. Example for the performance of basic extractors for fundamental harmonic extraction. ZXABE: Zero crossings analysis basic extractor, TABE: nonzero threshold analysis basic extractor. Signals: (1) original, (2) low-pass filtered at 6 dB/octave, (3) low-pass filtered at 12 dB/octave, (4) low-pass filtered at 18 dB/octave

extractor (TABE), and finally the TABE with hysteresis. The ZXABE sets a marker whenever the zero axis is crossed with a defined polarity. This requires that the input waveform has two and only two zero crossings per period. The TABE sets a marker whenever a given nonzero threshold is exceeded. The TABE with hysteresis acts like the normal TABE except that the marker is not set before a second (lower) threshold is crossed in opposite direction. This more elaborate device requires a lesser degree of low-pass filtering in the preprocessor.

The requirement of extensive low-pass filtering is one of two severe weak points of this otherwise fast and simple principle. For the ZXABE an attenuation of 18 dB/octave is necessary within the range of F_0 to be determined (McKinney, 1965; Erb, 1974). Accordingly, the amplitude of the signal at the basic extractor will vary by more than 50 dB due to the variations of F_0 alone. This dynamic range, increased by the intrinsic dynamic range of the signal (at least another 30 dB) is too much for the PDA to work correctly. The application of a ZXABE thus limits the possible range of F_0 and makes a manual switch or preset necessary. For the TABE the problem is not so acute, but the fact that the threshold must be adapted to the overall signal amplitude complicates the design of the PDA. - The second weak point is that this principle is a priori restricted to signals where the first harmonic is present. Low frequency band limitation, as occuring in telephone channels, causes these PDAs to fail systematically.

4.3 Simplification of the Temporal Structure. - The algorithms of this type take on some intermediate position between the principles of structural analysis and fundamental harmonic extraction. The majority of these algorithms (which are not too numerous) follow two principles: a) inverse filtering, and b) epoch detection. Both these principles deal with the fact that the laryngeal excitation function has a temporal structure which is much simpler and more regular than the temporal structure of the speech signal itself. The inverse filter approach cancels the transfer function of the vocal tract and thus reconstructs the laryngeal excitation function. If one is interested in pitch only and not in the excitation function itself, a crude approximation of the inverse filter is sufficient. Such an approximation is realized for instance when the analysis is confined to the first formant (Hess, 1976). The second principle, epoch extraction, is based upon the fact that at the beginning of each laryngeal pulse there is a discontinuity in the second derivative of the excitation function. Usually this discontinuity cannot be reliably detected in the speech signal due to phase distortions which occur when the signal passes the vocal tract. The signal is thus first phase shifted by 90° (applying a Hilbert transform). The squares of the original and the phase shifted signals are then added and yield a new signal which now shows a distinct peak at the time when the discontinuity in the excitation function occurs. The method works only when the spectrum of the investigated signal is flat to some extent. To enforce spectral flatness, the analyzed signal is for instance band-limited to high frequencies well above the narrow-band lower formants (Ananthapadmanabha and Yegnanarayana, 1975). Another way is is to analyze the LPC residual (Ananthapadmanabha and Yegnanarayana, 1979). The PDA by DeMori et al. (1977), finally, applies several independent epoch detectors, each of which covers a narrow band around at a different center frequency. At least one of these detectors is expected to operate in a region where the spectrum is sufficiently flat. This narrowband analysis works best when the center frequency of the individual epoch detector coincides with a formant.

The inverse filter approach has one weak point which occurs frequently with female voices. When F_0 is high, it may coincide with the

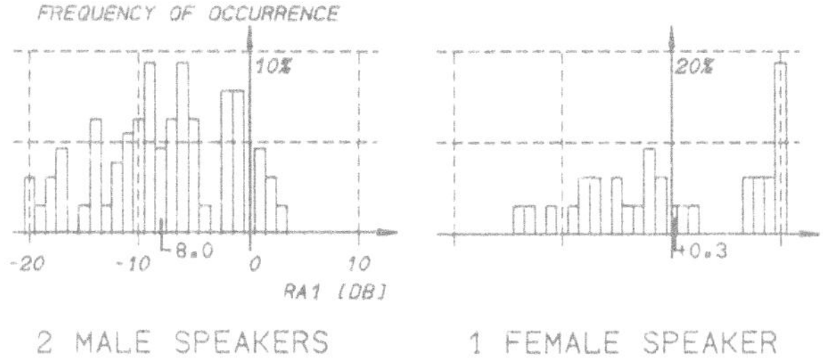

Fig.9. Distribution of the relative amplitude RA1 of the first harmonic for undistorted signals (three speakers)

first formant. If the inverse filter is not blocked, it removes the fundamental harmonic (which is extremely strong in this case) from the signal and brings the PDA into failure. The epoch detection principle, on the other hand, depends on the presence of a discontinuity in the second derivative of the laryngeal excitation function. This discontinuity is often weak, especially when speech is uttered with a soft or a falsetto voice. In the relevant publications on this subject, such signals are not investigated; the limitations of this principle thus have not yet been fully explored.

<u>4.4 Enhancement and reconstruction of the first harmonic</u>. If the principle of first harmonic extraction is to be applied to band-limited signals, the first harmonic must be enhanced or reconstructed, if missing. One way to do this is nonlinear distortion. In that respect, many proposals have been made from the beginning on (e.g., Grützmacher and Lottermoser, 1937; Risberg et al., 1960); much of this work has been reviewed by McKinney (1965). There is no uniform opinion, however, as to which nonlinear function (NLF) does this job in an optimal way. To further clarify this issue, the following experiment was carried out (Hess, 1979): Forty-eight isolated periods from 3 speakers (one female, two male) representing most of the vowels and sustained voiced consonants, were nonlinearly distorted by several NLFs: odd

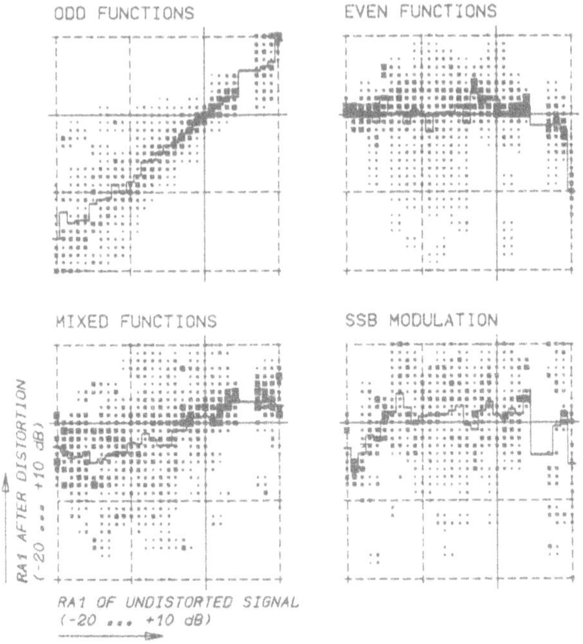

Fig.10. The effect of nonlinear distortion on the relative amplitude RA1 of the first harmonic

ones, even ones (such as full-wave rectification), "mixed" ones (neither even nor odd, e.g., half-wave rectification), and - finally - rectified SSB modulation (Risberg et al., 1960). The effect of these NLFs was then described measuring the relative amplitude RA1 of the first harmonic

$$RA1 := 20 \lg \frac{A(1)}{\max [\, A(i)\,]} \quad ; i = 2, 3, \ldots$$

In the experiment the frequencies of the higher harmonics were upward limited to 1200 Hz. Figure 9 shows the histogram of this parameter for the undistorted signal. There is considerable discrepancy between the data for the male and the female speakers. The results of the experiment are shown in Figure 10. The figure displays RA1 after distortion over RA1 of the undistorted signal. The odd functions, in general, show little effect. The even functions as well as the rectified SSBM greatly enhance the first harmonic when it is weak, but they attenuate it when it is dominant (for instance when a female voice is analyzed). The mixed NLFs are not optimal. Thus we can state that none of all these NLFs is able to always enhance the first harmonic of the signal in an optimal way.

4.5 Multi-channel approaches. Except for the algorithmic investigation of the temporal structure, most simple time-domain PDAs are restricted with respect to the range of F_0 or the type of signal to be processed. One way to increase the range or the reliability of these PDAs is to implement several of them in parallel and to perform some decision as to which one has the "correct" output. The partial PDAs may be identical in design, and each of them may process a subrange of F_0 (Léon and Martin, 1969; Dibbern, 1972). On the other hand, they may apply different principles without restriction of the frequency range. The PDA by Fant (1962), for instance, uses two, the one by Hess (1979) three nonlinear functions to enhance the first harmonic in different ways. The selection criteria in order to find the most likely channel are defined by a certain channel hierarchy (Léon and Martin, 1970), by a regularity check (Hess, 1979), or by syntactic rules (DeMori et al., 1977). The selection is continuously checked so that the PDA can change its choice at any time.

In summary, time-domain PDAs are mostly simpler and faster than short-term analysis PDAs. On the other hand, they are less reliable. Structural analysis algorithms are sensitive to phase distortions and short spurious signals, such as switching clicks. Many of the simpler PDAs can only process a limited range of F_0. The presence of the first harmonic is often required. A way out of this dilemma is the use of multi-channel PDAs where the advantages of several principles can be combined.

5. EVALUATION AND APPLICATION

If the performance of a measuring device is to be evaluated, one should have another instrument with at least the same accuracy. If this is not available, objective criteria are required in order to check and adjust the behavior of the new device. In pitch determination both these bases of comparison are difficult to generate, or they are even totally missing. There is no PDA which operates without errors, and there is such a variety of signals and marginal conditions that one cannot test any situation in advance. Thus it becomes understandable that designers of new PDAs have seldom provided detailed data on the performance of their algorithms. One of the rare studies which evaluates the performance of PDAs in detail was done by Rabiner et al. (1976) and McGonegal et al. (1977) in connection with a speech communication system. A reference pitch contour was obtained interactively using the semiautomatic PDA by McGonegal et al. (1975). Seven PDAs (two time-domain, five short-term analysis) were involved in this investigation. The main results read as follows.

1) None of the PDAs involved worked without errors, even under good recording conditions. Each PDA had its own "favorite" error; nevertheless, any error condition actually occurred for any of the PDAs.
2) The subjective evaluation does not match the preference of the objective evaluation. In fact, none of the objective criteria (number of gross errors, noisiness of the pitch contour, voiced/unvoiced errors) correlates well with the subjective scale of preference. Hence the question what errors in pitch determination are the really annoying ones for the human ear is still open.

The area of speech communication systems is one of the important application areas of pitch determination. There are other areas, however, which must not be neglected: a) phonetics and linguistics (includ-

Table 1. Profile of requirements for PDAs depending on the respective application. SPC: Speech communication; PHO: phonetics, linguistics, musicology; EDU: teaching aids, education; MED: medicine (phoniatrics). Ranking of importance: (1) mandatory, (2) important, (3) not negligible, (4) marginal, (5) insignificant

	SPC	PHO	EDU	MED
Accuracy	2	1–5	5	3
Quality and facility of display	5	1	1	1
Range of F_0	2	4	4	2
Robustness against signal degradation	1	3	5	5
Manual preset of range unnecessary	1	3	5	5
Cost of implementation	3	4	1	3
Real time performance	2	5–1	1	5
Ease of operation	5	2	2	3

ing musicology), i.e., the measurement of pitch contours as carriers of prosodic, phonetic, and musical information; b) education: training aids for the deaf as well as teaching aids for foreign languages; and c) the application in medicine as an aid in voice pathology and phoniatrics. Each of these applications has a different profile of requirements. Table 1 tries to compile and to weight some of these issues from the point of view of the respective application.

For all applications except speech communication, the display is a crucial question. For phonetics, on-line performance may be insignificant since the phonetician usually can wait for the results. There are some complementary tasks with respect to on-line performance and accuracy in this application area, however: analyses which need a great amount of data in a short time without a high accuracy required as well as analyses which need a high accuracy on a restricted amount of data (Lea, 1982). The student of a foreign language needs the results of his utterance at once; on the other hand, he doesn't need them very accurately. The same would apply to a device which is used as a visual aid for deaf patients*. The hardest requirements, however, are given in the speech communication area since the analysis is intended to be performed automatically, and since the human ear is a most critical judge. Hence most of the existing PDAs have been developed for this application, and most of the evaluations and reviews - including this one - have been made from this point of view. For applications in a speech communication system there are still open questions: none of all these principles works perfectly well for any situation, and it seems hard to believe that a new, revolutionary principle can completely solve the problem. The combination of several known principles with complementary behavior, however, together with a sophisticated control and selection logic, might form a new concept which would then permit research in pitch determination to advance one step further towards a general solution.

BIBLIOGRAPHY

ANANTHAPADMANABHA T.V., YEGNANARAYANA B. (1975): Epoch extraction of voiced speech. IEEE Transactions ASSP-**23**, 562-569

ANANTHAPADMANABHA T.V., YEGNANARAYANA B. (1979): Epoch extraction from linear prediction residual for identification of closed glottis interval. IEEE Transactions ASSP-**27**, 309-319

---

* This does not include some recent applications in this area: cochlear implants which provide deaf patients with information on pitch (Fourcin et al., 1978). In its profile of requirements this application is related to speech communication.

BAUDRY M., DUPEYRAT B. (1976): Analyse du signal vocal - Utilisation des extrema du signal et de leurs amplitudes - détection du fondamental et recherche des formants. Journ.Etud.Parole (GALF), vol. 7, 247-257

DeMORI R., LAFACE P., MAKHONINE V.A., MEZZALAMA M. (1977): A syntactic procedure for the recognition of glottal pulses in continuous speech. Pattern Recognition **9**, 181-189

DIBBERN U. (1972): Grundfrequenzmessung bei der menschlichen Sprache. In: Akustik und Schwingungstechnik. Berlin: VDE-Verlag; 345-348

DOLANSKY L.O. (1955): An instantaneous pitch period indicator. Journ.Acoust.Soc.Am. **27**, 67-72

DOLANSKY L.O., TJERNLUND P. (1968): On certain irregularities of voiced speech waveforms. IEEE Transactions AU-**16**, 51-56

DUBNOWSKI J.J., SCHAFER R.W., RABINER L.R. (1976): Real-time digital hardware pitch detector. IEEE Transactions ASSP-**24**, 2-8

ERB H.J. (1974): Ein Verfahren zur Bestimmung der Sprachgrundfrequenz in Echtzeit. Frequenz **28**, 23-28

FANT G. (1962): Speech analysis and synthesis: Technical (Final) Report. Stockholm: Speech Transmission Laboratory.

FILIP M. (1969): Envelope periodicity detection. Journ.Acoust.Soc.Am. **45**, 719-732

FLANAGAN J.L. (1972): Speech analysis, synthesis, and perception. Berlin, Heidelberg, New York: Springer.

FLANAGAN J.L., SASLOW M.G. (1958): Pitch discrimination for synthetic vowels. Journ.Acoust.Soc.Am. **30**, 435-442

FOURCIN A.J., ROSEN S.M., MOORE B.C.J., DOUEK E.E., CLARKE G.P., DODSON H., BANNISTER L.H. (1978): External electrical stimulation of the cochlea. In: Speech and Hearing - Work in progress. London: Dept. of Phonetics and Linguistics, University College; 3-32

FRIEDMAN D.H. (1977): Pseudo-maximum-likelihood pitch extraction. IEEE Transactions ASSP-**25**, 213-221

FRIEDMAN D.H. (1978): Multidimensional pseudo-maximum-likelihood pitch estimation. IEEE Transactions ASSP-**26**, 185-196

FUJIMURA O. (1968): An approximation to voice aperiodicity. IEEE Transactions AU-**16**, 68-72

GALAND C., ESTEBAN D., DUBUS F. (1976): Détection de la mélodie par autocorrélation non linéaire. Journ.Etud.Parole (GALF), vol. 7, 333-345

GOLD B. (1962): Computer program for pitch extraction. Journ.Acoust. Soc.Am. **34**, 916-921

GOLD B. (1977): Digital Speech Networks. Proceedings of the IEEE **65**, 1636-1658

GOLD B., RABINER L.R. (1969): Parallel processing techniques for estimating pitch periods of speech in the time domain. Journ. Acoust.Soc.Am. **46**, 442-448

GRÜTZMACHER M., LOTTERMOSER W. (1937): Über ein Verfahren zur trägheitsfreien Aufzeichnung von Melodiekurven. Akustische Zeitschrift **2**, 242-248

HESS W.J. (1976): A pitch-synchronous digital feature extraction system for phonemic recognition of speech. IEEE Transactions ASSP-**24**, 14-25

HESS W.J. (1979): Time-domain pitch period extraction of speech signals using three nonlinear digital filters. ICASSP-79, 773-776

HESS W.J. (forthcoming): Algorithms and devices for pitch determination of speech signals. Berlin, Heidelberg, New York: Springer.

IRWIN M.J. (1979): Periodicity estimation in the presence of noise. In: Proceedings of the 1979 Autumn Conference. Edinburgh: Institute of Acoustics; 55-58

KUBZDELA H. (1976): An analogue fundamental frequency extractor. In: Speech analysis and synthesis. edited by W.Jassem. Warsaw: Inst.of Fundamental Research, Polish Academy of Sciences; vol. 4, 269-279

LEON P., MARTIN PH. (1969): Prolégomènes à l'étude des structures intonatives. Paris, Montréal: Didier; Studia Phonetica, No. 2; 225 pp.

LEA W.A. (1982), this volume.

LIEBERMAN P. (1963): Some acoustic measures of the fundamental periodicity of normal and pathologic larynges. Journ.Acoust.Soc.Am. **35**, 344-353

MARKEL J.D. (1972): The SIFT algorithm for fundamental frequency estimation. IEEE Transactions AU-**20**, 367-377

McGONEGAL C.A., RABINER L.R., ROSENBERG A.E. (1975): A semi-automatic pitch detector (SAPD). IEEE Transactions ASSP-**23**, 570-574

McGONEGAL C.A., RABINER L.R., ROSENBERG A.E. (1977): A subjective evaluation of pitch detection methods using LPC synthesized speech. IEEE Transactions ASSP-**25**, 221-229

McKINNEY N.P. (1965): Laryngeal frequency analysis for linguistic research. Ann Arbor MI: Communic.Sciences Lab., Univ. of Michigan; Res.Rept. No.14; 340 pp.

MILLER N.J. (1975): Pitch detection by data reduction. IEEE Transactions ASSP-**23**, 72-79

MILLER R.L. (1970): Performance characteristics of an experimental harmonic identification pitch extraction system (HIPEX). Journ. Acoust.Soc.Am. **47**, 1593-1601

MOORER J.A. (1974): The optimum comb method of pitch period analysis of continuous digitized speech. IEEE Transactions ASSP-**22**, 330-338

MOSER M., KITTEL G. (1977): Rechnergestützte Tonhöhenbestimmung. Folia phoniatrica **29**, 119-126

NOLL A.M. (1967): Cepstrum pitch determination. Journ.Acoust.Soc. Am. **41**, 293-309

NOLL A.M. (1970): Pitch determination of human speech by the harmonic product spectrum, the harmonic sum spectrum, and a maximum likelihood estimate. In: Symposium on Computer Processing in Communication; edited by Microwave Institute. New York: The University of Brooklyn Press; vol. 19, 779-797

RABINER L.R. (1977): On the use of autocorrelation analysis for pitch detection. IEEE Transactions ASSP-**25**, 24-33

RABINER L.R., CHENG M.J., ROSENBERG A.E., McGONEGAL A. (1976): A comparative study of several pitch detection algorithms. IEEE Transactions ASSP-**24**, 399-413

REDDY D.R. (1967): Pitch period determination of speech sounds. Comm.of the ACM **10**, 343-348

RISBERG A. (1960): Voice fundamental frequency tracking. STL-QPSR, No.2

RISBERG A., MÖLLER A., FUJISAKI H. (1960): Voice fundamental frequency tracking. STL-QPSR, No.1, 3-5

ROSS M.J., SHAFFER H.L., COHEN A., FREUDBERG R., MANLEY H.J. (1974): Average magnitude difference function pitch extractor. IEEE Transactions ASSP-**22**, 353-361

SANCHEZ GONZALEZ F.J. (1977): Application of dissimilarity and a periodicity function to fundamental frequency measure of speech and voiced/unvoiced decision. In: Proc.of the 9th International Congress on Acoustics, Madrid 1977, paper I117, p.523

SCHROEDER M.R. (1968): Period histogram and product spectrum: new methods for fundamental-frequency measurement. Journ.Acoust.Soc. Am. **43**, 829-834

SCHULTZ-COULON H.J., ARNDT H.J. (1972): Tonhöhenschreibung in der phoniatrischen Praxis. Folia phoniatrica **24**, 241-258

SENEFF S. (1978): Real-time harmonic pitch detector. IEEE Transactions ASSP-**26**, 358-364

SONDHI M.M. (1968): New methods of pitch extraction. IEEE Transactions AU-**16**, 262-266

TERHARDT E. (1979): Calculating virtual pitch. Hearing Research **1**, 155-182

TILLMANN H.G. (1978): Bestimmung der Stimmperiode in der Zeitfunktion des digitalen Sprachschallsignals. IPSKMünchen, vol. 9, 207-213

TUCKER W.H., BATES R.T.H. (1978): A pitch estimation algorithm for speech and music. IEEE Transactions ASSP-**26**, 597-604

UN C.K., YANG S.C. (1977): A pitch extraction algorithm based on LPC inverse filtering and AMDF. IEEE Transactions ASSP-**25**, 565-572

WINCKEL F. (1964): Tonhöhenextraktor für Sprache mit Gleichstromanzeige. Phonetica **11**, 248-256

WISE J.D., CAPRIO J.R., PARKS T.W. (1976): Maximum likelihood pitch estimation. IEEE Transactions ASSP-**24**, 418-423

ZURCHER J.F. (1977): La mesure du fondamental par la détection de cretes. Techniques employés, résultats. Journ.Etud.Parole (GALF), vol. 8, 119-126

ACOUSTIC-PHONETIC DECODING AND ADAPTATION IN CONTINUOUS SPEECH RECOGNITION

Guy MERCIER

CNET 22301 - Lannion (France)

This is an overview of significant problems encountered in the attempt to automatically map continuous speech signals into phonemic labels. The first part is devoted to a general discussion of speech processing, feature extraction, units of recognition, coarticulation, segmentation, recognition strategies, training, normalization and speaker adaptation (chapter 2 to 9). In the second part, we describe how these early stages of analysis are performed in the KEAL speech understanding system, developed at CNET (chapter 10 to 14). The novelty of this phonetic recognizer mainly lies in the three following points : the strategy of recognition that takes into account different levels of recognition (syllables, phonemes, phones), the way by which segmentation and identification are combined, and the adaptation module which gives the phonetic processor leave to easily adapt to new speakers or to new tasks. Examples and results are given.

J.-P. Haton (ed.), Automatic Speech Analysis and Recognition, 69–99.

1 - INTRODUCTION :

In this communication, a general discussion of problems related to the process of transforming the continuous speech signal into a string or a lattice of discrete units will be presented.

We would like to emphasize that our main task is to recognize continuous utterances. The acoustic-phonetic processor is only one of the components of a more complex system. This component is an important one because the success of a general-purpose speech recognition system severaly depends upon the performance of the acoustic-phonetic processor.

This component can be completely integrated into the speech understanding system, as in the "I.B.M." , "Harpy", or Dragon systems where several knowledge-sources are introduced into a single network.

There are also many reasons why the acoustic-phonetic component should be conceived as a separate subsystem :

First, the knowledge sources involved in a speech understanding system are essentially different.

Secondly, conceiving the acoustic-phonetic component as an independent module allows more flexibility in design and calibration.

Thirdly, for limited speech recognition tasks, once this componenthas been calibrated, it can be integrated into a unified system.

The input speech to such a system is a time-varying waveform generally sampled and digitized by a speech encoder. The sampling rate may vary from 10 to 20 kHz. Our task is to find the boundaries between the consecutive speech sounds, and to assign them something like a phonemic label.

First we will try to set some of the problems we have to solve and secondly we will briefly present how this is done in our system.

2 - SPEECH REPRESENTATION AND ACOUSTIC PARAMETERS :

It is useful to divide an acoustic-phonetic recognizer into two parts : the acoustic processor and the phonetic analyzer.

For identifying speech sounds, it is possible to find some cues directly in the speech waveform. For instance, the quasi-periodic structure of certain portions of the speech signal indicate whether the segment is voiced or not.

However, the speech signal is usually further transformed in order to make other cues evident.

As we see in figure 1, various acoustic representations are possible. For all these representations, the speech waveform is analyzed for a short-time interval of about 20 ms and given a parametric representation each 10 ms. Within short intervals of the order of 10 ms the speech spectrum is assumed to be stationary, due to the inertia of speech articulators.

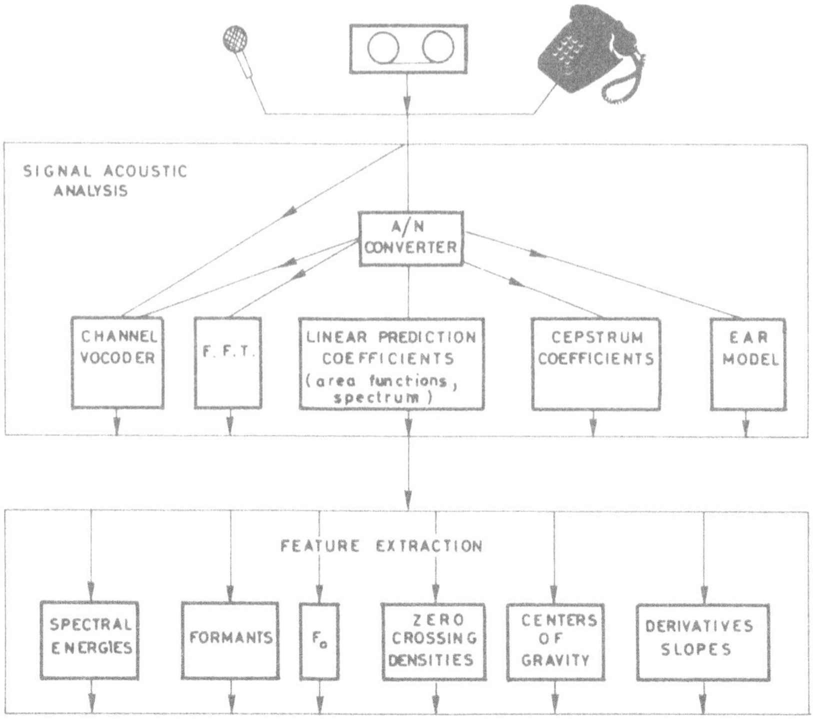

Fig. 1 - SIGNAL PROCESSING AND FEATURE EXTRACTION

Linear prediction analysis, F.F.T. and filter bank representations are most frequently used for speech recognition purposes ; homomorphic analysis (cepstrum coefficients) is also applied together with other methods.

All these different representations are of course closely related to each other, and some of them can be directly derived from others.

An interesting point in the cepstrum method is that the cepstral waveform decreases exponentially with time, and that only few coefficients are necessary to represent the speech sounds.

Another interesting viewpoint described by Davies and Mamelstein (1980) is to represent the spectrum in a mel-frequency scale (linear until 1000 Hz and logarithmic after) before applying the logarithmic transformation and the inverse F.F.T. or cosine-transform.

Mel-scale frequency cepstrum coefficients seem to be a good representation because excellent performances can be obtained with few coefficients and simple distance measurements, such as the euclidean distance (figure 2).

Many other important parameters can also be extracted and used for characterizing different speech sounds :

. The fundamental frequency of voice can be used as a voicing cue ; but its main significance is that the pitch contour carries prosodic information such as stress and intonation.

. The total energy curve is a good parameter for segmenting continuous speech into pseudo-syllables.

. The variations of the intensity in some frequency bands are also good cues for detecting boundaries between consecutive speech sounds.

. Formant frequencies, which are directly related to the vocal tract resonances, are of course among the most important parameters. It is known that vowels and sonorants are easily characterized by their formant frequencies, especially F1, F2, F3.

. The formant frequencies are now easily estimated when linear prediction analysis is used. This is also possible with an appropriate filter bank or with F.F.T., or cepstrum analysis (figures 3 and 4).

Vocabulary	Speech representation			
	L.P.C. coefficients	Frequency-mel coefficients		
	18 coef.	6 coef.	8 c.	10 c.
3	87,4 %	92,6 %	92,6 %	92,4 %
4	95,6 %	95,9 %	95,6 %	97 %
5	95,6 %	83,3 %	96,2 %	96 %

FIGURE 2 : COMPARING SPEECH REPRESENTATIONS AND WORD RECOGNITION ACCURACY.

Voc. n° 3 : 17 words /p/ /w/ a, /t/ /w/ /a/, /k/ /w/ /a/, ...
n° 4 : 9 words /a/ /r/ /p/, /a/ /r/ /d/, /a/ /r/ /m/, ...
n° 6 : 21 words /p//e/, /t//e/, /k//e/, ...

. One template per word
. Six repetitions in the recognition test.
. Dynamic programming comparison between words.

After M. COUVRAT and C. GAGNOULET.

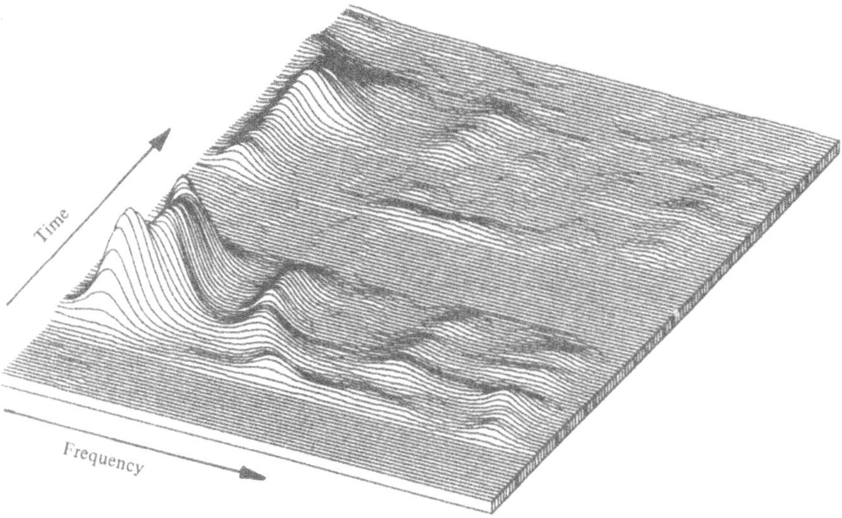

Figure 3 : Spectrum of the word "table" obtained by the cepstrum method.

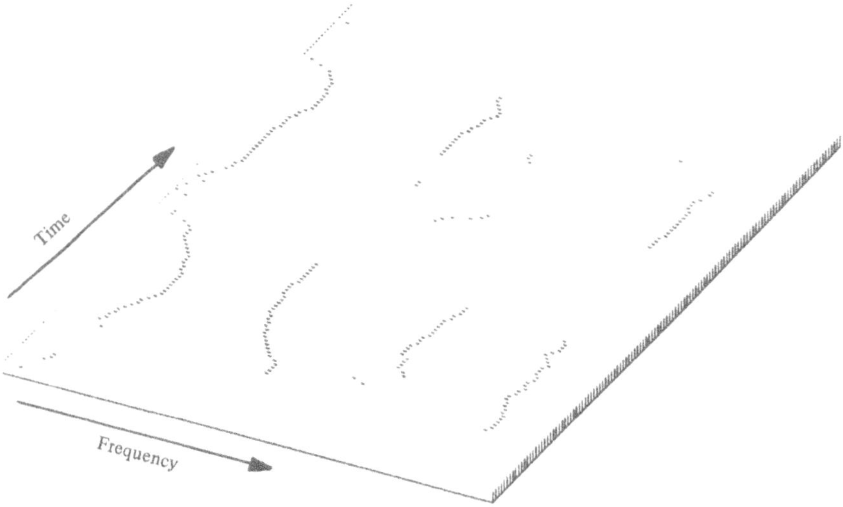

Figure 4 : Formants of the word "table" (cepstrum method)

. Zero-crossing densities are good parameters for detecting fricative sounds and are also used for estimating the formant frequencies.

. Time spectrum derivative may be useful for segmentation.

. Spectrum moments are used in some speech recognition systems (ZWICKER, 1978).

. Articulatory parameters have only recently been used in some speech recognition systems (SHIRAI, 1979 ; LOCHSCHMIDT, 1981).

. All of these parameters are used by the phonetic processing algorithms for segmenting, modeling of coarticulation and making phonetic categorical distinctions and labelling.

3 - UNITS OF RECOGNITION :

A basic question is : what are the primary units that must be first recognized to start the recognition process ? Should these units be words, syllables, diphones, clusters, phonemes, allophones, phones, acoustic events ... ?

June Shoup (1980) in the W. LEA's book : "trends in speech recognition" has discussed the possible advantages and disadvantages of each of these units. We have introduced some modifications and summarized them in figure 5.

For the recognition of limited vocabularies, words have been used with success as the units of recognition in many speech recognition devices. The main motivation probably resides in the three advantages pointed out in figure 5, but, when large vocabularies have to be recognized, a large set of references has to be created and shorter units have to be introduced in order to avoid learning and creating templates for each word separately.

. In recent speech recognition systems, there is a tendency to take the syllable as the basic unit (FUJIMURA, 1975 ; MERMELSTEIN, 1975 ; DE MORI, 1980 ; PERENNOU, 1981 ...).

. Good quality synthesis by diphones gives motivations for choosing diphones or demi-syllables as basic units for speech recognition. They are (or have been) used with success by RUSKE (1981), ROSENBERG (1981), WISWANATHAN (1981), MARIANI (1981) and others in different ways and different contexts.

RUSKE (1981), for instance, divides the words into syllables, syllables into demi-syllables, and these last units into consonant clusters and vowels, that are then time-normalized.

UNIT OF RECOGNITION	ADVANTAGES	DISADVANTAGES
WORD	. Avoids problems of coarticulation . Avoids problems of phonetic representation . Implies a simple speech recognition system.	. Extension to large vocabularies . Adaptation to new speakers is not easy
SYLLABLE	. The syllable nucleus is easy to locate . Much coarticulation information is included.	. The total number is relatively large . Precise boundaries are sometimes difficult to find.
DIPHONE	. Contains transitional and coarticulatory information	. Dictionary of diphones is relatively large (1000 items) . Problems of segmentation.
PHONEME	. The total number is small . Every word in the lexicon can be easily coded into a phonemic string.	. Some phonemes have various acoustic representations depending upon the context. . Some phonemes are not easy to locate. . Many rules or complex algorithms are needed for good segmentation and identification.
ALLOPHONE	. Can correspond to an acoustic segment easy to locate and to identify. . Avoids some coarticulation rules.	. The total number can be large . Some allophones are context-dependent.

FIGURE 5 : BASIC UNITS OF RECOGNITION : ADVANTAGES AND DISADVANTAGES (FROM JUNE SHOUP, 1979, PERSONAL MODIFICATIONS HAVE BEEN INCLUDED).

We come now to the most known basic units, the phonemes, which are perhaps the most difficult to identify. Advantages of taking these units as thebasic ones are obvious.

Unfortunately there is no straight forward link between a phoneme and its acoustic or articulatory representations. The difficulties of this problem have been pointed out by FANT (1973). They are summarized in the following schematic representation of sequential elements of speech (figure 6).

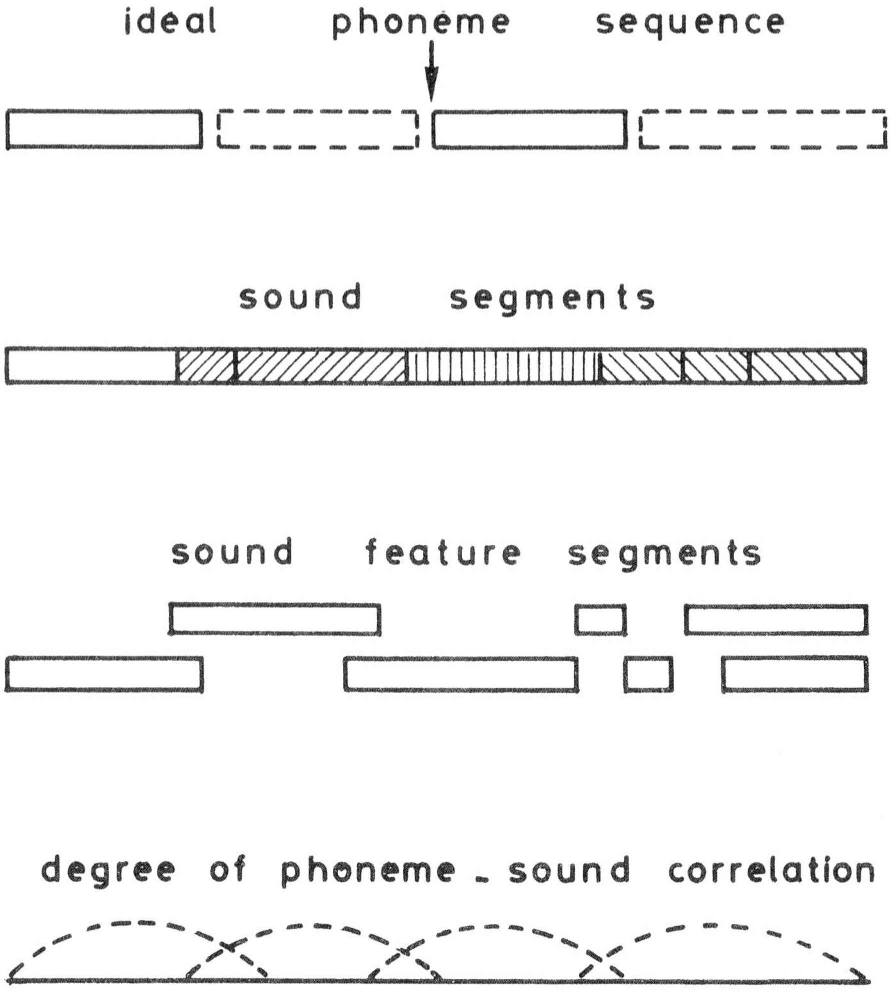

FIGURE 6 - SCHEMATIC REPRESENTATION OF SEQUENTIAL ELEMENTS OF SPEECH (after G. FANT).

We see that there are more sound segments than phonemes ; for instance, at least "silence + burst" for a stop consonant.

The distinctive features that characterize a phoneme are not necessarily present entirely in one acoustic segment, but they can overlap onto neighbouring segments or even disappear completely.

The manifestation of a phoneme in the physical signal is a continuous function, as shown in the last line of Figure 6.

These problems can be partially solved by introducing a larger set of units than phonemes (for instance sound segments or acoustic events), and by treating the different allophones of one phoneme as separate units. For the sake of simplicity we shall refer to all these units as "phones". A set of phones can also be found by clustering techniques (LELIEVRE, 1981).

Another method for solving the problem of coarticulation is to establish context dependent rules such as those used by DE MORI (1980) for classifying plosive and nasal sounds in CV utterances. This can be done through a hierarchical process in which the place of articulation of the vocalic contex is recognized before applying specific context-dependent rules. Formant frequencies are important cues for recognizing the place of articulation of vowels, and formant transitions are useful for modeling the coarticulation effects of plosives.

4 - SEGMENTATION :

This is a crucial point in many speech recognition systems. Two procedures can be used :

- Explicit segmentation : It consists in locating the basic units we have just defined such as syllables, diphones, phonetic segments, ..., in the continuous speech waveform.
  A broad preliminary segmentation is performed on the basis of changes in typical parameters such as energy, spectral derivative, Fo, centers of gravity, ...
  Usually, this segmentation has to be refined by means of hierarchical context-dependent strategies. The main advantage of this strategy is to recognize the segment as a whole.
- Implicit segmentation : In many systems, segmentation is obtained as a result of detecting and identifying a segment in some part of speech by using, for instance, dynamic programming techniques. We have an example of such a segmentation in figure 7.

Dealing with errors of segmentation : omission, insertion, merging, and spreading, is an important question. One way to approach this problem is to incorporate a "confidence measure" into the phonetic segment boundaries, with several possible decisions, and to build a segment lattice with overlapping hypotheses.

Another way is to have some probabilistic model of these errors (as in the I.B.M. system for instance) or to integrate them into the lexicon together with the phonological rules. In any case, the lexical analyzer must take into account all information provided by the phonemic lattice.

SAMPLE NUMBER	LABELS		SAMPLE NUMBER	LABELS		SAMPLE NUMBER	LABELS	
1	S F T		37	Y Z T		74	IN RI OI	
2	S F CH		38	Y Z T		75	U O1 OE	/l/
3	S F CH		39	G1 T1 D1	/k/	76	O1 U IN	[L]
4	S F CH		40		[K]	77	IN A1 O1	
5	S F CH		41			78	A1 A1 IN	/ɛ/
6	S F CH		42			79	A1 A1 IN	[AI]
7	S F CH		43	P Z K		80	A1 A1 IN	
8	S F CH	/s /	44	E O1 IN		81	A1 IN AI	
9	S CH F		45	P K S		82	Rl IN U	
10	S F CH	[S]	46	O1 E OE		83	T J UU	
11	S F CH		47	O1 E OE	/o/	84	F S CH	/s/
12	S F CH		48	O1 E OE		85	K CH T	
13	S F CH		49	O1 E OE	[O1]	86	S F T	[S]
14	S F CH		50	O1 E OE		87	S F CH	
15	S F T		51	M V G		88	F S T	
16	S F CH		52	M G G1		89		/t/
17	F S T		53	M V N	/m/	90		
18	T S F		54	M V B1	[M]	91		[T]
19	A1 IN AI		55	N M B1		92		
20	A1 A1 IN		56	N V N		93	Z G K	
21	A1 A1 EI		57	Z V		94	IN R1 EI	
22	A1 A1 EI		58	J Z		95	IN A A1	/ɛ̃/
23	EI AI A1	/e/	59	S S F		96	IN A E	[IN]
24	El A1 A1		60	S CH T	/s/	97	IN A E	
25	EI A1 U	[EI]	61	F F T		98	A IN E	
26	EI AI U		62	S S CH	[S]	99	A IN A1	
27	EI I U		63	S F F		100	IN A A1	
28	EI I U		64	S CH CH		101	IN A A1	
29	EI I U		65	S F CH		102	P R V	
30			66	T F S		103	IN A1 A	
31	EI I U		67	T CH F		104	IN A1 A1	
32	EI I U		68	T Y L				
33	EI I U		69		/e/			
34	EI I AI		70					
34	EI I AI		71					
36	EI I OU		72		[EI]			
			73					

FIGURE 7 - CENTISECOND LABELLING RESULT FOR THE FRENCH SENTENCE : "C comme Célestin" : /s/ /e/ /k/ /o/ /m/ /s/ /e/ /l/ /ɛ/ /s/ /t/ /ɛ̃/ together with an implicit segmentation.

5 - LEXICAL REPRESENTATION :

The representation of the lexicon, which is an important problem, is usually realized as a network or a non uniform coded string of phonetic units in order to provide all pronunciation variants.

Generative phonological rules allow expansion of the basic form of a word into different forms taking into consideration phenomena such as vowel reduction, duration, nasalization, voicing or devoicing, schwa (or mute /ə/ in french), deletion, juncture rules between words, stressed and unstressed syllables and so on...

6 - RECOCNITION STRATEGIES :

Let us now discuss the recognition strategies : it is possible to classify them into two categories, depending, to some extent, on the way the phonetic knowledge is represented.

Context-independent strategies include :

- the centisecond labelling procedure where, for instance, discriminant analysis can be used for labelling each sample.
- the first broad-phonetic classification procedures when feature detectors and specialized automata are often used.
- the template-matching procedures used for identifying phones, syllables, diphones and words, from sequences of acoustic samples or centisecond labels.

Context-dependent strategies include the hierarchical procedures with context-dependent rules. In this case, recognition is obtained by structural or syntactic methods.

These two strategies are often combined in a phonetic recognizer in order to have a system with feedback allowing a refinement of hypotheses.

For each of these strategies, various mathematical tools are available : pattern matching, dynamic programming, discriminant approaches, probabilistic automata, fuzzy automata, clustering techniques, etc...

7 - DISTANCES AND SCORING PHILOSOPHY :

Once one of these methods has been selected, distances between parametric representations have to be defined in order to obtain maximum discrimination between the pattern categories. One problem related to these distances is the way a score is assigned to a "segment-label" or to an optimal solution when searching this solution through a network. This is generally done by a weighted some of distances or by product of probabilities in the case of independent variables, or by using Markov transition probabilities. DE MORI (1980) and GUBRYNOWICZ (1981) use fuzzy algebra to assign degrees of plausibility to hypotheses. Each of these scores implies a training procedure in order to estimate each a priori probability or weight.

8 - TRAINING :

Parameters, thresholds, rules, network models, probability measures, etc..., necessary to represent the acoustic-phonetic knowledge and to identify the segments have to be learnt.

In order for this training to succed, it is necessary to choose representative data. Statistics and data analysis methods are useful because the estimation of a large body of data is necessary. Automatic or semi-automatic procedures for selecting representative examplesand extracting or adjusting the parameters have been studied (JELINEK, 1980), but a great care has to be taken in their use because bad segment boundaries or bad phonetic representations can lead to false phonetic knowledge and consequently to poor phonetic recognition.

9 - NORMALIZATION, ADAPTATION :

Other open problems concern speech normalization and adaptation :

. For amplitude, it is possible to use an automatic gain control circuit or a normalizing factor based, for instance, on the average amplitude or on the maximum formant amplitude values. The latter are compared to an empirical threshold.

. For duration, normalization can be achieved by segmentation, and by using non-linear time warping methods such as dynamic programming.

. Speaker normalization can be achieved by scaling the parameters : spectrum, formants, area functions and so on ... (see for instance the work of FANT (1973) for formant scaling and WAKITA (1977) for normalization based on vocal tract geometry.

It is also possible to use automatic or semi-automatic adaptation techniques to redefine or to adjust the phonetic knowledge : spectral templates, probabilities, thresholds, rules... A good discussion can be found in REDDY (1976), SAKAI (1980), JELINEK (1981).

10 - GENERAL ORGANIZATION OF THE KEAL ACOUSTIC-PHONETIC RECOGNIZER :

The KEAL acoustic-phonetic recognizer consists of components for carrying out three main tasks : acoustic analysis, labelling and training. These compoments are themselves composed of modules, each corresponding to an important step of analysis.

Each labelling module proceeds from the data or the hypotheses constructed during previous steps, refining them or building new ones in order to advance the phonemic recognition process.

The labelling procedure we have carried out consists in applying both an ordered sequence of logical rules (structuralor syntactic approach) and a set of decision functions (statistical discriminant approach) in order to distinguish phonemes.

The logical rules are used for segmentation and for recognizing distinctive features and rough-phonetic categories such as voiced, unvoiced, plosives, fricatives, nasals. Here is an example of such a logical rule : "When the proportion of energy between high and low frequencies is higher than a fixed threshold during a sufficiently long time, then a fricative segment is detected, with more or less high probability.

The ordering or "hierarchy" allows for taking into consideration the phonetic context that has been detected at the preceeding step.

The discriminant approach we use consists in representing the units to be recognized by points in a R^n space and then in separating them by means of hyperplanes (linear decision functions, figure 8).

This method is well-adapted for recognizing stationary acoustic segments like vowels, for which the coordinates of the vector representation in the R^n space are the parameters given by the acoustic analysis module.

These rules and functions represent the phonetic knowledge source of the system . Our logical rules have been established empirically by observing the acoustic signal in various sentences and different contexts. They leave room for improvements, which we hope will come with further empirical observation and the progress of basic acoustic phonetic research.

On the other hand, the parameters of the discriminant functions are computed automatically during a training phase ; they can be easily tuned to other speakers or to new recording conditions.

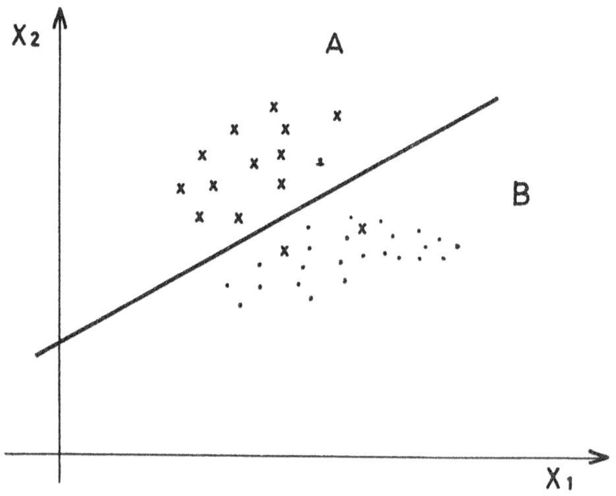

FIGURE 8 - SEPARATION OF THE CATEGORIES A AND B BY A LINEAR FUNCTION.

Figure 9 shows the general structure of the KEAL acoustic-phonetic recognizer. The arrows indicate how data and hypotheses are conveyed from one module to another. The end result is a phonetic lattice displaying alternative possibilities for segmentation and labelling (figure 12). This will allow us to hypothesize syllables and words.

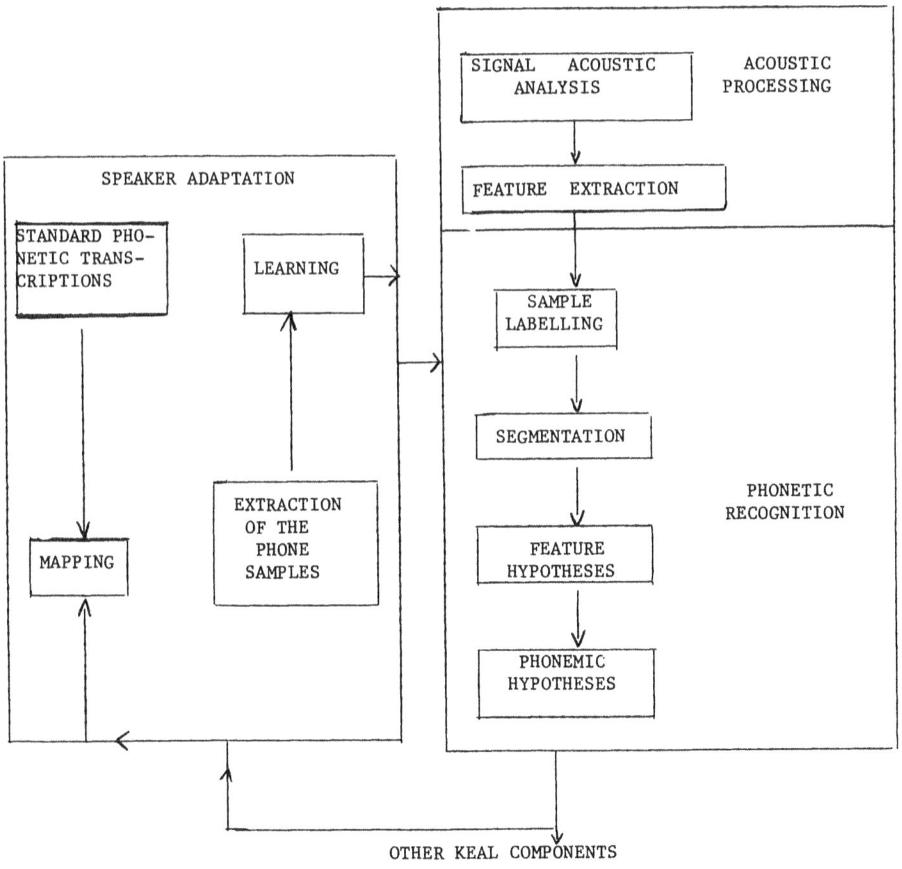

FIGURE 9 - GENERAL STRUCTURE OF THE PHONETIC ANALYZER, THE ACOUSTIC PROCESSOR AND THE ADAPTATION MODULE

In order to obtain the optimal phonetic lattice, we use "phones" as intermediate units because they seem to be more easily detectable from the speech signal. Some intermediate steps seemed necessary to achieve this goal :

. comparing speech analysis methods and selecting not only the best one but also the easiest one to implement and to interpret in terms of phonetic features.
. for each task of segmentation and identification, trying to find the optimal set of acoustic parameters.
. segmenting by taking into account the phonetic sample labels in addition to other features.
. completing the phonetic analyzer by adding new specialized modules such as :
    - recognition of the place of articulation of nasals and plosives
    - syllabic hypothesizer
    - vowel feature detector.
. eventually using clustering techniques for detecting allophones and selecting the set of phone references.
. modeling the time evolution of the speech signal within a segment by adequate functional approximations.
. automatic tuning of the parameters.
. using phonological constraints at the syllabic and lexical levels.

11 - ACOUSTIC ANALYSIS AND FEATURE EXTRACTION IN KEAL :

As in several other speech recognition systems, various parametric representations are available for KEAL (figure 1).

The first one is a 14-channel vocoder giving the short time spectrum every 13 ms in the frequency band 250-4200 Hz. For the moment, this is the acoustic analyzer which is actually used by KEAL.

Adjustment of the KEAL phonetic recognizer to the other acoustic analyzers we mention now is being developed in order to compare and to use them whenever possible.

A 16-channel analog vocoder has been designed (ZURCHER, 1980) whose frequency bandwiths are in artihmetic progression, giving better resolution in low frequencies. The total frequency band has also been extended to 5 kHz. Listening of nonsense monosyllables tests have shown an improvement of intelligibility from 80 to 90 % (ZURCHER, 1980).

With a 12-bit a/d converter and programs that compute F.F.T., linear prediction coefficients, cepstrum coefficients or cochlear model coefficients (CAELEN, 1979) every 10 ms, we obtain other possible speech representations.

KEAL also uses an analog pitch detector (ZURCHER, 1980) that detects and measures the fundamental frequency in the frequency-range 60-500 Hz.

The second block in figure 1 shows other acoustic feature parameters that are derived from the first series of parameters : formant frequencies, total energy, centers of gravity, spectral derivatives, slopes of the centers of gravity, etc...

All these parameters are valuable for segmentation and labelling.

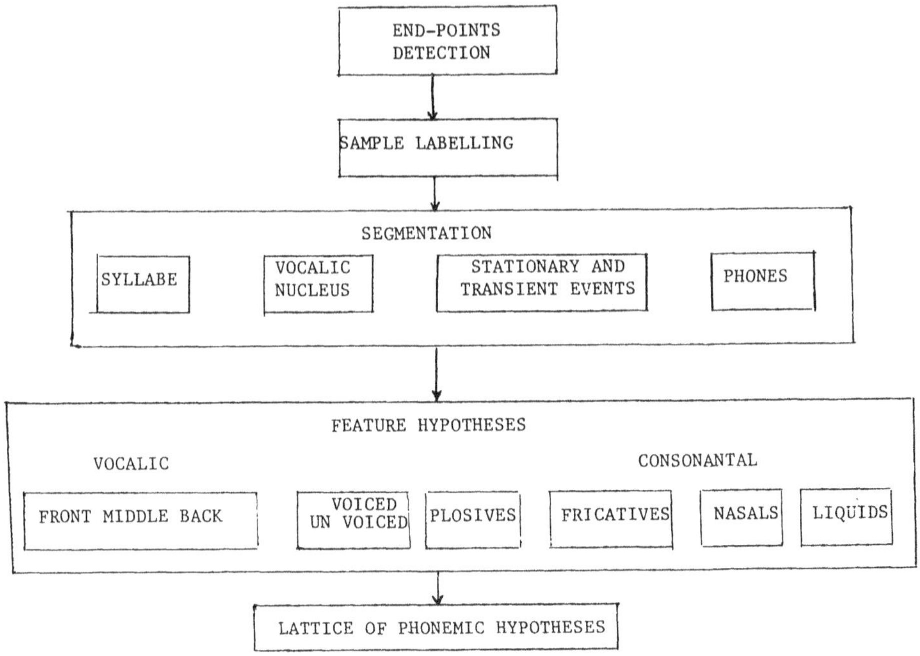

FIGURE 10 - GENERAL STRUCTURE OF THE PHONETIC ANALYZER.

12 - PHONETIC KNOWLEDGE REPRESENTATION AND PHONE-LABELLING IN KEAL :

At the moment, the phonetic recognizer contains the following levels (figure 10) : detection of the end points of a sentence, recognition of the features vocalic/non-vocalic, centisecond phone-labelling, segmentation into pseudo-syllables, detection of stable zones and transients between vocalic nuclei, assigment of broad phonetic labels to segments (stable zones and transient intervals) within each syllable, and generation of phonemic hypotheses.

12.1. End-points detection :

A first decision is obtained by means of three variables computed each centisecond : total energy, spectrum derivative and distribution of energy in frequency ; a linear function F(x) of these three parameters is then evaluated. The following logical rule is applied : "if F(x) is greater than a fixed threshold s_1 for at least 40 ms, then the beginning of a sentence is detected". A symetric decision rule is applied for detecting the end of a sentence with a threshold s_2 and a time duration of 140 ms. This end-point detection is refined after the centisecond-phone labelling has been performed, in order to adjust the boundaries.

An algorithm that detects energy peaks in some fixed frequency bands has been added. It is followed by an algorithm based on fuzzy algebra to evaluate whether the sentence end-points (beginning or ending) belong to one of the peak intervals.

Through these various "filters", the true boundaries of the analyzed sentences are obtained in almost all cases.

Another good idea probably would be to include certains types of background noise in the lexicon of phones or words (TUFFELLI, 1981).

12.2. Vocalic - non vocalic labelling :

The assigment of vowel or consonant label to each centisecond spectrum starts with the computation of the following parameters : energies and centers of gravity of the energy distribution in the low, middle, high and total frequencies. From these acoustic parameters a set of logical variables are derived allowing us to assign the vocalic or non-vocalic label to the analyzed sample in terms of the value of a boolean function of these variables. At the same time, the label "silence" is assigned to samples with very little energy. The use of fuzzy algebra or probabilities would allow us to take a less rough decision.

12.3. Centisecond "phone" labelling :

A "phone" label is assigned to each centisecond spectral sample, by means of linear decision functions the coefficients of which have been computed during the learning stage ; these functions are defined as follows : each centisecond sample is represented by a vector X(t) the first coordinates of which correspond to the differences of energies between consecutive vocoder-channels ; the other following coordinates correspond respectively to the total energy of the spectrum, the center of gravity of the total energy distribution, the energies defined inside four frequency sub-bands and the center of gravity of the energy distribution in these sub-bands.

Within each category (vocalic, non vocalic, silence) defined in the preceeding paragraph it is possible to distinguish among a number m of different phones (for the category "silence" there is obviously only one phone, making unecessary the assigment of a supplementary label to a "silent" sample). The number m is determined either heuristically, or by means of clustering techniques (LELIEVRE, 1981).

Each phone usually has a compact representation in the space where the 10 ms vector sample points : $X(t) = (X_1^t, \dots X_n^t)$ are represented.

The labelling algorithm makes use of a separation of this space by hyperplanes (figure 8), which are determined by "fitting" to the clusters an equilateral polyhedron with m vertices. The best fit is determined by a least square procedure (MERCIER, 1981).

Now suppose that the least squares procedure yields (w_{ij}) $(i = 1, \dots, n)$ as the coefficients specifying the orientation of the jth hyperplane decision surface.

In order to label a sample X(t), the m-1 following functions $h_j\ (Y(t))$ are evaluated

$$h_j\ (Y(t)) = \sum_{i=1}^{n+1} w_{ij}\ Y_i^t \qquad j = 1, \dots, m-1$$

with $Y(t) = (Y_n^t, \dots, Y_{n+1}^t) = (X_1^t, \dots, X_n^t, 1)$

Let us now define

$$H\ (Y(t)) = (h_1(Y(t)), \dots h_j(Y(t)), \dots, h_{m-1}(Y(t)))$$

Then it is possible to compute the m following scalar products $F_i(Y(t)$ with

$$F_i(t) = F_i(Y(t) = a_i^T.H(Y(t) \qquad \text{for } i = 1, \dots, m.$$

with the $a_i = (a_{i1}, \ldots, a_{ih}, \ldots, a_{im-1})$ corresponding to the vertices of a polyhedron.

The m $F_j(t)$ values are compared. The sample X(t) is assigned the three phone labels corresponding to the three greatest values of $F_i(t)$ in decreasing order (figure 7).

12.4. Segmentation into pseudo-syllables :

The objective of this task is to locate the vocalic nucleus of each syllable.

There are three steps :

. First the curve E(t) representing variations of the energy as a function of time is split into consecutive segments, each containing an energy maximum.
. These segments are then investigated in order to see whether or not they contain a vocalic nucleus. Nine acoustic parameters representing wide-band energies, formant-positions and the number of vocalic samples are used for making this decision.
. Third, the boundaries of the pseudo-syllables are determined. This is usually done by taking the stationary portion between two vocalic nuclei which contains the least energy, and designating its first sample as the boundary. This boundary seems to correspond well enough to the beginning of the phonological french syllable.

Segmentation is about 95 % correct when sentences are uttered normally. For the moment, results are less good for rapid speech, but we believe they could be improved by appropriate tuning.

At the end of this process, the length, the intensity of the syllable and the vowel localization are computed, allowing us to further use this information.

Special procedures allow us also to detect in some cases the french mute /ə/ at the end of a word.

12.5. Segmentation into "phones" :

Vowels are located approximately within the stationary zone surrounding the maximum of energy of each syllable. The most important parameter for locating stationary and transient events is the derivative P(t) of the spectrum envelope. The outputs of this segmentation may be questioned by the consonant identification process, which eventually modifies this preliminary segmentation.

12.6. Identification of the segments :

12.6.a : Vowel recognition :

Two alternative vowel recognition proceduresare available. The first one, which is very simple, consists in computing m mean values G(i) on the sequence of samples representing the vowel segment, with :

$$G(i) = \frac{1}{\ell} \sum_{t=t_1}^{t_\ell} F_i^t \qquad i = 1, \ldots, m$$

where ℓ is the number of samples in the segment and F_i^t is the function defined in the § 12.3.

The phonetic labels that correspond to the highest values of the function G(i) are generated and stored in the phonetic lattice.

This function seems to be too simple when the segment is not stable and it needs to be refined in order to deal with transients segments. A polynomial approximation for modeling segments is under way.

As far as this procedure is concerned it is important to note that some vowels such as /a/, /ɔ/, /i/, or /y/ have to be represented by two or more "phones" in order to improve the recognition.

In this procedure, we also compute, for each possible phone detected within a segment its mean position along this segment. This allows us, in some cases, to detect whether or not there are two vowels or a diphtong in the segment (figure 11).

The second process is based on the notion of distinctive features and corresponds to an hierarchical process that detects these features (ROSSI, 1981). A vowel recognition rate higher than 80 % is obtained.

For the moment, only the output of the first process is taken into consideration.

12.6.b : Consonant-recognition :

An hierarchical process is performed (figure 10). Voiced consonants are separated from unvoiced consonants by means of the following parameters : outputs of the pitch detector, energies and differences of energy in the low frequency channels (250-650 Hz), and proportions of phonetic labels corresponding to voiced phonemes that have been assigned to each centisecond sample by the sample-labelling module.

The feature "interrupted" is detected if the segment consists of a zone of silence or of little low-frequency energy (buzz-bar) followed by a strong burst and a sudden variation of the spectrum parameters (that is, essentially energies and centers of gravity or formants).

The feature "fricative" will be located if the segment is composed of enough samples with a frequency center of gravity higher than a fixed threshold.

SAMPLE NUMBER N° Echantillon	1	2	3	4	5	6	7	8	9	10	11	12	13	14	15	16	17	18	19	20	21	22	23	24
Codage CENTISECOND	UU	M	M	N	M	Y	O	E	O	O	A	IN	IN	IN	OE	OE	IN	IN	IN	AI	ES	EI	V	G
Phonetique LABELLING	M	UU	UU	G	G	R	OE	O	A	A	IN	A	OE	OE	IN	IN	OE	OE	AI	IN	AI	AI	G	V
des échantillons	N	N	G	M	N	M	E	OE	IN	IN	O	OE	A	A	A	A	A	A	OE	OE	O	O	Y	N
Treillis phonétique / PHONETIC LATTICE			M											IN										
			N											OE										
		UU										A												
										O														
																				AI				
time							E															temps		
N° Segment			1					2a				2							2b					
N° Syllabe			1									1												

Figure 11

<u>Etiquetage centiseconde pour le mot "moins" avec détection possible de la diphtongue</u> : O IN ; /O/ /$\tilde{\varepsilon}$/

Centisecond labelling of the word "moins" together with the phonetic lattice and a possible detection of the french diphtong /o/ /$\tilde{\varepsilon}$/

A glide (l) or (r) will be located after a plosive sound if there is at least one maximum on the energy curve together with enough energy in the low frequencies.

Recognition of the place of articulation of plosives is under way. Low frequency, middle-frequency and high-frequency centers of gravity during the burst, and formant or center-of-gravity transitions are the two sets of parameters used. The decision algorithm takes into account the place of articulation of the following vowel.

Recognition of the three french nasal consonants (m), (n), (ɲ) is also under development ; it will be based upon the values and the transitions of the three first formants, together with the position of the first antiformant ; it also has to take into account the vocalic context (DE MORI and al. 1979 ; CAELEN, 1979 ; D. DUBOIS, J. MARTINAGE, 1981). Preliminary results give a recognition score of 90 % when trying to separate /m/ from /n/.

The phonetic labels which are assigned to segments may have different degrees of depth. A segment can remain labelled just non-vocalic or non vocalic-interrupted if evidence about other acoustic features is not good enough.

After this hierarchical procedure has taken place, the same discriminant approach as defined in the § 12.6.a is applied to the consonant segment. A refinement of the preceeding label and the generation of the final phonemic hypotheses is provided by this approach.

Here again, for applying this discriminant approach it is necessary to have more phones than phonemes ; especially it is essential to have at least two phones for each plosive sound, corresponding respectively to the silence or the buzz-bar segment and to the burst segment followed eventually by the transitional segment to the vowel.

12.7. Building the phonemic lattices :

The output of the phonetic analyzer is represented in the form of a sequence of segments sharing the following information (figure 12) : rank number in the sequence, reliability value, beginning and ending time of the segment, several candidate phonemes ranked by their degree of confidence, number of the syllable to which the segment belongs. Each syllable also has a likelihood score. This lattice is then given to a word spotter able to make hypotheses or verifications about the possible sequence of words in a utterance (VIVES, 1981).

12.8. Training and tuning :

In order to evaluate the coefficients of the linear discriminant functions used to separate the phoneme classes, the following procedure has been designed (figure 13).

PHONETIC LATTICE	SEGMENT NUMBER		1	2	3		4	5	6		7	8
	SEGMENT WEIGHT		1	1	1		1	0,5	1		1	1
	SYLLABE NUMBER		1	2	2		2	3	3		4	4
	SEGMENT BEGINNING		4	18	24		37	49	53		72	77
	SEGMENT ENDING		14	22	36		47	53	57		77	85
MAPPING	PHONEMIC OUTPUT		o	z k j	e y		b d n g	k p t	e ɛ		z t j	ɔ̃ ɔ
	STANDARD TRANSCRIPTION	p	o	z	e	y	n	k	ɛ	s	j	ɔ̃
MAPPING RESULT	SEGMENT BEGINNING	NO	4	18	24		37	49	53	NO	72	77
	SEGMENT ENDING	MAP	14	22	36		47	53	37	MAP	77	85

FIGURE 12 - PARTIAL OUTPUT OF THE PHONEMIC LATTICE AND MAPPING RESULT FOR ADAPTATION.

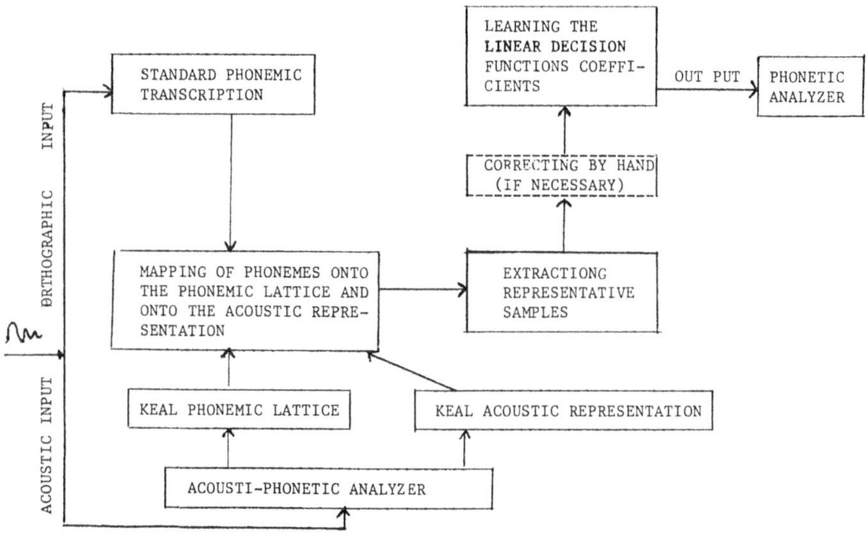

FIGURE 13 - ADAPTATION AND LEARNING.

The phonetic segment references necessary for building the learning set are automatically extracted by using a matching program (VIVES, 1979). This program maps the acoustic phonetic lattice given by KEAL onto an ideal phonemic transcription of the syllables, words or sentences composing the reference data set and onto their corresponding acoustic representation. It then extracts automatically the corresponding acoustic phone-like segment.

In case of mapping errors, the limits of the segments have eventually to be adjusted (or deleted) by hand, then providing the correct acoustic samples.

A stochastic approximation program which minimizes a least-square criterion on this set of acoustic-phone references, supplies the phonetic analyzer with the optimal coefficients W of the linear discriminant functions. Intermediate results of the stochastic approximation program are kept in memory.

At the moment, this facilitates the adjustment of coefficients to a new speaker and makes it possible to adjust on the basis of a partial reference set. We also have in mind the eventual possibility of automatic adaptation to new speakers. An example of functioning is given in figure 14.

12.9. Results :

Presently, the program is able to identify a maximal number of 17 vowel-like phones and 26 consonant-like phones. Recognition results are given in tableI.Four speakers (1 female, 3 males) have taken part in four training stages.

The differences in results between speakers are induced by many factors which are not easy to control : recording conditions, number of references, errors of segmentation, kind of linguistic units making up the data set (syllables, isolated words, sentences) from which the acoustic segments are extracted.

13 - CONCLUSION :

We have noted significant problems related to phoneme recognition and some of the techniques used to solve them. We have then described how problems of coarticulation, segmentation, labelling, training and adaptation are approached in the KEAL system.

Much effort is still needed to increase our phonetic knowledge and improve present performance : the following directions of research are suggested :

- Establishment of well-defined context-dependent rules for recognizing the most context-dependent phones.
- Improvement of the decision rules by introducing unsupervised learning models in the system, by computing consistent statistics and by using perceptual distances.

/f	ɛ	R	m	e	l	a	p	a	R	ɑ̃	t	ɛ:	z/	PHONEMIC CODING
F	AI	R	M	EI	L	A	P	A	R	AN	T	AI	Z	CODAGE STANDARD / STANDARD CODING
K	OE		M	EI	L	OE	P	A	R	AN	T	AI	Z	TREILLIS PHONEMIQUE / PHONEMIC LATTICE
T	IN		UU	AI	D	A	K	O			K	OE	G	
P	A		Y	OE	UU	AI	T	E			D	EI	W	
F	AI		Y	U	Z	IN	D	IN			P		N	
2	2	0	1	1	1	2	1	1	1	1	1	1	1	N° DE CADRAGE / MAPPING NUMBER
2	7		15	20	24	27	37	39	45	46	60	63	76	DEBUT / BEGINNING
6	13		18	24	25	31	39	44	49	52	63	73	79	FIN / ENDING

LE RESULTAT DU CADRAGE VOUS SEMBLE-T-IL CORRECT ? OUI

IS MAPPING CORRECT ? YES

FIGURE 14 - AN EXAMPLE OF MAPPING THE STANDARD TRANSCRIPTION ONTO THE CORRESPONDING PHONEMIC LATTICE

Speaker	I	II	III	IV
Number of differents phones	10 vow. 16 cons.	10 vow. 16 cons.	10 vow. 16 cons.	10 vow. 16 cons.
Total number of references	266	262	262	218
Correct-recognition percentage	67	59	59	61
Recognition percentage within hypotheses 1 + 2	80	72	72	76
Recognition percentage within hypotheses 1 + 2 + 3	86	82	80	85
Kind of utterances	isolated words	isolated words	isolated words	continuous speech

TABLE 1 - PERCENTAGE OF PHONETIC RECOGNITION

- Automatic speaker normalization and adaptation.
- Improvement of the time normalization.
- Use of perceptual and linguistic knowledges for building up a well-defined phonetic recognition model.
- Improvement of the clustering techniques for obtaining the optimum number of different phones.
- Taking into account prosodic cues (such as micromelody) and phonological knowledge, and introducing them into the lexicon.
- Adapting the form of the phonemic output lattice to make its information more useful to higher stages of processing.

ACKNOWLEDGEMENTS :

I would like to express many thanks to Dr S. ISARD, F. CLERMONT and B. DELGUTTE, for fruitful discussions on these topics and for correcting the English form of this text.

I am grateful to Prs. R. DE MORI and M. ROSSI for useful discussions and for having taken part in this work.

I wish to thank C. GAGNOULET, J. LE BRAS, C. SORIN and J. VAISSIERE for reviewing this paper, and R. CHEVREUL for drawing the figures.

I am also grateful to our secretary J. LE DENMAT and to M. TOUDIC for typing this manuscript, and to colleagues of the speech group who have taken a great part in the work described here.

REFERENCES :

CAELEN J., 1979 : "Un modèle d'oreille, analyse de la parole continue, reconnaissance phonémique" ; Thèse d'état, TOULOUSE.

CASTELAZ P.F., and NIEDERJOHN R.J., 1978 : " A comparison of linear prediction, FFT, zero-crossing analysis techniques for vowel recognition", Conference record of ICASSP, pp. 541-545.

CHAPLIN W.G., LEVADI V.S., 1967 : "A generalization of the linear threshold decision algorithm" to multiple classes. Computer and information sciences II. Tou A.P. NEW-YORK - London.

COUVRAT M., GAGNOULET C., 1981 : "Optimisation en reconnaissance par mots. Note technique interne CNET.

DAVIES S.B. and MERMELSTEIN P., 1980 : "Comparison of parametric representations for monosyllabic word recognition in continuously spoken sentences", IEEE Transactions on A.S.S.P. Vol. ASSP-28 N° 4, p. 357 à 366.

DE MORI R., 1980 : "Automatic phoneme recognition in continuous speech : a syntactic approach", Spoken language generation and understanding, D. REIDEL, Publishing Company, J.C. Simon editor - pp. 191 à 220.

DUBOIS D., MARTINAGE J., 1981 : "Identification des consonnes nasales dans la parole continue", rapport de stage CNET, ISEN.

EMERARD F., 1977 : "Synthèse par diphones et traitement de la prosodie". Thèse de doctorat de spécialité en linguistique, Université de Grenoble.

FANT G., 1967 : "Auditory pattern of speech". In W. Wathen-Durin (Ed.), Models for the perception of speech and visual form. CAMBRIDGE Mass. M.I.T., Press.

FANT G., 1973 : "Speech sounds and features", CAMBRIDGE, MA, M.I.T. Press.

FLANAGAN J.L., 1972 : "Speech analysis, synthesis and perception", 2nd edition,Springer Verlag, NEW-YORK.

FU U.S., 1974 : "Syntactic methods in pattern recognition", Academic Press, NEW-YORK, London.

FUDJIMURA O., 1975 : "The syllable as a unit of speech recognition". IEEE, Trans on A.S.S.P. Vol. 23 p. 82 à 87, Feb. 75.

GAGNOULET C., GILLET D., MERCIER G., VAISSIERE J., VIVES R., 1981 : "Le système de compréhension de la parole KEAL". Note technique interne CNET, 1981.

GUBRYNOWICZ R., 1981 : "Mapping acoustic cues into phonetic features". The fourth F.A.S.E. symposium on acoustics and speech, 1981 - VENEZIA.

HATON J.P., SANCHEZ C., 1981 : "Méthodes synchrones et asynchrones en reconnaissance phonétique de la parole", Séminaire en codage et décodage phonétique, Toulouse.

JELINEK F., MERCER R.L., BAHL L.R., 1980 : "Continuous speech recognition : statistical methods". Summer School on Speech Recognition, CISM-IBM, Italia.

KLATT D., 1980 : "SCRIBER and LAFS : two new approaches to speech analysis". Trends in speech recognition, W. LEA, editor, Prentice Hall, pages 529 à 555.

LEA W.A., 1980 : "Speech Recognition, Past, Present and Future", Trends in speech recognition p. 39 à 98, Wayne LEA Editor, Prentice Hall.

LELIEVRE A., 1981 : "Classification hiérarchique automatique pour la création de phones ou de mots de référence en reconnaissance de la parole". 12èmes J.E.P. - MONTREAL.

LIENARD J.S., TEIL D., CHOPPY C., RENARD G., 1977 : "Diphone synthesis of French : Vocal response unit and automatic prosody from the text". IEEE-ICASSP, Hartford, 1977.

LOCHSMIDT, B.F., 1981 : "Acoustic phonetic analysis using an articulatory model", Nato Advanced study institute, juin, juillet 1981 BONAS (GERS), FRANCE

MARIANI J., 1981 : "Reconnaissance de la parole continue par diphonèmes", Séminaire encodage et décodage phonétique, Toulouse.

MASSARO D.W., 1975 : "Understanding language : An information processing analysis of speech perception, reading and psycholinguistics", A. Press, NEW-YORK, London.

MERCIER G., GERARD M., 1981 : "L'apprentissage des paramètres de reconnaissance phonétique dans un système de reconnaissance de la parole continue". Congrès Rec. Formes et intelligence artificielle. NANCY Septembre 1981.

MERCIER G., 1981 : "L'analyse acoustique, l'analyse phonétique et l'apprentissage dans le système de reconnaissance de la parole continue KEAL II", 12èmes JEP - MONTREAL .

MERMELSTEIN P., 1975 : "A phonetic-context controlled strategy for segmentation and phonetic labelling of speech - IEEE Trans., or A.S.S.P. Vol. 23 p. 79-82, February 75.

MICLET L., GRENIER Y., LEROUX J., 1981 : "Reconnaissance phonémique par prédiction linéaire, l'expérience de l'ENST" 12èmes JEP - MONTREAL .

PERENNOU G., DE CALMES M., 1981 : "Le décodage au niveau phonologique dans Arial II". Séminaire encodage et décodage phonétique, Toulouse.

PISONI and SAWUSCH, 1979 : "Structure process in speech perception" p. 16-36 Cohen and Nooteboom editor Springer Verlag.

ROSENBERG A.E., RABINER L.R., LEVINSON S.E. and WILPON J.G., 1981 : "A preliminary study on the use of demisyllables in automatic speech recognition", Proc. ICASSP-81 pp. 967-970

ROSSI M., LECORRE C., MERCIER G., 1978 : "Indice de détection de formants sur analyse spectrale par canaux, 9èmes JEP - LANNION p. 105-114.

ROSSI M., NISHINUMA Y., TREVARAIN O., MERCIER G., 1981 : "Reconnaissance des voyelles par les indices et les traits", Séminaire encodage et décodage phonétique, Toulouse.

ROTH M.T., 1981 : "Etude de la voyelle dans les monosyllabes en français". Thèse de 3ème cycle - Université de Haute-Bretagne - RENNES II.

ROTH M.T., 1981 : "Etude de la voyelle dans les monosyllabes en francais". Thèse 3ème cycle de linguistique générale, Université de Haute-Bretagne - RENNES II.

RUSKE G., SCHOTOLA T., 1981 : "The efficiency of demisyllable segmentation in the recognition of spoken words. Proc. ICASSP-81, p. 971 à 974.

SAKAI T., 1980 : "Automatic mapping of acoustic features into phonemic labels". Spoken language generation and understanding, J.C. SIMON, editor, pages 147 à 190, Reidel Publishing Company.

SHIRAI K. and HONDA H., 1978 : "Feature extraction for speech recognition based on articulatory model", Proc. of the 4th IJCPR pp. 1064-1068.

SHOUP J., 1980 : "Phonological aspects of speech recognition". Trends in speech recognition, W. LEA, editor, Prentice Hall page 125 à 138.

SORIN C., 1981 : "La perception de la parole", Cours ENST BREST Février 1981.

STELLA M., 1980 : "Comparaison de différents coefficients de prédiction linéaire pour la reconnaissance des mots isolés". NT/LAA/TSS/36 CNET-LANNION.

TUFFELLI D., 1981 : "Constitution d'une base de données acoustiques", Séminaire encodage et décodage phonétique, Toulouse.

VAISSIERE J., 1980 : "Speech recognition programs as models of speech perception". In the cognitive representation of speech T. MYERS, J. LAVER and O. ANDERSON, editors, North Holland Publishing Company, in press.

VIVES R., 1980 : "Utilisation de l'information phonémique et syllabique pour la reconnaissance des mots prononcés isolément ou dans les phrases". Recherches acoustiques Vol. VI, CNET-LANNION, FRANCE, pages 117 à 133.

VIVES R., 1979 : "Un programme de cadrage pour l'adaptation du locuteur en reconnaissance automatique de la parole" CNET - Note technique NT/DAS/SST/24.

WHITE G.M., and NEELY R.B., 1976 : "Speech recognition experiments with linear prediction, bandpass filtering and dynamic programming", IEEE Trans. on ASSP Vol. 94 p. 183-188.

WISWANATHANV.R., MAKHOUL J., SCHWARTZ R., 1981 : "Medium and low bit rate speech transmission", Nato advanced Study Institute on automatic speech analysis and recognition, Bonas, Gers, FRANCE.

WOLF J., WOODS W., 1980 : "The HWIM speech understanding system". Trends in speech recognition, W. LEA, editor, Prentice Hall, pages 316 à 339.

ZURCHER F., 1980 : "Le vocodeur à canaux : une nouvelle jeunesse ?". Recherches/Acoustique, Volume VI, CNET. Pages 23 à 40.

ZUE V., 1980 : "Acoustic processing and phonetic analysis" in trends in speech recognition, pages 101 à 124, Wayne LEA editor. Prentice Hall signal processing series.

ZWICKER E., TERHARDT E., and PAULUS E., 1979 : "Automatic speech recognition using psychoacoustic models. JASA 65, 487 - 498.

ACOUSTIC-PHONETIC KNOWLEDGE REPRESENTATION: IMPLICATIONS FROM SPECTROGRAM READING EXPERIMENTS

Victor W. Zue

Department of Electrical Engineering and Computer Science
Massachusetts Institute of Technology
Cambridge, Massachusetts 02139, U.S.A.

ABSTRACT

This paper presents a summary of several spectrogram reading experiments designed mainly to uncover the amount of phonetic information that is contained in the speech signal. The task involved identifying the phonetic contents of an utterance only from a visual examination of the spectrogram. The results generally support the notion that there is a great deal of phonetic information in the speech signal that can be extracted by the proper application of phonetic rules. From these results, it is argued that phonetic recognition in speech recognition systems can be improved substantially, and that improved phonetic recognition will lead to speech recognition systems of greatly increased complexity and sophistication.

1. INTRODUCTION

Speech recognition is a research topic that has fascinated many speech scientists over the past several decades. Prior to the 1970s, research efforts were primarily directed towards isolated word recognition (Hyde, 1968). Since 1970, we have witnessed a flourish of research activities in continuous speech recognition. Such activities in the U.S. were at least in part due to the 5-year, multi-site effort initiated by the Advanced Research Project Agency of the U. S. Department of Defense. The ARPA project terminated abruptly in 1976, with some of the contractors meeting the original goal. [For a detailed treatment of the initial ARPA goal and a review of the final systems, see Newell et al. (1971) and Klatt (1977), respectively]. Since then, research

J.-P. Haton (ed.), Automatic Speech Analysis and Recognition, 101–120.

in speech recognition has, with few exceptions, taken on a different emphasis. These more recent systems generally operate in a speaker-dependent mode, and recognition is accomplished by recognizing isolated words or phrases using mostly pattern recognition techniques (Lea, 1980). Efforts to generalize such techniques to connected words and phrases have been met with limited success.

Almost all the state-of-the-art recognition systems, whether for isolated-word or continuous speech recognition, depend heavily for their success on the training of the system to a particular speaker's voice. In almost every case, the speaker has to train the system on each word or phoneme. The recognition is performed by matching the input speech to the stored templates and picking the template that is the closest to the input, using some distance measure. While such techniques have been quite successful for a system with a limited vocabulary and for a given speaker, they do not come close to human performance on virtually unlimited speaker and vocabulary.

There are two general types of pattern matching approaches. The first involves storing individual patterns for each word in the dictionary. Such an approach becomes increasingly impractical as the dictionary grows, both in storage and computational requirements. Furthermore, such an approach leads to little flexibility in handling word boundary effects, which can cause major distortion of a given word compared to its form if enunciated in isolation.

The second approach is to store a dictionary of templates deemed to be representative of all of the phonemes of the language. Each word is then entered as a sequence of such templates. This approach is better able to handle word boundary effects, by incorporating appropriate rules into the dictionary structure. However, subtle but specific variations in a given phoneme which may yield important clues to the identity of its neighboring phonemes are generally totally ignored, and in fact would probably serve to lower its score against a "canonic" form stored in the dictionary. As an example, we offer /s/ in an /sk/ environment, which often shows a concentration of energy near 1800 Hz in addition to the usual high frequency energy. This acoustic cue, when present, is extremely reliable in identifying the following velar consonant. In a template matching scheme, such an /s/ would probably still score best against an /s/ template, but there would be no mechanism for utilizing the 1800 Hz peak to aid in the identity of the /k/.

Another problem with phoneme template matching schemes is that they generally weigh all of the information equally regardless of the spoken speech sound. In the phoneme /u/ of American English, the first formant will always be low in frequency, but the second formant can be almost anywhere in the F2 region. One solution would be to store a whole sequence of dictionary entries for /u/ covering the entire F2 range. It would however be better to focus on F1 for identifying /u/, and use the location of F2 to aid in the identity of adjacent phonemes. Pattern recognition has no place for such decision strategies.

There are at least two other reasons why template matching alone can not be expected to give adequate performance as the systems grows in complexity and usage. First, regardless of the particular type of representation of the speech signal (raw spectrum, LPC coefficients, etc.), there does not exist a set of theoretically motivated distance metrics. As a consequence, acoustic information related to the linguistic (or phonetic) identity of a given sound is confounded with variability related to speaker specific characteristics, and the same distance metric is applied indiscriminately for the different types of speech sounds. Second, the templates tend to be highly speaker-dependent. The adaptation of a system to new speakers is likely to be a time consuming and painful process. The alternative approach, the one that we are proposing, is to incorporate our knowledge of the acoustic characteristics of speech sounds, and how these characteristics are modified by the environments, in order to provide a representation that is more linguistically based.

It is our belief that acoustic-phonetic knowledge representation is a major roadblock in the design of advanced speech recognition systems that are meant to approach human performance. In particular, we believe that the ability of humans to perceive speech in hostile acoustic environments, spoken by familiar as well as unfamiliar speakers, cannot be duplicated easily by machine without adequately incorporating these knowledge sources. Our belief is at least in part due to the results of recent spectrogram reading experiments which have shown that the acoustic signal is rich in phonetic information.

The purpose of this paper is to summarize several spectrogram reading experiments conducted over the past three years, describe the factors contributing to such a positive result, and speculate on the implication of our research on the design of speech recognition systems.

2. SPECTROGRAM-READING EXPERIMENTS

Ever since the invention of the sound spectrograph some thirty years ago (Koenig et al., 1946), the spectrogram has been the single most widely used form of display for speech. The popularity of the spectrogram is at least partly due to the fact that it is relatively easy to produce, and that it displays the temporal and spectral characteristics that are known to provide the segmental cues to speech sounds. To be sure, a speech spectrogram sometimes introduces distortions to the acoustic structure of speech and often does not provide adequate information on certain linguistically relevant cues, such as stress and intonation. Nevertheless, a speech spectrogram gives a good description of the segmental acoustic cues of speech, and it has been an invaluable tool in the development of our understanding of speech production and perception.

A speech spectrogram of the utterance, "The soldiers knew the battle was won," is shown in Figure 1. As Figure 1 reveals, a spectrogram provides a display of the energy in the speech wave in terms of frequency -- along the vertical axis; time -- along the horizontal axis; and intensity -- by the darkness of the markings.

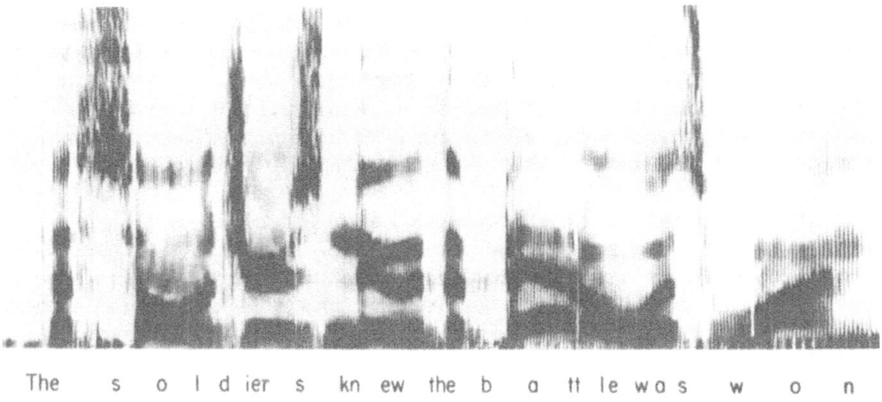

Figure 1 Spectrogram of the utterance "The soldiers knew the battle was won." The sentence was spoken by a male speaker and the spectrogram was made on a Voiceprint Laboratory Sound Spectrograph.

Is it possible to "read" a spectrogram? With sufficient practice, can one examine the spectrogram of an unknown utterance and determine what was said? The common belief is that it is not possible (Fant, 1962; Liberman, 1968; Lindblom and Svensson, 1973) and that no amount of training will allow one to do so. The results of past spectrogram reading experiments by Potter et al. (1947), Klatt and Stevens (1973), and Svensson (1974) are consistent with such a belief. In the classic book **Visible Speech**, Potter and his colleagues reported an experiment in which they attempted to teach normal as well as hearing-impaired subjects to communicate with each other using the Direct Translator, a real-time spectrographic display. The best subject, a deaf engineer, was able to acquire a vocabulary of 800 words after 200 hours of training. Vocabulary acquisition appeared to be a steady linear function of the amount of training; a word took about 15-20 minutes of practice to acquire. Since the visual patterns were recognized at the word level, it is difficult to assess their results in terms of phonetic recognition.

Klatt and Stevens (1973), serving as their own subjects, attempted to phonetically label a set of 19 spectrograms of unknown English sentences. In order to minimize the possibility of recognizing words in spectrograms, a "mask" was placed over each spectrogram so that only 300 msec of speech was visible at one time. In addition, each spectrogram was read in a single pass from left to right. Klatt and Stevens correctly identified approximately 33% of all segments, and provided a correct partial specification (e.g. "plosive", "nasal") on an additional 40%.

As part of his doctoral dissertation, Svensson (1974; see also Lindblom and Svensson, 1973) asked students who had taken a spectrogram-reading course to phonetically label spectrograms of unknown utterances. The subjects were provided with a flow chart describing a set of binary decisions to use when reading spectrograms. In addition, they were told to provide only one label per segment, and were given an hour to interpret each spectrogram. The results, according to Svensson, were disappointing, with performance ranging from a low of 22% to a high of 51% with an average of 38% of the segments correctly labeled.

In this section we report the results of several spectrogram-reading experiments that we have conducted over the past three years. These experiments collectively address several questions. First, how well can unknown spectrograms be "read". More specifically, how much phonetic information can be extracted from the acoustic signal. Second, how can the incorporation of other knowledge such as syntax, semantics and pragmatics improve the speed and accuracy of spectrogram reading. Third, how (and if) such a skill can be learned by others. Since the speed at which spectrograms can be read is of little concern to speech recognition researchers, the second question will not be addressed in this paper.

2.1 How Accurately Can Spectrograms Be Read?

The first experiment, designed to uncover the amount of phonetic information that can be extracted from a spectrogram, was reported by Cole et al. (1980). The task involves the phonetic transcription of unknown English utterances by visual examination of the spectrogram. In this experiment, spectrograms were made from 23 English utterances spoken by two male speakers. Each utterance is approximately 2 sec in length and contains an average of 6.4 words. The utterances consisted of normal English sentences, as well as phonetically, syntactically, and semantically anomalous sentences. In addition, spectrograms of 45 words in the carrier phrase, "Say --- again," were also prepared. This second set of spectrograms was intended to uncover to what extent word boundary information may influence performance.

In order to measure the accuracy of the reader's transcriptions, we asked three trained phoneticians to transcribe the utterances by listening to them. Figure 2 compares the transcription produced by the reader with that produced by each transcriber (T) for the sentence shown in Figure 1.

In order to evaluate segmentation accuracy, we arbitrarily adopted the criterion that a segment is assumed to exist when two

---

VZ: ð ə s oʷ ǰ ɚ̂ z n ü ð ə b ɛ ɾ ə w ʌ z w ɔ n
č ɜ̂ yu v v æ l̩ l ə

T1: ð ə s o l ǰ ɚ̂ z n u ð ə b æ ɾ l̩ w ɨ z w ʌ n

T2: ð ʌ s o l ǰ ɚ̂ z n u ð ə b æ ɾ l̩ w ʌ z w ɔ n

T3: ð ə s o l ǰ ɚ̂ z n u ð ə b æ ɾ l̩ w ə z w ʌ n

Figure 2 Transcriptions produced by the reader while reading the spectrogram, and by the three Ts who listened to the speech. The utterance is the one shown in Figure 1.

---

or more Ts produced a segment label. According to this criterion, there were 499 segments in the 23 utterances, of which the reader identified the existence of 485, or slightly more than 97%. It should be noted that segmentation is not an unambiguous process. In fact, 20 alternative segment paths were introduced, of which 13 were correct. It is interesting to note, however, that no extra segments were introduced by the reader.

The reader used a single label to identify a segment 52% of the time, two labels 35% of the time, and three labels 6% of the time. On the remaining 33 segments, the reader produced 17 partial transcriptions, 13 optional segments, and provided no label in three cases. If we exclude optional labels from consideration, and count each partial transcription as three labels, then the reader produced an average of 1.53 labels to each segment. When more than one label was given, they were almost always rank ordered so that it was possible to score the results for first, second, and third choices. We arbitrarily decided to score all partial transcriptions as equivalent to a third choice.

Table 1 summarizes the results on labeling. It can be seen that segment labels produced by the reader agreed with at least one T on 425 of 499 segments, or 85%. Approximately two out of three

TABLE 1
Agreement on Segment Labels between the Reader and Any T

	Proportion	%
First Choice	335/499	67
Second Choice	64/499	13
Third Choice	26/499	5
TOTAL	425/499	85

times the agreement is in the first choice. Closer examination indicates that there is a tendency for better agreement in consonants than vowels. This measure of labeling accuracy might seem, at first glance, quite generous, since the reader's labels were compared individually with that by each transcriber. However, it must be kept in mind that phonetic transcription is an **interpretation** imposed upon a continuously varying signal, and that transcribers rarely agree completely on the transcription of a given utterance. In this experiment, the agreement among all three transcribers is approximately 85%.

A detailed analysis of the reader's performance revealed a number of insights into his methods. Perhaps most interesting, higher-level knowledge about the syntactic and semantic structure of English was rarely used to interpret spectrograms. Performance was significantly **better** on sequences of words and nonsense words (92%) than on English sentences (83%), suggesting the use of phonological compensation by the transcribers based on contextual information.

The results on words in a known carrier phrase represents substantial improvement over that of normal and anomalous sentences. The existence of all 201 segments was correctly identified by the reader. The labels produced by him agreed with at least one of the Ts 93% of the time.

2.3 Can Spectrogram Reading Be Taught?

The results cited in the previous section are encouraging in that they illustrate the richness of phonetic information in the acoustic signal. These results in themselves are of little scientific interest, however, if it can not be demonstrated that such skill can be represented by explicit knowledge and thus can be taught to others. With respect to the design of advanced speech recognition systems, the ability to transfer such knowledge to others must first be demonstrated before any claims can be made regarding improved machine performance.

Several attempts have been made to teach the spectrogram reading skill to others. One such attempt was reported by Cole and Zue (1980) in which five students who enrolled in a graduate course in speech production and perception were asked to label phonetically five unknown spectrograms. The students' performance was compared to the transcriptions produced by two experienced phoneticians who listened to the sentences. Working as a group, the students identified the existence of approximately 94% of the segments. Averaging over all classes of sounds, the students agreed with either of the transcribers on the first, second, and third choices, respectively, 51, 24, and 8% of the time.

In a separate experiment, Seneff (1979) reported on the results of a spectrogram reading experiment that involved phonetic labeling and word hypothesizing for a set of 49 sentences spoken by one male speaker. Prior to the reported experiment, Seneff estimated herself to have studied about 30 to 40 spectrograms of continuous speech. During the experiment, she was provided with feedback immediately following the completion of each spectrogram. The feedback included not only the correct answer, but also instructions regarding acoustic cues or phonological processes. Phoneme identification scores improved steadily from 59% for the first ten sentences to 69% for the last ten sentences for an average of 64%. It should be noted that Seneff allowed herself only one label per segment.

The results reported by Seneff are more significant than those reported by Cole and Zue (1980), in this author's opinion, for several reasons. First, the corpus of the Seneff study (49 sentences and 1153 segments) is considerably larger than that reported by Cole and Zue (5 sentences and 104 segments). Second, it is very difficult to infer from Cole and Zue the performance of individual students, since the students worked as a group throughout ther experiment, whereas in the second experiment Seneff was the only subject. Finally, the experiment by Seneff was more controlled both in style and content, and a careful log of the learning experience was kept, thus allowing a more direct assessment of the transferability of this skill.

In addition to the two experiments described above, the spectrogram reading course has been given by this author on at least three other occasions to groups of students ranging in number from 8 to 13. The students varied significantly in their background, ranging from seasoned speech researchers to naive subjects with no exposure to acoustic phonetics. The course usually started with a brief description of the acoustic theory of speech production and the acoustic characteristics of consonants and vowels in stressed consonant-vowel (CV) environment. This is then followed by a description of the coarticulatory and phonological processes that govern the concatenation of segments into larger units. By the end of the course, which is approximately 40 hours in duration, the students usually have been exposed to spectrograms of approximately 50 sentences. Although a formal test of the subjects was not performed, it is estimated conservatively that the phonetic recognition accuracy on sentences is about 50%.

3. PHONETIC KNOWLEDGE REPRESENTATION

The conclusion to be drawn from the experiments described in the previous section is that the acoustic signal is rich in phonetic information, that such information can often be extracted from a spectrographic representation of the signal, and that the process of extracting this information often involves the application of explicit rules that can be taught to others. The results are all the more interesting considering that past attempts at spectrogram reading have met with much less encouraging results. These results are also significantly better than those reported in the literature on phonetic recognition in speech recognition systems (Klatt, 1979). What, one must logically ask, is the cause of such improved performance?

One of the most important factors that contributed to the improved performance in spectrogram reading is our improved understanding of the acoustic characteristics of fluent speech. To be sure, there has been ongoing research on the acoustic properties of speech sounds over the past few decades, and a great deal of knowledge has been acquired. However, with few exceptions, these research efforts have been focused on the acoustic properties of consonants and vowels in stressed consonant-vowel syllables. It was not until the past decade that researchers began to focus on the acoustic characteristics of speech sounds in continuous speech. We now have a much better understanding of the properties of speech sounds in different phonetic environments [see, for example, Umeda (1974), Kameny (1975), Klatt (1975), Zue (1976), Umeda (1977), and other citations in the following sections]. Furthermore, we are beginning to develop a quantitative understanding of the phonological processes governing the concatenation of words [see, for example, Oshika et al. (1975), Cohen and Mercer (1975)]. In fact, a few of the effects have been studied in some detail (Zue and Laferriere, 1979; Zue and Shattuck-Hufnagel, 1980). Therefore, our improved ability to interpret spectrograms is primarily due to the collected research efforts of many researchers. In addition, as a consequence of studies on the properties of speech sounds and of the auditory responses to speech-like sounds (see, for example, Kiang, 1980), we are gaining better insights into how the speech signal is processed in the auditory system, what portions of the signal are carrying the principal information concerning the distinctive phonetic dimensions, and what portions show more variability with respect to these dimensions. For example, the role of the burst spectra and amplitudes and of rapid onsets and offsets in identifying place of articulation and other features for consonants has been documented (Zue, 1976; Blumstein and Stevens, 1979).

To summarize, it is our feeling that our ability to extract a great deal of phonetic information from the acoustic signal is primarily a reflection of our improved understanding of the factors that contribute to the phonetic identities of speech sounds and their acoustic correlates. Spectrogram reading is nothing more than a demonstration of the proper utilization of our gained knowledge of how the acoustic cues for phonetic contrasts are encoded in the speech signal. Native speakers of a language demonstrate this ability whenever they communicate with each other by voice.

In the next section, we will attempt to identify the various ways phonetic knowledge is encoded in the speech signal and show how these sources of knowledge are utilized in spectrogram reading. We will argue that a phonetically-based speech recognition system must incorporate these knowledge sources.

3.1 Acoustic Phonetic Knowledge Sources

If the ultimate goal of a speech recognition system is to reach the performance of human speech perception, i.e., extremely high recognition on nearly all speakers, it would be important to know how acoustic phonetic knowledge is being utilized by humans. After all, the task of phonetic recognition faced by a computer is not unlike that faced by the native speaker of a language when trying to decode the utterance spoken by another native speaker. Aside from knowing the inventory of the speech sounds used in a language, i.e., the phonetic segments and their acoustic correlates, there are several levels at which acoustic phonetic knowledge is utilized heavily. We shall address each one of these knowledge sources in turn.

The first of these sources involves the constraints imposed by the language on the allowable sequences of speech sounds. In American English, there are only some forty possible consonant clusters that can start a word. Thus, for example, a native speaker knows that "vnuk" is not an English word. As another example, if we specify that the third phoneme of a three-element, word-initial cluster is /l/, then the remaining two phonemes (/s/

and /p/, as in "splash") are uniquely determined **without** the need of further acoustic evidence. The utilization of sequential constraints is not restricted to the word-initial position. In American English, for example, nasal-stop clusters almost always follow the homorganic rule requiring that the nasal and the stop have the same place of articulation (as in "camp", "tent"). The proper identification of one element of the cluster again determines the other element without the need of further acoustic evidence. The knowledge of constraints on allowable phoneme sequences enables the listener to **fill in** phonetic details that might be distorted or even absent in the acoustic signal. A speech recognition system must incorporate this knowledge such that it can also infer phonetic identity of speech sounds in the absence of clear acoustic cues.

When speech sounds are connected to form larger linguistic units, the canonic acoustic characteristics of a given speech sound will change as a function of its immediate phonetic environment. As an illustrative example, consider the utterance, "Tom Burton

Figure 3 Spectrogram of the sentences, "Tom Burton tried to steal a butter plate," spoken by a male speaker. The spectrogram illustrates the various acoustic realizations of the phoneme /t/.

---

tried to steal a butter plate," shown in Figure 3. Every word, except "a", in this sentence contains a single occurrence of the phoneme /t/. However, depending upon the immediate phonetic environment and stress pattern, the underlying /t/'s are realized alternatively as an aspirated /t/ ("Tom"), an unaspirated /t/ ("steal"), a retroflex /t/ with extended aspiration ("tried"), an unreleased /t/ ("plate"), a flap ("butter"), or a glottal stop ("Burton"). The acoustic characteristics of these realizations are seen to be drastically different.

The modification of the acoustic properties of speech sounds as a function of the phonetic environment is not a phenomena that is restricted to within a word. When words are concatenated to form phrases and sentences, significant acoustic changes can result, as evidenced in the following example. Figure 4 shows a spectrogram of the seven words "did", "you", "meet", "her", "on", "this", and "ship", spoken in isolation as well as in a sentence,

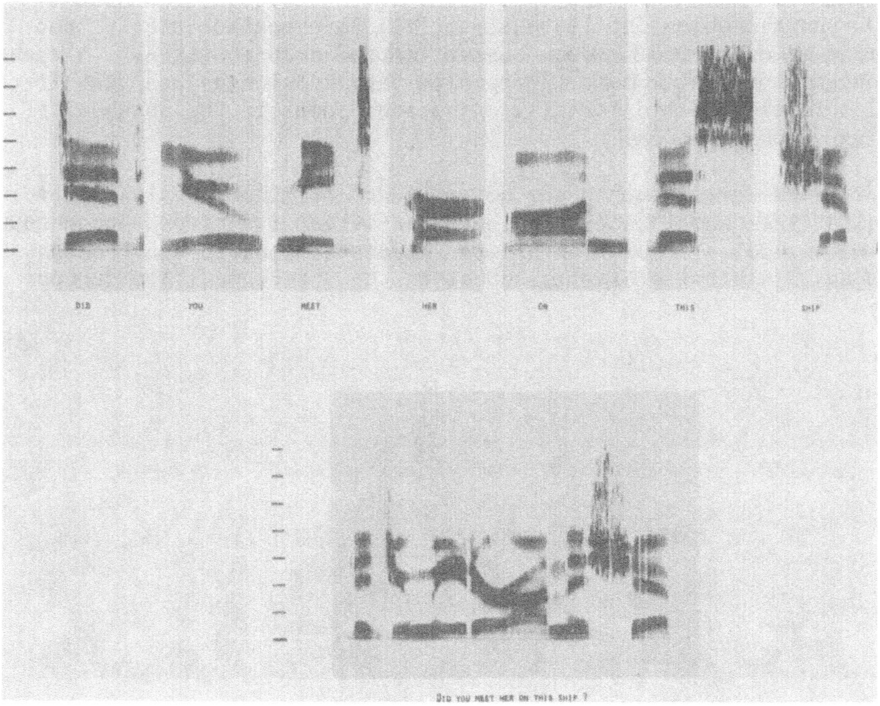

Figure 4 Spectrogram of the words "did", "you", "meet", "her", "on", "this", "ship", spoken in isolation and in a sentence by a male speaker. The spectrogram illustrates the phonological processes such as palatalization and flapping that can operate at word boundaries.

"Did you meet her on this ship?" We can see, for example, that the word-final /d/ and the word-initial /y/ in the word pair "did you" are realized acoustically as a single /ǰ/; the word-final /t/ and the word-initial /h/ are realized as a single flap; and the word-final /s/ and the word-initial /š/ in the word pair "this ship" are realized as a single, long /š/. These kinds of phonetic changes at word boundaries, particularly when there are adjacent word-final and word-initial consonants, are extremely common in normal American English. In order to properly perform lexical access, the nature of these phonological rules must be understood.

Whereas it is true that the acoustic realizations of phonetic segments are highly context-sensitive, most of the variations, such as the ones illustrated in Figures 3 and 4, are, in fact, systematic and can be captured by explicit rules (e.g. /t/ becomes a glottal stop [?] when preceded by a stressed vowel and followed by a syllabic nasal [n̩], as in "Burton"). Over the past decade, research in fluent speech has enabled us to gain a good understanding of the nature of these rules and how they interact. Although our present knowledge of the inventory of such rules is still incomplete, such knowledge, however fragmented, must be incorporated into a speech recognition system such that words can be recognized from the seemingly ambiguous acoustic signals.

Another factor affecting phonetic knowledge representation is the fact that a given phonetic contrast often has multiple acoustic cues. For example, the voicing characteristic of English stop consonants can manifest itself in many ways (Klatt, 1975; Zue, 1976). Some of the more salient acoustic cues include: a) the prevoicing during closure, b) the duration of the voice-onset time (VOT), c) the extent of the first formant transition, d) the amplitude of the burst, e) the fundamental frequency contour, and f) the duration of the preceding vowel. While it is often true that these multiple acoustic cues provide redundant phonetic information, there are also situations where only selected cues are available to aid recognition. For example, if a stop consonant is not released (as the /t/ in "basketball"), then those cues related to the release will not be present, thus rendering a decision algorithm based on VOT or the burst spectra useless.

We have demonstrated the fact that phonetic contrasts often have multiple acoustic cues. On the other hand, a given acoustic cue could also have multiple causes. For example, the duration of phonetic segments is influenced by a multitude of factors ranging from phonetic and phonological (such as inherent duration, stress, segmental interaction) to syntactic, semantic, and discourse knowledge (Klatt, 1976). Therefore, the application of acoustic phonetic rules must not be binding so that the effects of other factors can be taken into account when additional knowledge about the utterance becomes available.

One other factor that is important in the representation of acoustic phonetic knowledge is the fact that acoustic phonetic information is encoded differentially in the speech signal. Speech perception experiments (Cutler and Foss, 1977), spectrogram-reading experiments (Klatt and Stevens, 1973; Cole et al., 1980), and analysis of early speech recognition systems (Lea, 1980) all seem to indicate that humans and machines can perform better recognition for vowels in stressed syllables and consonants preceding stressed syllables, suggesting that the acoustic information is more robust around these syllables.

3.2 Example of Acoustic-Phonetic Knowledge Utilization in Spectrogram Reading

Having discussed the multitude of factors that influences the representation of acoustic phonetic knowledge, we will now turn to an example of how these knowledge sources interact during spectrogram reading. We want to stress the fact that the spectrogram-reading aspect of our example is only a method to **demonstrate** the knowledge interaction. We do not feel strongly that phonetic recognition should necessarily be performed on such a representation of the speech signal. In fact, as we shall see, some acoustic information is poorly represented in a spectrogram.

Figure 5 shows the spectrogram of an unknown utterance spoken by a male speaker. The spectrogram was generated digitally, and the display included the speech waveform and several parameters that, we believe, complement a conventional spectrogram. The quality of the digital spectrogram is, in our opinion, comparable to its analog counterpart. We will attempt to walk through this utterance from left to right, illustrating the process of applying acoustic-phonetic knowledge.

There is a well-enunciated glottal stop at the beginning of the sentence (t=0.17 sec) suggesting that the sentence began with a vowel, with no preceding consonant. The glottal stop appears as an irregularity in the fundamental frequency, which is visible both in the time waveform and in the spectrogram. (Another cue for the absence of a preceding consonant is the absence of noticeable rising F1 transition, typical from a consonant to a vowel. Since F1 and F2 are both high (around 700 Hz and 1800 Hz, respectively) in the middle of the vowel, the most likely choice for the vowel is /æ/. (/ɛ/ may be a second choice, but the long duration of the vowel makes it a distant second choice.)

Following the vowel is a region of low acoustic output except in the low-frequency portion. Since the energy dropped abruptly from the preceding vowel, the phoneme is probably a voiced stop. The identity of the stop can be sought by a process of elimination. If the stop were a labial stop (/b/), one would expect all formants to fall sharply at the end of the vowel. If it were a velar stop (/g/), then one would expect F2 and F3 to move towards each other. Since there is little motion in F2 or F3 at the end of the vowel, the most likely candidate is a /d/.

It might be supposed that the burst of energy following the stop gap (t=0.38 sec) is the burst of the /d/. However, it is

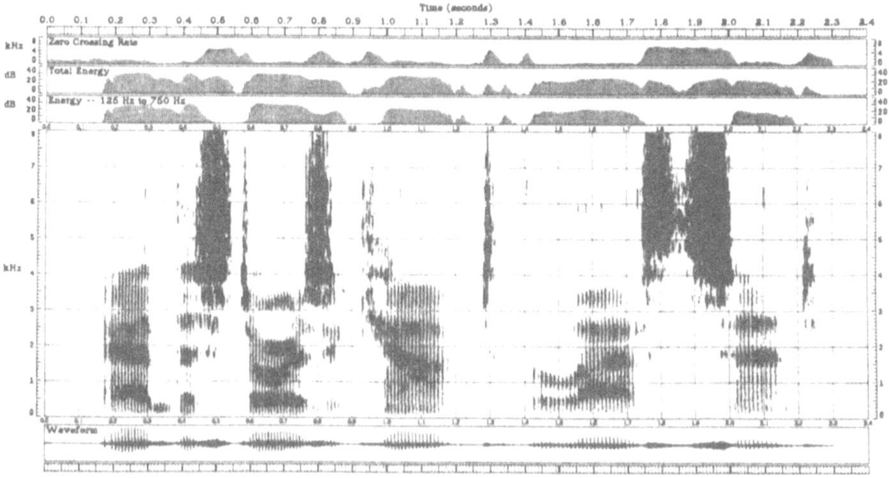

Figure 5 Digital spectrogram of an unknown utterance which is used to illustrate the application of acoustic-phonetic knowledge. In addition to the spectrogram, the waveform and several parameters are also displayed to complement the information available on a spectrographic display.

instead associated with the voiced fricative /ð/, which is often realized in a stop-like manner when preceded by a consonant. The stop is therefore unreleased. A true /d/ burst would be stronger in amplitude and lower in frequency than the frication noise generated for /ð/. (See, for example, the stop burst at t=0.57 sec.)

Following the phoneme /ð/ is a short, moderately front, vowel, possibly an unstressed /ɪ/. After the short vowel is a strong fricative (a definite /s/) followed by a short gap and a burst of energy (t=0.57 sec) which exhibits frequency characteristics typical of a dental stop (/d,t/). Since there is little aspiration following the burst, a /d/ or an unaspirated /t/ are the likely choices. The stop could therefore either be a /d/ with a boundary preceding it (since /sd/ is not an allowable English cluster), or a /t/ in a cluster with the /s/ (/t/ is not aspirated when it appears in a cluster with /s/).

After the stop is a vowel with sizeable second formant motion. Since F3 is quite low, especially for the second-half of the vocalic segment, a postvocalic /r/ is proposed. The vowel counterpart /ɝ/ is ruled out because F2 is much too low. After taking into account the formant transitions due to the preceding alveolar stop and the following /r/, we propose a low back vowel such as /ɔ/ or /o/.

After the /r/ is another strong alveolar fricative which must be either /s/ or /z/. Since it is shorter in duration than the first /s/ and since there are some clear pitch striations (especially at the onset and below 4 kHz), the phoneme is probably a /z/. The low-frequency energy parameter also indicates that this fricative has substantial low-frequency energy, further corroborating that it is a /z/.

Following the /z/ is a very short vowel which is easily overlooked. This vowel, based on duration and formant frequencies, can only be a front reduced vowel /ɨ/. After the /ɨ/ is another stop gap followed by a clearly aspirated release. The second and third formant transitions during the aspiration and during the preceding schwa clearly indicate a velar stop, i.e., a /k/.

The vowel following the /k/ is rather long with a very high F1. The high F1 suggests /a/ or /æ/. However, /æ/ is ruled out because F2 is too low (compare with the word "add"). Notice that there is also quite a bit of motion in F2, suggesting the possibility of a diphthong. However, F1 seems to be quite steady, although somewhat diffuse and weak in energy. An experienced reader will recognize the diphthong /aw/ in a nasalized environment. The additional nasal pole and zero in the low-frequency portion of the vowel spectra cause the appearance of an unusually high F1. Both the nasal formant and the vowel formant are visible in this digital spectrogram, since the analyzing filter is narrow enough to resolve them.

When there is a nasalized vowel there is usually a nasal. However, in this case the nasal is very hard to find. In fact, a phonological rule involving homorganic nasal-stop clusters must be invoked. The rule states that the duration of a homorganic nasal is greatly reduced when it is followed by a **voiceless** stop (Zue and Laferriere, 1979). Since F2 is rising slightly at the end of the vowel, /nt/ is a likely choice.

The stop gap ends in a strong burst with spectral characteristics indicating an alveolar place of articulation. It is tempting to associate the burst with the /t/ in the /nt/ cluster. However, the stop gap is quite long in duration, particularly unusual for a stop in a cluster. Thus a more appropriate hypothesis would be that there is another /t/ following the /nt/ cluster and that the first /t/ is, in fact, unreleased.

The release of the /t/ (t=1.28 sec) is quite long, to be followed by yet another consonant which is low in amplitude. Phonotactic considerations rule out the possibility of this /t/

forming a cluster with another consonant. Thus, the aspiration of the /t/ probably included a devoiced schwa, a common acoustic realization of an unstressed "to" in continuous speech.

The weak consonant following the aspirated /t/ is voiced, as evidenced by the periodicities in the waveform. The only candidates are the two voiced fricatives /v,ð/ or the voiced stop /b/. (/d/ and /g/ are not probable due to the relatively weak release.) Given our previous discussion on the realization of /ð/ following another consonant, /ð/ appears to be the most logical choice.

The consonant is followed by a schwa, which is followed by an intervocalic sonorant. The slow formant transitions suggest that the sonorant is either a /w/ or an /l/. /l/ is the more likely candidate due to the distinct separation of F1 and F2 and the rather marked discontinuity leading into the following vowel.

Following the /l/ is a relatively long, fairly steady-state vowel with a reasonably high F1 and mid range F2. The vowel /æ/ is the only vowel to fit this category. Note that the second formant for this /æ/ is somewhat lower than that of the first /æ/ we encountered. This is due to the coarticulatory effect of the /l/, which tends to lower the second formant of front and central vowels.

After the /æ/ is another /s/, extremely long in duration, with a fading out of energy mid way through. The length alone suggests that there must be more than one segment. In fact, this pattern is characteristic of the /sts/ cluster, with the /t/ closure never completely formed.

The final vowel is clearly nasalized, and, as in the previous nasalized vowel, no distinct nasal murmur is observed. The spectral characteristics of the release enable us to propose another /nt/ cluster. The first formant for the vowel is very difficult to locate, a problem often encountered with nasalized vowels. In such situations one learns to rely mainly on F2. The only vowels in English that can have such a second formant value are /ɛ/ and /æ/. Since the vowel is not very long, especially for the last segment of an utterance, /ɛ/ is probably a better choice than /æ/.

The string of phonemes that we have proposed thus far is shown in Figure 6. Even with an open lexicon, the number of words that can be proposed is quite small. After ruling out several potential word strings on syntactic or semantic grounds, the reader will probably accept "Add the store's account to the last cent" as the most likely candidate for a complete sentence.

ʔ æ d̚ ð̆ ɪ s t ɔ r z ɨ k aw n t̚ tʰ ð̆ ə l æ s t s ɛ n t
ɛ z d o æ

Figure 6 Proposed phoneme string for the utterance shown in Figure 5.

4. IMPLICATIONS TO SPEECH RECOGNITION

The design philosophy of most speech recognition (or understanding) systems in the past decade has been based on the belief that the acoustic signal does not provide sufficient information to identify the linguistic content of an utterance. In fact, the change in terminology from "speech recognition" to "speech understanding" reflects a departure from the view that speech can actually be "recognized" by machine, and the acceptance of the view that syntactic, semantic, and pragmatic knowledge must be used heavily to recognize an utterance (Reddy, 1976). While we are in no way disputing the importance of higher-level knowledge sources in the process of speech recognition by humans and machine, the results from spectrogram reading experiments clearly show that there is a great deal more phonetic information in the speech signal than was previously believed, and that such information is often explicit and can be captured by rules.

As we have stated earlier, it is our belief that acoustic phonetic recognition is the major road-block in the design and implementation of advanced speech recognition systems with capabilities approaching that of humans. We believe that developing phonetic recognizers with significantly better performance than those reported in the literature should be possible, and such improved phonetic recognition will enable us to tackle problems with relaxed constraints imposed by syntax, semantics, and pragmatics, thus enabling us to approach problems of increased complexity and sophistication.

Just exactly how the phonetic knowledge should be captured and represented is a matter that must be tackled jointly by researchers in diverse disciplines. The speech scientists and acoustic phoneticians must specify the phonetic rules of a given language in quantitative detail. Parameters capturing the relevant acoustic characteristics must be determined and extracted from the speech signal. Researchers in artificial intelligence must specify how such knowledge should be represented in the computer such that information can be accessed and digested in a cohesive manner. We are far from solving the problem, but the spectrogram reading experiments do provide us with reenforced conviction that the problem may be solved in the not so distant future.

REFERENCES

Blumstein, S.E. and Stevens, K.N. (1979) "Acoustic Invariance in Speech Production: Evidence from Measurements of the Spectral Characteristics of Stop Consonants," J. Acoust. Soc. Am., Vol. 66, No. 4, 1001-1017.

Cohen, P.S. and Mercer, R.L. (1975) "The Phonological Component of an Automatic Speech Recognition System," in **Speech Recognition: Invited Papers Presented at the 1974 IEEE Symposium**, ed. D.R. Reddy, 275-320, (Academic Press, New York).

Cole, R.A. and Zue, V.W. (1980) "Speech as Eyes See It," Chapter 24 in **Attention and Performance VIII**, ed. R.S. Nickerson, 475-494 (Lawrence Erlbaum Asso., Hillsdale, New Jersey).

Cole, R.A., Rudnicky, A.I., Zue, V.W., and Reddy, D.R. (1980) "Speech as Patterns on Paper," Chapter 1 in **Perception and Production of Fluent Speech**, ed. R.A. Cole, 3-50 (Lawrence Erlbaum Asso., Hillsdale, New Jersey).

Cutler, A. and Foss, D.J. (1977) "On the Role of Sentence Stress in Sentence Processing," Language and Speech, Vol. 20, 1-10.

Fant, G. (1962) "Descriptive Analysis of the Acoustic Aspects of Speech," Logos, Vol. 5, 3-17.

Hyde, S.R. (1972) "Automatic Speech Recognition: A Critical Survey and Discussion of the Literature," in **Human Communication: A Unified View**, edited by E.E. David and P.B. Denes (McGraw-Hill, New York).

Kameny, I. (1975) "Comparison of Formant Spaces of Retroflexed and Nonretroflexed Vowels," IEEE Transactions on Acoustics, Speech, and Signal Processing, Vol. ASSP-23, 38-49.

Kiang, N. S.-Y. (1980) "Processing of Speech by the Auditory Nervous System," J. Acoust. Soc. Am., Vol. 68, 830-835.

Klatt, D.H. (1975) "Voice Onset Time, Frication and Aspiration in Word-Initial Consonant Clusters," J. Speech and Hearing Research, Vol. 18, 686-706.

Klatt, D.H. (1976) "Linguistic Uses of Segmental Duration in English: Acoustic and Perceptual Evidence," J. Acoust. Soc. Am., Vol. 59, No. 5, 1208-1221.

Klatt, D.H. (1977) "Review of the ARPA Speech Understanding Project," J. Acoust. Soc. Am., Vol. 62, No. 6, 1345-1366.

Klatt, D.H. and Stevens, K.N. (1973) "On the Automatic Recognition of Continuous Speech: Implications from a Spectrogram-Reading Experiment," IEEE Transactions on Audio and Electroacoustics, AU-21, 210-217.

Koenig, W., Dunn, H.K., and Lacey, L.Y. (1946) "The Sound Spectrograph," J. Acoust. Soc. Am., Vol.18, 19-49.

Lea, W.A. (1980) **Trends in Speech Recognition**, (Prentice-Hall, Englewood Cliffs, New Jersey).

Liberman, A.M., Cooper, F.S., Shankweiler, D.P., and Studdert-Kennedy, M. (1968) "Why Are Speech Spectrograms Hard to Read?" American Annals for the Deaf, 1968, Vol. 113, 127-133.

Lindblom, B.E.F. and Svensson, S.G. (1973) "Interaction between Segmental and Nonsegmental Factors in Speech Recognition," IEEE Transactions on Audio and Electroacoustics, AU-21, 536-545.

Newell, A., Barnett, J., Forgie, J.W., Green, C.C., Klatt, D.H., Licklider, J.C.R., Munson, J., Reddy, D.R., and Woods, W.A. (1973) **Speech Understanding Systems: Final Report of a Study Group** (North-Holland/American Elsevier, Amsterdam).

Oshika, B.T., Zue, V.W., Weeks, R.V., Nue, H., and Aurbach, J. (1975) "The Role of Phonological Rules in Speech Understanding Research," IEEE Transactions on Acoustics, Speech, and Signal Processing, Vol. ASSP-23, 104-112.

Potter, R., Kopp, G., and Green, H. (1947) **Visible Speech**, (van Nostrand, New York).

Seneff, S. (1979) "A Spectrogram Reading Experiment," Term paper submitted for a Graduate Course on Sound, Speech, and Hearing, Massachusetts Institute of Technology.

Svensson, S.G. (1974) **Prosody and Grammar in Speech Perception**, Monographs from the Institute of Linguistics, University of Stockholm, (MILOS), Vol. 2.

Umeda, N. (1975) "Vowel Duration in American English," J. Acoust. Soc. Am., Vol. 58, 434-445.

Umeda, N. (1977) "Consonant Duration in American English," J. Acoust. Soc. Am., Vol. 61, 846-858.

Zue, V.W. (1976) "Acoustic Characteristics of Stop Consonants: A Controlled Study," Sc.D. Thesis, Department of Electrical Engineering and Computer Science, Massachusetts Institute of Technology; Also published by the University of Indiana Linguistic Club.

Zue, V.W. and Laferriere, M. (1979) "Acoustic Study of Medial /t,d/ in American English," J. Acoust. Soc. Am., Vol. 66, No. 4, 1039-1050.

Zue, V.W. and Shattuck-Hufnagel S. (1980) "Palatalization of /s/ in American English: When is a /š/ not a /š/?" J. Acoust. Soc. Am., Vol. 67, S27.

PHONEMIC RECOGNITION BY LINEAR PREDICTION: EXPERIENCES AT ENST

MM. L. MICLET, Y. GRENIER, J. LE ROUX

Ecole Nationale Superieure des Télecommunications
46 Rue Barrault, 75634 Paris Cedex 13

The introduction of identification methods, stemming from System Theory and in particular the Linear Prediction has recently given rise to the development of new methodologies in signal analysis, ans especially in speech. The idea underlying these methods considers the vocal system as a filter belonging to a certain family parametric. According to this approach, the analysis of speech signals within this established mathematical framework enables the identification of the characteristic parameters of the signal. During a short period, the speech signal represented by the numerical value of these parameters. The inversion of the model permits checking the correctness of the hypotheses made, by producing a synthetic signal, the quality of which measures the validity of the model.
For the past ten years, large advances have been made in the techniques of speech modelling, due to the ease of designing algorithmes on a computer.

Apart from applications in analysis, compression and synthesis, the study of speech signals by identification methods suggests techniques for its recognition : if the parameters extracted are pertinent, one should be able to use them to compare a speech segment thus described with stored prototypes, and with the help of metric, choose the sound which it most resembles.

J.-P. Haton (ed.), Automatic Speech Analysis and Recognition, 121–137.

The recognition of speech signals based on their analysis by linear prediction has been developed at ENST for more than 5 years. The purpose of this article is to review the different studies which have been made, and to propose certain subjects, the study of which seems indispensable if one wishes to progress in this domain. After the introduction of the basic notions, the principle of the "adaptive channel vocoder" is recalled : this being a bank of inverse filters, each one being matched with a phoneme (or sub-phoneme) of the speaker. The passage of the signal in the "vocoder" designates as the phoneme articulated at a certain moment the one which corresponds to the channel for which the error of prediction is lowest. This system of recognition works directly on the autocorrelation of the input signal, at a rate of 100 HZ.

The main problem of such a system is of course the training of each channel model. A procedure based on a method of clustering has been developed at ENST, which enables one to determine in a semi-automatic fashion, a coherent series of channels.

Another aspect is that of adaptation to the speaker. We introduce a method, stemming from the Canonical analysis of correlations, which automatically enables, starting with the channel adaptive vocoder for a speaker, the calculation of that for another speaker. The projection function is learned from a series of key words pronounced by the two speakers : it is thus a procedure for automatic adaptation to the speaker.

So the reconstruction of phonemes can be explained based on recognition factors stemming from the "Vocoder". Automata associated with the channels perform local smoothing, and the series of segments is consequently treated in order to derive a classic "phoneme lattice".

The strengths and weaknesses of such a system of phonenic recognition can be analysed, and certain research axes are hence proposed in order to improve the performance of this system. The latter are already promising.

1. SPECTRAL ANALYSIS WITH LINEAR PREDICTION PARAMETERS

Here we consider the problem of Pattern Recognition using linear prediction parameters.
Otherwise, more general identification methods can be used in Pattern Recognition /4/ but the nature of the speech signal and its characteristic traits make it

a very special case. In this article we propose to describe the methods of phoneme recognition used in ENST, on the basis of the analysis of the speech signal by linear prediction. We shall begin by introducing some notation.

Let S(t) be the signal ; by sampling at fixed frequency (typically, every10 ms), we have the series : $S_n = S(nT)$.
The evolution of S_n is modelized by a linear autoregressive model :

$$S_n = \frac{1}{a_o} \left(e_n - \sum_{i=1}^{p} a_i S_n - i\right)$$

where p is the prediction order.
The minimization of $E = \sum_{n=1}^{N} {}_n^2$ over N samples leads to the optimal values of $\{a_i\}$, which define the model of the signal.
Let $\hat{p}(\omega)$ be the spectrum of this model, i.e. the spectrum of the output for white noise input. Then we also have :

$$E = \frac{T}{2\pi} \int_{-\frac{\pi}{T}}^{+\frac{\pi}{T}} \frac{p(\omega)}{\hat{p}(\omega)} d\omega$$

where $p(\omega)$ is the spectrum of the signal.

Let the cepstrum of the signal be the inverse Fourier transform of the logarithm of the spectrum of the signal ; the cepstral coefficients can be computed as :

$$C_n = \frac{1}{2\pi} \int_{-\pi}^{\pi} \log\left(\hat{p}(\omega)\right) e^{jn\omega} d\omega$$

It can be shown that :

$$\begin{cases} C_1 = a_1 \\ jC_j = j\, a_j - \sum_{k=1}^{j-1} k\, C_k\, a_{j-k} \qquad j = 2 \ldots p \end{cases}$$

These equations allow the estimation of the soectrum and cepstrum, starting from linear prediction coefficients.

2. RECOGNITION WITH AN "ADAPTIVE CHANNEL VOCODER"

2.1. BASIS OF THE METHOD

The linear prediction model suggests a method for signal recognition.
Let us assume we have to "recognize" a signal S i.e. assign it

to one of C signal classes, each one being represented by its linear prediction coefficients $(a_i^j)\ \begin{matrix} i=1, p \\ j=1, C \end{matrix}$

At each sampling instant, it is possible to compute the prediction errors of S_n, relative to all of the models :

$$e_n^j = -\sum_{i=1}^{p} a_i^j \, S_{n-i}$$

The spectral representation on a window of width N, is :

$$E^j = \frac{T}{2\pi} \int_{-\frac{\pi}{T}}^{\frac{\pi}{T}} \frac{p(\omega)}{\hat{p}(\omega)}$$

where $\hat{p}_o(\omega)$ is the spectrum of the jth model.
The value E^j will be minimum for the "closest" signal to S, for j = 1, C. This kind of decision technique has been intensively used for several years :/5/, /6/, /7/, /8/, /9/, /10/.

2.3. SIMULATION

In a real process, the computation will not be performed at each new sample, but on a window of N samples (typically, N = 100). The prediction error for the jth model can also be written as :

$$E^j = (A^j)^T \, V \, (A^j)$$

where

$$A^{jT} = (a_o^j \ \dots \ a_p^j)$$

and

$$V = \begin{pmatrix} v_o & v_1 & \cdots & v_p \\ v_1 & & & \vdots \\ \vdots & & & v_1 \\ v_p & \cdots & v_1 & v_o \end{pmatrix}$$

where $v_i = \sum_{n=p}^{N} S_n \, S_{n-i} = v_{-i}$

This can also be written :

$$E_j = (\alpha^j)^T \, v$$

where

$$v^T = (v_p \ v_{p-1} \ \dots \ v_1 \ v_o \ v_1 \ \dots \ v_{p-1} \ v_p)$$

and

$$\alpha^{jT} = (\alpha_{-p}^j \ \alpha_{-p+1}^j \ \dots \ \alpha_{-1}^j \ \alpha_o^i \ \alpha_1^j \dots \ \alpha_p^i)$$

where

$$\alpha_i^j = \sum_{k=o}^{k=p-|i|} \alpha_k^j \quad \alpha_{k+i}^j$$

These formulas give a simple simulation method for recognition, which can be called "adaptive channel vocoder" ; the model of each channel is the set of parameters

$$(\alpha_i^j) \begin{array}{l} i=1,P \\ j=1,C \end{array}$$

The class to which the signal S is assigned is the kth class, such that :

$$j,k \in [1,C], \forall k \neq j \; (\alpha^k)^T v < (\alpha^j)^T v$$

The design of a "adaptive channel vocoder" is :

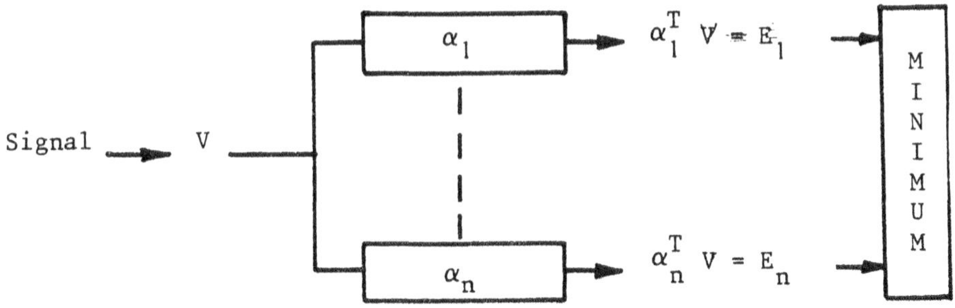

Thus a recognition decision is provided at each instant for which the correlation V of the signal S (typically 10 ms) has been calculated.

This type of distance utilises, in a very direct and simple way analysis by linear prediction. It can be compared to other distances (calculated on the linear prediction coefficients) on the training data bases in order to measure its properties ; in particular, for the euclidean distance between the cepstral coefficients the effectiveness of which is well known /9//10/. One conclusion that we have drawn from these comparisons is that the adaptive channel vocoder acts as a good discriminator based on scope of linear prediction parameters, and that the economy of calculation which it represents, compensates for the slight degradation, relative to the spectral distance already mentioned.

3. UTILISATION OF AUTOMATIC CLASSIFICATION METHODS AS AN AID TO TRAINING

3.1. STANDARD TRAINING PROCESS

As with all problems related to Pattern Recognition, the question of the quality of the training process is crucial and must he linked to the choice of distance used in decision. The very definition of the adaptive channel vocoder provides a very simple method of training.

First of all suppose that one has established a supervised training set that means knowing both the signal representations and their phonemic labels of a set of speech samples. Under the hypotheses of linear prediction, one must presume that the elements of the training data are stationary, and thus correspond to stable parts of phonemes. This working hypothesis is plausible for voiced sounds and fricatives, and very debatable for plosives and non-voiced sounds. This observation is evident in the recognition results.
Staying within this framework, we are now going to work on the phoneme level, supposed stationary, and not on that of the transitions (diphonemes). The training of the adaptive channel vocoder to continuous speech presumes some other treatment of the transitions, which we shall discuss in paragraph 5. Under these hypotheses, we have then defined a training set composed of labelled parts of the signal. In practice this is done by an examination of the signal on an interactive graphic console which allows segmentation and storing the labelled parts and phonemic labels.

Standard training consists in calculating one channel (and only one) of the adaptive channel vocoder for each phonemic category processed a priori. In practice, this numbed is fixed at 32. The calculation of the channel parameters is carried out by as a criteria the minimisation of the average of the output from the channel for all the labelled parts corresponding to this channel. In other words, if the phonemic class # i possesses M_i "points", that is M_i instants of 10 ms whose correlation has been calculated, one calculates α_i in order to minimise :

$$E_i = \frac{1}{M_i} \sum_{k=1}^{M_i} (\alpha_i)^T V_k$$

where the (V_k) k = 1, M_i are all the points of the training data labelled with phoneme # i.

Then :

$$E_i = (\alpha_i)^T V c_i$$

where

$$V c_i = \frac{1}{M_i} \sum_{k=1}^{M_i} V_k$$

Thus in this training process we calculate the parameters α_i of the channel by applying the algorithms of linear prediction to the center of gravity of the correlation vectors of the base points with the same phonemic label as the channel.

This type of training presumes, in principle, that the phonemic categories imposed a priori are really those of the speaker studied and that the errors of labelling while creating the training set, though inevitable are rare.
The shifting of the center of gravity by aberrant points can effectively induce considerable errors in discrimination between the phonemic categories. In spite of these weaknesses, which the semi-automatic aid which are presently going to expose tries to remedy, this method of learning provides a good support in evaluating the quality of the adaptive channel vocoder. Let us give an example here : the considered case is composed of parts of signals extracted from words pronounced by only one speaker. It consists of 1700 points (or 17 seconds of total speech) and it is divided into 32 phonemic classes.

The 32 corresponding channels of the channel adapted vocoder are determined by the method of centers of gravity. Then all of the data is input to this vocoder ; the rank of a given point is determined (by increasing prediction errors) according to which channel has the same phonemic label as this point (the ideal being of course that it be in the first rank) : Taking all the base points and using a 15th order predictor one obtains typically /9/ :

percentage of points recognised in the first rank	66.0
percententage of points recognised in the first four rankes	88.6

These results are quite satisfactory, and seem to validate the hypotheses made, especially since the majority of errors arise from non voiced plosives and sounds which are essentially transitory (certain phonemes /R/, or semi vowels).

3.2. IMPROVEMENT

One way to remedy the "blind" aspect of the above method is to study the less well known phonemic categories with the object of modeling them by two or more channels ; or even to create a channel whose label consists of two phonemes, indicating in this way that it modelises a union of intermediate level with respect to the analysis carried out. If we then use a finer description we need non supervised training methods, that is to say automatic classification. The experiment has indeed proved to us that the direct examination of base point "clouds", by factorial analysis, was very biased because of the distance used in recognition by the adaptivechannel vocoder method. On the contrary the non-supervised methods, such as "dynamic clusters" /11/, allows for the use of the same metric in training as in recognition.

In this case we want to look in the union of points, which we can presume to be labelled by the same phoneme # i, for the N_i optimal sub-channels. The value of N_i must a priori be fixed ; the sub channels are given by the partition of the union C ; of points which minimise :

$$J = \sum_{j=1}^{N_i} \alpha_j \sum_{k \in C_j} Vk$$

where C_j is the sub-class # j.

The "dynamic cluster" algorithm uses as "kernel" the adapted sub-channels ($\alpha_1 \ldots \alpha_{n_i}$). The partition is given with the distance $E(v,\alpha) = \alpha^T v$.

The partition having being found, the sub channel is defined with the aid of the center of gravity of each category found in the partition.

The method of "Dynamic clusters" consists here in carrying out alternativly a recognition through adapted channels (that is to say with the "distance" of the linear prediction) on a non labelled population ; then a training using the center of gravity of each category thus separated.

The convergence in a finite number of steps is assured by the properties of the metric thus chosen.

In practice, the training method is semi-automatic, and takes place in the following manner : the standard channels for a speaker are calculated ; the training data is transferred to these channels, the more confusable channels, the least discriminatory in recognition are looked at. The data points of the training which corresponds to these channels are isolated, and an automatic classification through the "Dynamic clusters" algorithm is carried out, the number of channels being greater than two. A certain number of interactive procedures allow us at this stage to eliminate the aberrant points, which are very destructive in automatic classification and to look for the number of the best adapted categories. Afterwards one puts the data in to the modified adaptive channel vocoder, and the improvement made is evaluated. Then we continue to look for the sub channels in the mixed phonemic zones. As an example, for another speaker, let us give here the same figures as above (order of prediction : 15).

	First Rank	Four First Rankes
Standard channel adapted vocoder 32 channels	60.1	85.6
Modified channel adapted vocoder 73 channels	75.6	87.2

One sees then a noticable improvement in the quality of recognition - quite stable if the four best channels are used.

This method of training gives in fact a fine description of the training space by defining the sub phonemes which "match" better the true space studied.

The interactive and semi-automatic aspect of the training thus achieved allows an elaborate manipulation of data, and leads to a precise definition of a adaptive channel vocoder associated with a speaker.

4. ADAPTATION TO THE SPEAKER

One of the characteristics of the adaptive channel vocoder and generally of methods of speech recognition by linear prediction is the great sensitivity to the speaker. The recognition of a data base stemming from speaker L_2 on the channel adapted vocoder trained on speaker L_2 is indeed diminished One can quote for example the following figures /12/ (under the same conditions as above) :

		Training L_1	Speaker L_2
Recognition Speaker	L_1	97.1	70.0
	L_2	55.0	93.1

This degradation is not necessarily a handicap in constructing a multi speaker system : it translates simply the strong dependence of an adaptive channel vocoder on the phonemic space of the speaker to which it was adapted ; if it is possible to find the correspondance between speakers in the form of a transformation which the channels undergo, it can on the contrary be an advantage : rather than looking for an adaptive channel vocoder reflecting an "average" speaker, one tries and find a mathematical expression of the transformation from one speaker to another, from the training data ; better still, this transformation can be found by examining a series of words or identical phrases pronounced by the two speakers, which shall allow us to relate the different instances of the same phonemes.

The method which has been adopted comes from Data Analysis, where it is known under the name of Analysis of canonic correlations. It is a statistical procedure whose aim is to put into correspondence closely possible the two "phonemic universes" of the speakers calculating a double projection of one on to the other. A first experiment consisted in calculating these operations of projections from the training data of the speakers /12/ On obtains then, for the same speakers

	L_1	L_2
L_1	97.1	94.3
L_2	90.4	93.1

These figures allow for the verification of the theoretic validity of the method however, a real experiment of adaptation must take place automatically : the second experiment consisted in extracting the projections, not from the training data , but from identical words pronounced by the two speakers. The "space" between the two speakers necessary in order to relate their identical phonemes, is achieved by dynamic comparison on the energy of the signal. Afterwards, the adaptive channel vocoder of the second speaker is calculated from that of the first, the latter being determined from a training data. The second speaker is thus parameterised in an entinely automatic way. In these conditions, we find, for the same speakers /13/.

	L_1	L_2
L_1	97.1	89.4
L_2	81.6	93.1

The experiments carried out in automatic adaptation seem encouraging enough, and this is a crucial point for the achievement of high quality phonemic prototypes in a speech recognition system.

5. DECODING IN PHONEMES

The system of recognition by adaptive channel vocoder provides (cf figure 1) a fixed rate decision, each speech interval of 10 ms being decoded under the list of nearest phonemes ; the segmentation in phonemes and the passage to a more classical phonemic lattice can be made at this stage, that is to say that the segmentation is done here after the recognition (one can imagine naturally the inverse process, which is more classical). The interest in this approach is in providing a fine description of the signal in the phonemic space of the adaptive channel vocoder through which it was analysed. In order to extract the phonemes, the idea is to "smooth" the phonemes recognised in time, and to fall back on the energy of the signal to aid this segmentation on the recognition data. The chosen procedure is to associate to each channel an automaton, whose role is to follow the behaviour of this channel (in relation to the others) in recognition. A good rank ordering of channels (deducted from the associated errors of prediction) during a sufficient length of time assures that the phoneme associated with the channel must be "near" during the length of the phoneme really emitted. An automaton indicates then the beginnings and the ends of segments where the associated phoneme is present and attributes to it a score calculated from the corresponding errors of prediction.

All the results thus obtained are then analysed and the segmentation signs (indications) produced by the automaton are confronted with those of another automaton, which segments in accordance with the energy. The segmentation boundaries are merged in order to mark the limits of phonemes, ans the segments thus obtained make up the desired phoneme lattice (see Figure 2).

Such a system affords great flexibility of implementation since a different automaton can be associated with each phoneme (for example, for a plosive, the decision delay must be shorter than for a fricative). One can also adapt these automata to the global characteristics of the speaker (slowness of articulation, or systematic co articulations). But the adjustment of these parameters requires training on a very large mass of data ; their optimal determination could, there again, be helped by automatic training methods. At the present, the characteristics of the phonemic decoder have been fixed in a simple way independent from the speaker since enough data has not yet been processed in order to have a precise idea about a fine adjustment of the parameters.

6. STRENGTHS WEAKNESSES AND FUTURE WORK

Several years of experience in analysis and speech recognition by linear prediction has taught us the vast possibilities of these techniques, and their "fragility" at the same time, we understand the care with which training must be conducted before synthesis or the recognition can be achieved.

As for the strengths, a valuable characteristic of linear prediction methods is their algorithmic simplicity both in simulation and in real implementation. The realization of an adaptive channel vocoder is indeed very easy, and its efficacy can be tested at a very low computing cost. The very principal of this recognition device tends toward an on-line treatment of the signal and possibly in real time with a fast correlator.

The theoretical advantages of these modeling methods are well known /3/, /1/. From a practical view point, in light of the quality of recognition of most of the phonemes (including a careful training process - see below) the close adequation of a speaker to his adapted channel vocoder and the non-theoretic possibility to carry out an adaptation for each speaker.

However certain precautions must be taken in the analysis. A pre-emphasis (high pass filtering) of the signal is necessary ; it is carried out by a simple double differentiation of the correlation. It is calculated every 10 ms, with a window of 25.6 ms. This can sometimes induce a "resonance" on the pitch of the speaker.

As for the weaknesses of the method, an obvious short coming is the impossibility of precise recognition of non-voiced plosive phonemes. They are effectivly characterised by short temporal phonemes, and their recognition by adaptive channel vocoder is rare, and of little significance. It seems that a detection for these phonemes should be envisaged, then a recognition by other methods. On the contrary, the nasal vowels, which in theory are badly described by "all pole" models, are in practice well recognised. The most delicate point in the development of a system of recognition by adaptive channel vocoder seems to be definitely that of channel training. One notices a great sensitivity of results for the relatively slight variations of the channel parameters. This is due to the fact that the decision carried out is taken on a criteria of "distance" which is badly controlled : the parameter space is of a high dimension (typically, 15) and the methods of data analysis and projections in the sub-spaces do not give a correct representation. Slight errors in learning can induce a considerable bias ; this is the reason why the method described in paragraph 3, or any other approach allowing the verification of the validity of the training phase, seems necessary.

For future research directions undertaken in this domain, let us quote the following axes :

- study and comparison of spectral distances stemming from linear prediction (generalisation of adaptive channel vocoders) ;

- variable determination of the order /9/ ;

- study of a system of low bit rate transmission by recognition on on a very complex adaptive channel vocoder ;

- use of large data bases in order to fix the parameters of the system.

On this last point particularity, a lot of works remains to be carried out (some of which is indebted to collaboration with the CNET-Lannion teams). The experience already acquired allows us to hope for further progress in these domains.

Phoneme corresponding at the channel recognized in first, second, ... Signal Energy

E	Y	J-	EE	I	ON	AI	R	.010133
E	Y	L	EU	EE	AI	N	AU	.015160
E	Y	J-	EU	I	EE	AI	R	.018555
Y	E	J-	I	EU	EE	AI	R	.010079
Y	E	J-	EE	N	I	EU	AI	.002033
Y	N	EE	E	J-	EU	I	AI	.002538
L	U-	Y	N	EE	E	EU	AI	.002367
L	N	U-	EU	E	Y	AU	AI	.003892
L	N	M	U-	EU	AI	AU	Y	.006950
L	N	EU	U-	AI	AU	A	Y	.011926
L	EU	N	AU	AI	E	A	Y	.018637
L	EU	W-	EE	AU	O	N	AI	.018803
W-	O	OU	L	ON	EE	EU	AU	.016706
W-	OU	O	ON	L	R	EE	N	.014269
OU	W-	O	ON	R	L	EE	N	.012131
W-	OU	O	ON	R	EE	L	N	.010858
W-	OU	O	ON	R	B	G	EE	.011307
W-	OU	O	ON	R	B	EE	G	.013187
W-	O	OU	ON	R	B	EE	G	.017232
W-	O	OU	ON	R	AU	EE	L	.025201
O	W-	ON	R	OU	AU	EE	L	.041955
W-	O	R	AU	ON	OU	EU	EE	.062543
W-	AU	R	EU	AN	ON	AI	EE	.056222
EU	AU	EE	AI	IN	A	L	AN	.041285
EU	AI	AU	A	L	E	N	Y	.046832
AI	EU	E	L	A	N	U-	Y	.049599
E	AI	EU	L	U-	A	Y	N	.043428
E	AI	Y	L	U-	EU	A	N	.034469
E	Y	AI	U-	L	EU	N	A	.026383
E	Y	L	U-	AI	EU	N	AU	.012976
L	N	M	U-	W-	O	EU	E	.012364
L	N	M	U-	EE	W-	O	EU	.014129
L	N	M	U-	W-	OU	O	EU	.015281
N	L	M	U-	W-	O	EU	AU	.015642
N	M	L	U-	OU	W-	O	EU	.015855
M	N	L	U-	OU	W-	EE	O	.016129
U-	OU	EE	W-	M	G	N	ON	.015419
OU	V	EE	R	J-	ON	I	E	.014983
U-	EE	J-	V	R	ON	I	G	.015370

Figure 1

Recognition with an adaptive channel vocoder
Each line is 10 ms.
Sentence to be recognized :
/E/L/W-/A/N/ ......

534	E	Y	J-	EE	I	ON	AI	R	.010133	
535	E	Y	L	EU	EE	AI	N	AU	.015160	
536	E	Y	J-	EU	I	EE	AI	R	.018555	Best phoneme
537	Y	E	J-	I	EU	EE	AI	R	.010079	for the segment
538	Y	E	J-	EE	N	I	EU	AI	.002033	
539	Y	N	EE	E	J-	EU	I	AI	.002538	
540	L	U-	Y	N	EE	E	EU	AI	.002367	
							END OF SEGMENT 45★★★★			Y E
541	L	N	U-	EU	E	Y	AU	AI	.003892	
542	L	N	M	U-	EU	AI	AU	Y	.006950	
543	L	N	EU	U-	AI	AU	A	Y	.011926	
544	L	EU	N	AU	AI	E	A	Y	.018637	
545	L	EU	W-	EE	AU	O	N	AI	.018803	
546	W-	O	OU	L	ON	EE	EU	AU	.016706	
							END OF SEGMENT 46★★★★			L·EU N
547	W-	OU	O	ON	L	R	EE	N	.014269	
548	OU	W-	O	ON	R	L	EE	N	.012131	
549	W-	OU	O	ON	R	EE	L	N	.010858	
550	W-	OU	O	ON	R	B	G	EE	.011307	
551	W-	OU	O	ON	R	B	EE	G	.013187	
552	W-	O	OU	ON	R	B	EE	G	.017232	
553	W-	O	OU	ON	R	AU	EE	L	.025201	
554	O	W-	ON	R	OU	AU	EE	L	.041955	
555	W-	O	R	AU	ON	OU	EU	EE	.062543	
556	W-	AU	R	EU	AN	ON	AI	EE	.056222	
							END OF SEGMENT 47★★★★			W- OU O
557	EU	AU	EE	AI	IN	A	L	AN	.041285	
558	EU	AI	AU	A	L	E	N	Y	.046832	
559	AI	EU	E	L	A	N	U-	Y	.049599	
560	E	AI	EU	L	U-	A	Y	N	.043428	
561	E	AI	Y	L	U-	EU	A	N	.034469	
562	E	Y	AI	U-	L	EU	N	A	.026383	
563	E	Y	L	U-	AI	EU	N	AU	.012976	
							END OF SEGMENT 48★★★★			E EU AI
564	L	N	M	U-	W-	O	EU	E	.012364	
565	L	N	M	U-	EE	W-	O	EU	.014129	
566	L	N	M	U-	W-	OU	O	EU	.015281	
567	N	L	M	U-	W-	O	EU	AU	.015642	
568	N	M	L	U-	OU	W-	O	EU	.015855	
569	M	N	L	U-	OU	W-	EE	O	.016129	
570	U-	OU	EE	W-	M	G	N	ON	.015419	
571	OU	V	EE	R	J-	ON	I	E	.014983	
572	U-	EE	J-	V	R	ON	I	G	.015370	
							END OF SEGMENT 49★★★★			N L M

Figure 2 : Example of phonemic decoding.

BIBLIOGRAPHY

/1/ C. GUEGUEN
Introduction à l'analyse de la parole
7° journées d'Etude sur la Parole
(GACF). Nancy Mai 1976.

/2/ J. MAKHOUL
Linear Prediction : A tutorial review
Proc. of IEEE Vol 63 n° 4 Apr. 75.

/3/ J. MARKEL, A. GRAY
Linear Prediction of Speech
Springer-Verlag 1976.

/4/ C. GUEGUEN
Sur une approche "système" de la Reconnaissance des Formes

Congrès AFCET/IRIA
Reconnaissance des Formes et Traitement des Images
Chatenay-Malabry Fév. 78.

/5/ H. WAKITA
Residual energy of Linear Prediction applied to vowel and speaker recognition
IEEE Trans. on ASSP Vol Juin 76.

/6/ C. GUEGUEN, G. CARAYANNIS, T. FARJAUDON, F. LE CHEVALIER
Un vocodeur à canaux adaptés
L'Onde Electrique Vol 55 n°7 (1975).

/7/ F. ITAKURA
Minimum prediction residual principle applied to speech recognition
IEEE Trans. on ASSP Vol 23 1975.

/8/ Y. GRENIER, L. MICLET
Reconnaissance de signaux par prédiction linéaire

Congrès AFCET/IRIA
Traitement d'images et Reconnaissance des Formes
Chatenay-Malabry Fév. 1978.

/9/ L. MICLET, A. NEHME
Expériences en Reconnaissance de la Parole
par Prédiction Linéaire.

2° Congrès AFCET/IRIA
Reconnaissance des Formes et Traitement des Images
Toulouse 12-14 Sept. 1979.

/10/ J.J. MARKEL, A. GRAY
Distance Measures for Speech Processing
IEEE Trans. on ASSP Vol 24 n° 5 Oct. 76.

/11/ E. DIDAY
Analyse Factorielle Typologique
Rapport Laboria n° 27 1973
INRIA (Rocquencourt).

/12/ Y. GRENIER, J.C. MAURIN
Adaptation au locuteur par analyse canonique des corrélations

2° Congrès AFCET/IRIA
Reconnaissance des Formes et Intelligence Artificielle
Toulouse Sept. 79.

/13/ M. MICHEL
Adaptation d'un système de reconnaissance
Rapport de stage projet de 3° Année
ENST 1979.

/14/ J. LE ROUX, L. MICLET
Transcription orthographique-phonétique et synthèse en temps réel de la parole par prédiction linéaire

10° journées d'Etudes sur la Parole du GALF
Grenoble Mai 1979.

ACOUSTIC-PHONETIC ANALYSIS BASED ON AN ARTICULATORY MODEL

Bernd Lochschmidt

Research Institute of the Deutsche Bundespost
Darmstadt, W.Germany

ABSTRACT
In the application of automatic speech recognition in new calling services there are still many open questions related to frequency band limitation, speaker independence and the need for a relatively large vocabulary.
MOSES is a speech recognition research project aimed at the development and study of the procedures required for the above-mentioned application in an overall system. Essential prerequisites for this are a flexible and modular system structure and a high computer performance.
The main feature of this speech recognition system, at the moment only used for isolated words, is an acoustic-phonetic analysis assisted by an articulatory model to achieve a microsegment description of the speech input.

1. INTRODUCTION

Extensive research activities in various branches of science have contributed, in recent years, to gaining detailed information on the p h e n o m e n o n o f l a n g u a g e . Hence, it is expected that speech recognition systems will be applied in many fields by the middle or end of this decade. Special interest is taken in applications in public inquiry services for which some promising beginnings are already visible /1/. This, however, should not obscure the fact that many problems of automatic speech recognition have not been solved so far, in particular, with respect to its use in the public telephone network. Apart from the difficulties caused by frequency band limitation and all kinds of interference, this holds, above all, for the problems arising from the need for speaker independence and an adequately large vocabulary as well as the requirement for user acceptability and a sufficiently robust procedure.

J.-P. Haton (ed.), Automatic Speech Analysis and Recognition, 139–151.

In view of such requirements, it is hardly reasonable to use recognition procedures with a purely technical approach which process, for example, a word as a whole pattern. In this global signal processing, essential features are frequently obscured or rendered unintelligible. This is partly due to signal portions attributable to speaker-typical properties and interference. Hence, it is necessary to develop more sophisticated systems which adequately account for the properties of the human speech production process and adapt themselves, by learning, to different speakers and channels. However, this requires a more phonetically-oriented approach in which the recognition accounts for the sound structures and the physiological constraints of articulation.

A modular speech recognition system has been designed for investigating all necessary details of a sufficiently effective acoustic-phonetic analysis. In addition to the components used for signal analysis and parameter extraction, this system also includes, in particular, those for dividing the signal into microsegments, for extracting articulatory parameters and for normalizing speaker differences. The system is implemented on a digital processor. Fig. 1 is a block diagram of the MOSES recognition system which contains a verification stage in addition to the complex for the acoustic-phonetic analysis. Details of the individual components and their functions will be given in the following.

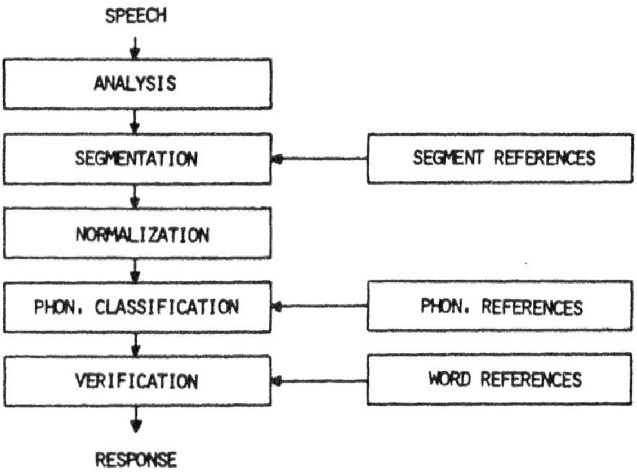

Fig. 1 Block diagram of the MOSES system

2. ANALYSIS

As a general rule, the analysis serves for achieving a considerable reduction of the data flow and for the conversion of the speech signal into a parametric form of representation which largely facilitates the feature extraction and classification performed thereafter. The parameters obtained from the analysis are intended to yield a very accurate and effective description of the essential features of speech in an economical way. This has to be done in consideration of the large tolerances of utterances, feature variations resulting from coarticulation effects and the differences existing between various speakers.

Since an insufficient feature extraction can hardly be corrected afterwards, even by making use of syntactic and semantic information, it is extraordinarily important for the whole recognition process that the features are extracted with great care. The following sections show the way in which the features can be efficiently extracted by means of the LPC analysis and with the inclusion of an articulatory model.

2.1 LPC analysis

The procedure of 'linear prediction (LPC)' is very important for digital speech processing because it offers the advantage of a very compact representation of speech signals. The theory, solution algorithms and properties of the LPC analysis are described in detail, for example, by Markel and Gray /2/.

In this context it is necessary to mention a very important property of linear prediction: it offers a simple way of determining, from the calculated filter coefficients a_i , the reflection coefficients k_i and from these the *area function* of the acoustic tube model.
The following relation holds for the ratio between the areas A_i and A_{i-1} of two adjacent tube segments

$$A_{i-1} / A_i = (1+k_i) / (1-k_i) \quad \text{with } i = 1,\ldots,10. \tag{1}$$

It has been shown that, under properly chosen boundary conditions, this model closely resembles the human vocal tract, especially with respect to vowel sounds /3/. Other important parameters are the residual energy (RE) and the normalized residual energy (NRE) which are used for the segmentation.
(The following input and analysis conditions have been selected for the acoustic-phonetic analysis: sampling frequency = 8 kHz, bandwidth of the filter = .15 ... 3.8 kHz, filter order = 10, preemphasis factor = .98, window = 30 ms and frame interval = 10 ms).

2.2 Articulatory parameters

As indicated above, the LPC analysis offers the possibility of calculating the area ratios A_{i-1}/A_i of the acoustic tube model.
In this form, however, the results of the analysis are still far too complicated for a further evaluation. Much more useful results can be achieved if the restrictions of articulation, which are attributable to the anatomical and physiological constraints involved in the production of speech, are directly included in the model, i.e., if a speech production model is used /4/.

An articulatory model describes the vocal tract very efficiently by giving details of the shape and position of the articulatory organs. The most important characteristics, apart from the width of the lip opening, are the shape and the position of the tongue. As few as three articulatory parameters are sufficient for obtaining a good model, above all, of the production of vowels. The decrease in the number of parameters is very advantageous, but even more important is the fact that the articulatory parameters make it relatively easy to account for dynamic processes and feature variations arising from coarticulation effects. Finally, it is possible to normalize the length of the vocal tract and, hence, to compensate for anatomical differences between various speakers (see section 4).

In the mathematical formulation of the model, the shape of the vocal tract is described by specifying the widths w(i) of the tube segments, the tube segment beginning at the lips having the index $i = 0$ and the tube segment beginning behind the glottis having the index $i = M$ (where $M = 8$). The value of w for $i = 0$ is indicated by the first articulatory parameter, the width $z = w(0)$ of the lip opening. The width w(i) measured in the range $1 \leq i \leq M-1$ between the palate or pharynx and the surface of the tongue is mainly determined by the tongue contour resulting from the shape and position of the tongue. The formulation of the equation is as follows:

$$w(i) = p(i)x + q(i)y + \overline{w}(i) \quad \text{with } i = 1,\ldots,M-1 \text{ and } M = 8 \qquad (2)$$

where x and y are the two other articulatory parameters of the model, which, in this case, constitute weighting factors of two special tongue contour functions p(i) and q(i), whereas $\overline{w}(i)$ holds for the contour of the relaxed tongue.
Eq.(2) can be written by means of the following vectors:

$$\underline{W} = \underline{P}x + \underline{Q}y + \underline{\overline{W}} \qquad (3)$$

where $\underline{W} = |w(i)| = |w_i|$ and, accordingly, $\underline{P} = |p_i|$, $\underline{Q} = |q_i|$, $\underline{\overline{W}} = |\overline{w}_i|$.
The vectors $\underline{P}$ and $\underline{Q}$ are the eigenvectors which - like the neutral contour vector - can be obtained from a principal component analysis performed on the basis of different sounds and speakers. As a consequence, the vectors take account of the special anatomical and physiological constraints and peculiarities of the human vocal tract.

This requires, however, the knowledge of the actual widths of the vocal tract which, in the final analysis, can only be obtained from the evaluation of a very large number of X-ray photographs. This paper is based on the results achieved by R. Harshman and P. Ladefoged /5,6/ which were modified for the articulatory model (see Fig. 2).

Starting from Eqs. 2 or 3 and the eigenvectors (Fig. 2), the individual components and the resulting tongue contour for a certain pair of articulatory parameter values (x=1, y= -1) can be shown in an anatomical and linearized representation (Fig. 3).

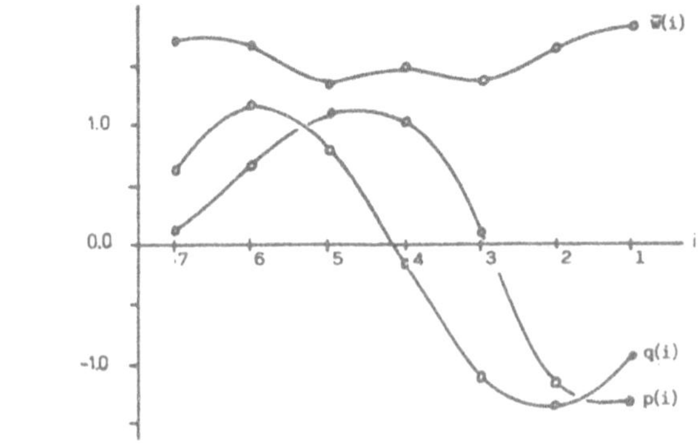

Fig. 2: Eigenvectors P, Q and W of the articulatory model

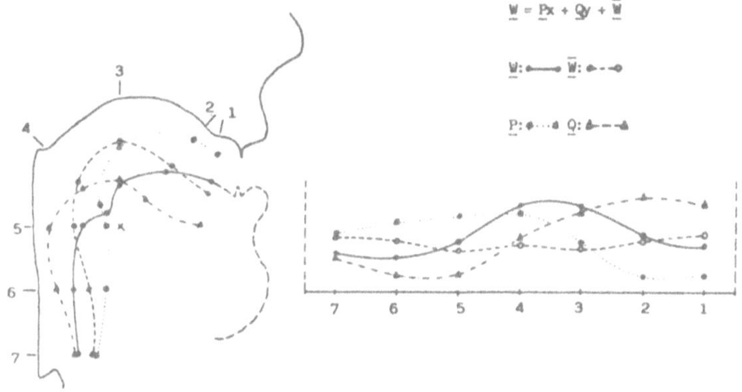

Fig. 3: Components and resulting tongue contour of the articulatory model for x=1 and y= -1 (anatomical/linearized)

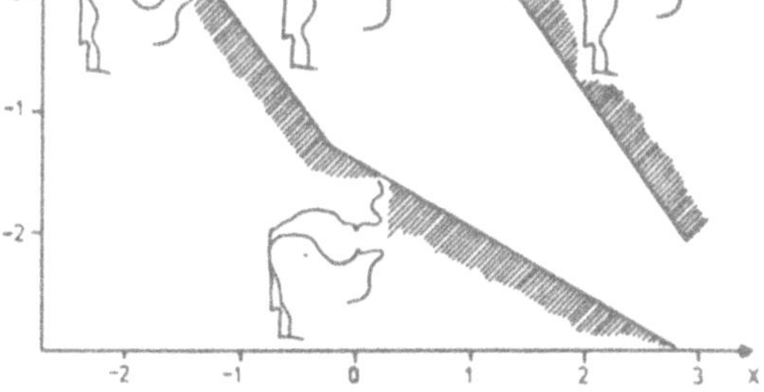

Fig. 4: Range of the potential values of articulatory parameters in the x-y plane and some extreme contours of the model

The fact that the width of the vocal tract cannot decrease below zero (constriction!) results in the boundary condition

$$w_i \geq 0 \quad \text{for } i=1,\dots,7 \tag{4}$$

The range of the potential values of the articulatory parameters x and y is limited by this boundary condition to the area outlined in Fig. 4. This figure also shows the tongue contours for some extreme positions.
The articulatory model is basically defined by Eq. (3) in connection with the parameter sets obtained for $\underline{P}$, $\underline{Q}$ and $\overline{\underline{W}}$. In the signal analysis performed for the purpose of speech recognition, the articulatory parameters z, x, and y are approximated. In this approximation, the area function

$$A_i^* = c_i w_i^a \qquad i = 1,\dots,7 \tag{5}$$

which depends on the tongue contour of the articulatory model (c and a are used as special transformation constants of the model) is optimally adapted, step by step, to the area function obtained from the LPC analysis. The following error criterion is used:

$$F = \sum_{i=1}^{M-1} \frac{(A_i^* - h\tilde{A}_i)^2}{\min(A_i^*, \tilde{A}_i)} \quad . \tag{6}$$

$\tilde{A}_i$ are the areas of a tube model with M=8 segments, calculated, in the normalization (section 4), by interpolation from the areas A_i of the LPC analysis (Eq. (1)). With the aid of the normalization factor, the two area functions to be compared are normalized to the same vocal tract volume.
Fig. 5 shows the way in which the articulatory analysis is performed.

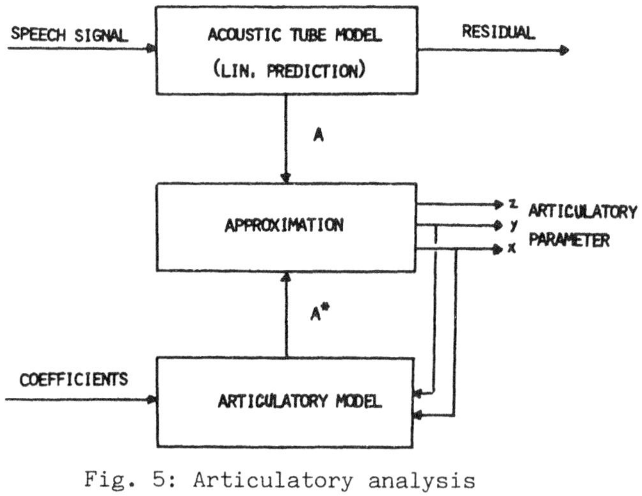

Fig. 5: Articulatory analysis

Fig. 6 shows the clusters obtained in the articulatory analysis for eight characteristic vowels of the German language. It is evident from the figure that the tongue contour parameters x and y alone do not allow these clusters to be sufficiently separated (left diagram). However, if the articulatory parameter z indicating the width of the lip opening is included, the clusters can be sufficiently separated in the three-dimensional cluster space (right diagram).

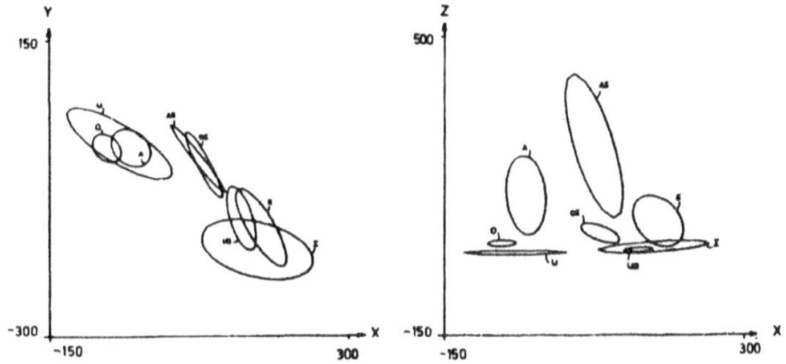

Fig. 6: Articulatory clusters of eight German vowels

2.3 Extraction of further parameters

Apart from the parameters derived from the 'classical' LPC analysis, there are several others which can be used for speech recognition. Some of them can be directly obtained from the speech signal. In the case discussed here, these parameters are the signal energy (E) and the zero crossing rate (ZCR) /7/.
In addition, there exists another group of parameters which are important for further processing and for the classification and which can be extracted from a combination of known parameters.
In the MOSES speech recognition system, these are the parameters 'back to total ratio (BTR)' and 'signed front to back ratio (SFBR)' /8/.

3. SEGMENTATION

The segmentation of the signal according to characteristic features is an important process for deriving phonetic information. In the MOSES system, the separation is performed in such a way that homogeneous segments are obtained. The criteria used for this purpose are both the results of a preclassification of individual signal sections (static segmentation) and a measure M of the variations of the signal as a function of time (dynamic segmentation). A segment boundary is marked whenever the class changes or M exceeds a threshold. It is typical of this procedure that single sounds are described by more than one segment. Since these segments reflect the microstructure of an utterance, they may appropriately be termed m i c r o s e g m e n t s . The procedure of a combined static and dynamic segmentation is shown in Fig. 7 and will be described in detail in the following.

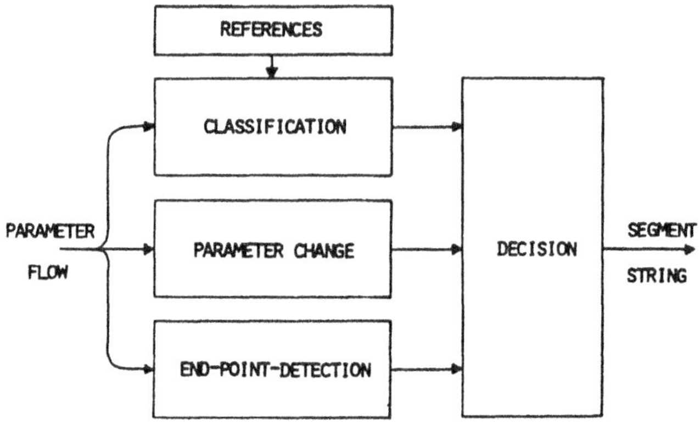

Fig. 7: Block diagram of static and dynamic segmentation

3.1 Static segmentation

The process of speech production offers possibilities for a preclassification of sounds into the groups vowels, nasals, voiced fricatives, voiceless fricatives and 'silence'. This preclassification is performed for all frames. At the moment, the seven parameters E, ZCR, NRE, k1, k2, BTR and SFBR are used for classifying the signal frames in order to obtain sequences of signal frames belonging to the same class. The distribution of these parameters is discussed in detail in /9/. Tests have shown that, for this application, the classification according to the probability-density distance yields the best segmentation results (7.3 percent errors).

3.2 Dynamic segmentation

The static segmentation groups all signal frames into certain classes. Whenever the feature vector travels from one cluster into another one, there is a transition into another class, and a segment boundary is defined. This crude classification, however, is not able to detect changes in sound quality within the same sound class. This requires a dynamic criterion which evaluates the change of a feature within the segment obtained from the preclassification and defines additional segment boundaries, if necessary. The measure used for this purpose is the change of the log-area ratios:

$$LAR(i) = \log (A_i / A_{i-1}) \quad . \qquad (7)$$

The change in the areas of the acoustic tube model from the last segment boundary marked at t_n to t_{n+K} is calculated and an additional segment boundary is defined, if a given threshold is exceeded.

This combined segmentation splits the signal into a sequence of microsegments and labels the segments according to the preclassification. Fig. 8 gives a word example showing the result of this segmentation procedure.

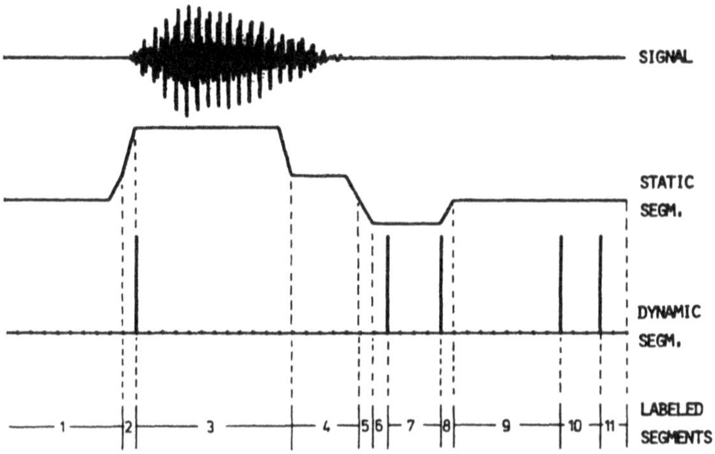

Fig. 8: Result of segmentation of the word /zɛks/

4. NORMALIZATION

The speech signal contains a large number of superimposed items of information. Since only the linguistic information is relevant for speech recognition, the speech velocity, influences of signal transmission and, above all, speaker-typical features must be regarded as disturbing factors the effects of which have to be minimized in an adequate way. The extraction of normalized features from these parameters, i.e., features which are largely cleared of such influences, is extraordinarly important for speaker-independent speech recognition.

The time normalization of different speech velocities is largely achieved by the formation of feature-dependent segments of variable length.

To compensate for different or varying signal levels, the moving average of the maximum levels is taken in the acoustic-phonetic analysis and each new speech input is normalized to this value. In this way, problems of an analog gain control can be largely avoided.

By far the most difficult problem for speaker-independent systems, however, is the s p e a k e r n o r m a l i z a t i o n . Feature variations are caused by different dimensions, above all, by different lengths of the vocal tracts of different speakers. Fant /10/ has shown that the vocal tract lengths can be derived, in a first approximation, from the formants of the speech spectrum. This is done by marking steady-state vowel frames for every word and calculating the corresponding formants. The vocal tract lengths (VTL) derivable from the formant structures are averaged for every word and learned in a non-supervised process /11/.

This is followed by the normalization of the area function (see section 2) with the aid of the learned vocal tract length. At the same time, an interpolation is performed to obtain an area function with a fixed number of 8 sections. The articulatory model uses this normalized area function for calculating the articulatory parameters. Although this linear scaling does not fully account for the differences between male and female speakers, this procedure is nevertheless able to achieve, in a first approximation, a certain normalization of anatomical differences of vocal tracts.

5. CLASSIFICATION

After the analysis, segmentation, and normalization of the signal, it is necessary to classify the microsegments obtained. For this purpose the *mean feature vector* $\underline{x}$ is derived from the parameter sets of the frames belonging to a microsegment. Then it is possible to determine the distances between the segment represented by the vector $\underline{x}$ and the different sound classes within the preclassified sound group, i.e., if a segment has been identified, for instance, as 'nasal', only the distances to all classes of this group (/m/, /n/ and /ŋ/) are determined. The smallest distance decides into which class the segment belongs. This classification uses the Euclidian distance

$$v_i = ((\underline{x} - \underline{m}_i)^T \Sigma_i^{-1} (\underline{x} - \underline{m}_i))^{1/2} \quad , \tag{8}$$

where the mean vectors $\underline{m}_i$ and covariance matrices Σ_i have to be previously determined for all classes i.

For the sake of completeness, it should be mentioned that, in order to reduce the effects of misclassifications, both the best candidate and the second one are identified. Using the probabilities obtained from the different distances, misclassifications can be easily corrected by applying phonetic knowledge. Since this approach is also recommendable for the preclassification, there are, in total, four candidates and their corresponding probabilities for each segment.

The result of the *acoustic-phonetic analysis* obtained in the signal analysis, segmentation and classification is given in Fig. 9 which lists the microsegments of the word example /zɛks/ shown in Fig. 8.

```
*******************************************************************************
 SEG  STF NFR :PA:FL:FH:NA:VO:                                      I DAUER    I
-------------------------------------------------------------------------------
   1    1   9 :   :**:  :  :  : S   71   F   49   *  100   *    0  0I********* I
   2   10   1 :   :   :**:  :  : SH  99   Z   33   AE: 98   E:  10  2I*        I
   3   11  13 :   :   :  :  :**: AE: 98   E:  17   SH  98   V   47  0I**********I
   4   24   4 :   :   :**:  :  : SH  96   V   53   CH  71   X   61  0I****     I
   5   28   1 :   :**:  :  :  : CH  40   S   11   *  100   *    0  0I*        I
   6   29   1 :**:  :  :  :  : *  100   *    0   S   18   CH   8  6I*        I
   7   30   4 :**:  :  :  :  : *  100   *    0   *  100   *    0  0I****     I
   8   34   1 :**:  :  :  :  : *  100   *    0   S   43   CH  24  2I*        I
   9   35   8 :   :**:  :  :  : S   15   CH  10   *  100   *    0  0I******** I
  10   43   3 :   :**:  :  :  : CH  81   X   41   *  100   *    0  0I***      I
  11   46   2 :   :**:  :  :  : F   18   S    2   *  100   *    0  0I**       I
*******************************************************************************
```

Fig. 9: Result of the acoustic-phonetic analysis (Table of microsegments)

6. VERIFICATION

To complete the recognition, it is necessary to assign the word to be identified to one word of the known vocabulary or to reject it. This process, termed *verification*, consists of two steps.

To begin with, several *hypotheses* for the most probable combinations of segment sequences have to be derived from the table of microsegments shown above. For this purpose, those segments of the microsegment list are systematically combined which are most likely to occur as segment sequences. According to the probabilities, a hypothesis list is compiled from these segment sequences.

A special procedure of *dynamic programming* is then applied to check the agreement of the hypotheses with the words of the given vocabulary. Insertions, deletions and substitutions are evaluated according to cost functions and the costs are added /12/. The decision is made in favour of the hypothesis involving the lowest 'cost'. The corresponding word of the vocabulary is read out and passed on for conversational processing. If the cost exceeds a certain threshold, the hypothesis is rejected.

The dynamic programming procedure applied here requires a special *transcription* of the vocabulary given in orthographic writing into microsegments corresponding to the hypothesis list. Moreover, it is necessary to specify the cost of sound substitions. For this purpose, it is recommendable to use a *sound substitution matrix* (with the elements W).

The distances

$$D(i,j) = \mathrm{Min}\begin{cases} D(1,j-1) & + \; W(i,j) \\ D(i-1,j-1) & + \; 2W(i,j) \\ D(i-1,j) & + \; W(i,j) \end{cases} \tag{9}$$

are then calculated. The distance D(M,N) represents the 'overall distance' (sum of all costs) between a pair of segment sequences to be compared, which finally is adequately normalized.

The verification procedure is illustrated in Fig. 10.

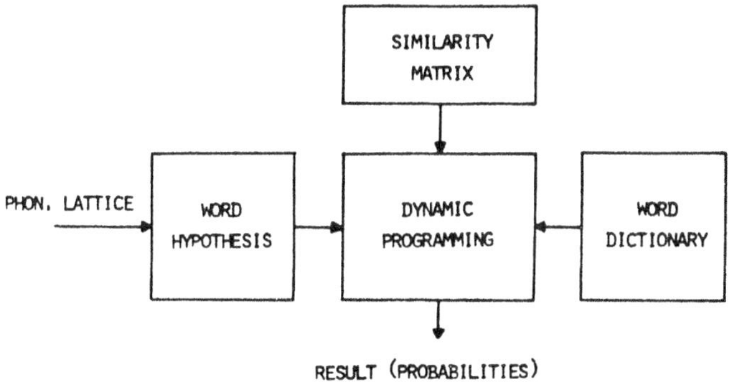

Fig. 10: Block diagram of the verfication procedure

7. SUMMARY

The performance of the acoustic-phonetic analysis applied in the modular speech recognition system MOSES is tailored to the particular requirements of automatic inquiry services: recognition of speech limited to telephone bandwidth, speaker independence and large vocabulary.

The speech recognition system is implemented as a program system with all necessary components on a 16-bit processor (PDP 11/34). Time-consuming operations like analysis or classification are performed by an attached array processor (AP-120B). Since the computing time has not yet been fully minimized, the real-time factor is approximately 90.

The results so far obtained have shown that the acoustic-phonetic analysis based on a special articulatory model is a valuable tool for speech recognition. However, to increase the recognition rate, it will be necessary to make use of much more phonetic knowledge for the recognition process, if speaker independence is regarded as the most important target. This holds, in particular, for improved decision criteria based on extensive statistical research and for the application of special rules in the conversion of the table of microsegments into phonetic symbols. Moreover, if specific characteristics of the vocabulary are accounted for by a modified verification procedure, a considerable increase of the recognition rate can be expected. Appropriate procedures are being developed.

REFERENCES :

/1/ LEA, W.A.: "Trends in Speech Recognition", Prentice Hall Inc. 1980

/2/ MARKEL, J.D., GRAY, A.H.: "Linear Prediction of Speech", Springer Verlag, 1976.

/3/ WAKITA, H.: "Estimation of Vocal Tract Shape by Optimal Inverse Filtering and Acoustic/Articulatory Conversion Method" SCRL, Santa Barbara, California, SCRL Monograph No. 9, 1972.

/4/ SHIRAI, K.: "Vowel Identification in Continuous Speech Using Articulatory Parameters", Proceedings of ICASSP 81, Atlanta, Vol. 3, 1981, pp. 1172-1175.

/5/ HARSHMAN, R.: "Factor Analysis of Tongue Shapes", Jasa, Vol. 62, No. 3, Sept. 1977, pp. 693-707.

/6/ LADEFOGED, P.: "Generating Vocal Tract Shapes from Formant Frequencies", Jasa, Vol. 64(4), Oct. 1978, pp. 1027-1035.

/7/ ATAL, B.S., RABINER, L.R.: "A Pattern Recognition Approach to Voiced-Unvoiced-Silence Classification with Application to Speech Recognition", IEEE, ASSP-24, No. 3, June 1976, pp. 201-212.

/8/ KASUYA, H. ; WAKITA, H. : "An Approach to Segmenting Speech into Vowel- and Nonvowel-Like Intervals", IEEE Vol.ASSP-27, Vol. 4, 1979, pp.319-327

/9/ LOCHSCHMIDT, B. : "Segmentation of Band-Limited Speech Signals Using the Parameters of the Linear Prediction Method", Frequenz 34, Vol.5, 1980, pp.136-141

/10/ FANT, G. : "A Note on Vocal Tract Size Factors and Nonuniform F-Pattern Scalings", Speech Sounds and Features, MIT Press, Cambridge, 1973, pp.84-93

/11/ WAKITA, H. : "Normalization of Vowels by Vocal-Tract Length and Its Application to Vowel Identification", IEEE, Vol.ASSP-25, N° 2, Apr. 1977, pp.183-192

/12/ SAKOE, H. ; CHIBA, S. : "Dynamic Programming Algorithm Optimization for Spoken Word Recognition", IEEE, Vol. ASSP-26, N° 1, Feb. 1978, pp.43-49

THE EFFICIENCY OF DEMISYLLABLE SEGMENTATION IN THE RECOGNITION OF SPOKEN WORDS

Günther Ruske and Thomas Schotola

Technical University of Munich, Fed. Rep. of Germany

Abstract. The efficiency of syllabic segmentation and recognition is demonstrated in an experiment using three different word recognition systems and a vocabulary of 1000 words. In each system the preprocessing is carried out by a special loudness analyzer which yields 22 specific loudness functions. The first system avoids any segmentation and the total word pattern is time normalized to a constant length. In the second system syllable nuclei are detected and used as segment boundaries; the segments are time normalized and the resulting word pattern classified. The third system classifies each demisyllable using vowels and consonant clusters as decision units. For increasing vocabulary sizes the demisyllable recognition system shows less decay in recognition performance than the other systems and is significantly advantageous for a vocabulary consisting of 1000 words.

1. INTRODUCTION

Several experiments were carried out in order to investigate the influence of different segmentation procedures as well as the application of different decision units in automatic word recognition. In many commercial word recognition systems the words them-

---

This article is based on "The efficiency of demisyllable segmentation in the recognition of spoken words" by the above authors appearing in IEEE Int. Conf. on Acoustics, Speech and Signal Processing, Copyright © 1981 IEEE. This research has been accomplished within the SFB 50 "Kybernetik", Munich, which is supported by the "Deutsche Forschungsgemeinschaft" (DFG).

J.-P. Haton (ed.), Automatic Speech Analysis and Recognition, 153–163.

selves are successfully used as decision units. Here, recognition consists of a similarity measurement of the complete word pattern with a set of reference templates representing the "vocabulary" of the system. However, if a large vocabulary has to be processed the number of those words which are very similar and whose only difference in the end is a single phoneme will, of course, increase. In this case segmenting the speech signal into smaller units does offer certain advantages. The introduction of smaller units such as syllables, parts of syllables, diphones or phonemes means that the classification procedure does not have to discriminate the complete vocabulary directly, but that the words are composed of smaller units. Therefore the efficiency of segmentation must be evaluated as a function of vocabulary size. For this purpose two extreme systems are compared in the following: a word recognition system which does not perform any segmentation at all and a demisyllable recognition system which uses consonant clusters and vowels as decision units.

The consideration of the syllable structure takes place in many systems after primary classification. Since segmentation and classification are both nonlinear operations (throwing away information) it is important to decide which operation should be applied first. In several systems the syllable structure is taken into account within the signal domain. In (1) syntactic methods are applied in order to localize syllable nuclei and process consonantal parts. Those methods which achieve the segmentation of syllables by evaluating the loudness contour or an energy function are reported in (2,3).

In our own previous experiments (4,5,6) the segmentation into demisyllables was performed using a loudness function obtained by reproducing human loudness sensation with a model. In this system some recognition errors result from the problems which arise when dividing the consonantal part between two consecutive syllable nuclei into one final and one initial demisyllable. Another problem may arise when a diphtong has to be cut in half in the syllable nucleus. In fact, uncertainties and errors during segmentation will often result in final recognition errors. Due to these complex relations it is not easy to explain the differences in the performance of the demisyllable system and the simple word recognition system. So that the efficiency of syllabic segmentation alone could be investigated a third system was developed which performs a syllabic segmentation but uses the complete words as decision units as is done in the first system.

The following experiments were carried out using isolated words in order to evaluate the recognition performance at the acoustic-phonetic level. That means that other knowledge sources such as syntax or semantics are excluded; only lexical information is used to decide whether a word is a member of the vocabulary or not. An important parameter is the vocabulary size which varies from 125 to 1000 words in the following experiments.

2. METHOD

Fig. 1 shows a block diagram of the three word recognition systems described in this paper. They start from the same processing of the speech signal. This is to ensure equal conditions for all systems as far as possible. Effective preprocessing is obtained using a hardware model which simulates the human loudness sensa-

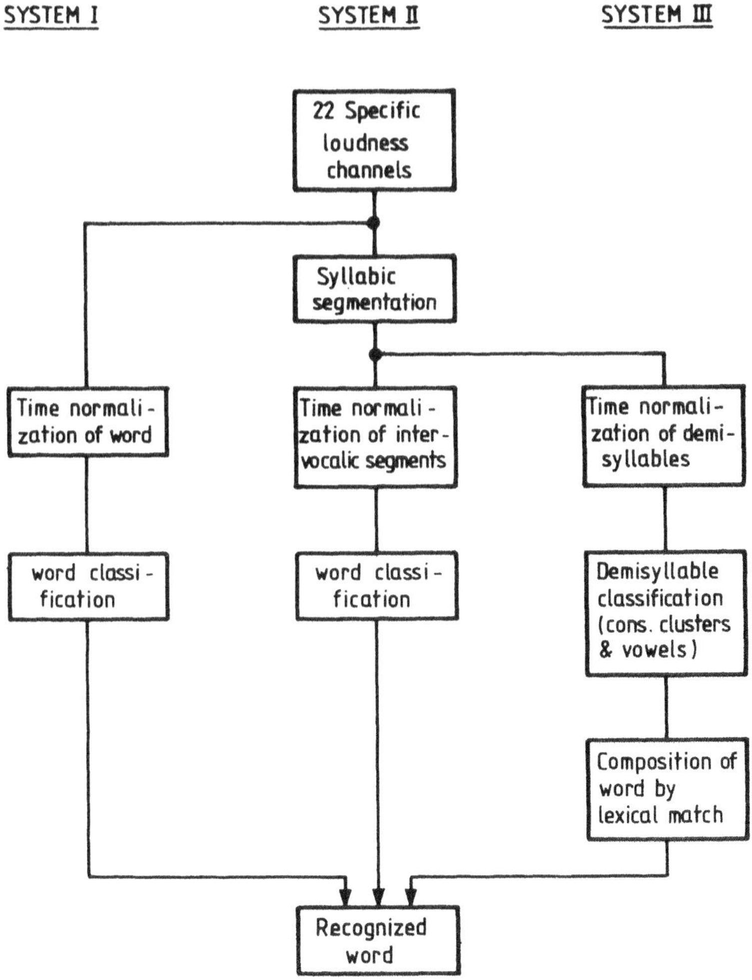

Fig. 1 Block diagram of the three systems

tion thereby reflecting one of the basic auditory functions of hearing. The model yields 22 specific loudness functions covering a frequency range of 50 Hz to 8.5 kHz arranged on a critical-band--rate scale (6). The sampled data of these functions represent the input patterns (as a series of "loudness spectra") of the spoken words. Fig. 2b shows the specific loudness representation of the German word "kommen" (/komən/). In all systems the similarity measurement is carried out by calculating the "city-block" distance between the loudness spectra of the unknown word and the reference templates.

The temporal coordination of corresponding spectra can be effectively performed applying "dynamic pattern matching" methods. Our own previous experiments have shown that a similar performance is achieved using a "dynamic interpolation" procedure which yields patterns of a constant length. Dynamic interpolation means that the trace of consecutive loudness spectra in the 22-dimensional pattern space is divided into a constant number of new interpolation points which are distributed equidistantly along this trace (4). In this way the procedure effects a nonlinear time normalization. In order to allow a comparison of the different systems, the average number of loudness spectra representing the templates is the same in all systems.

System I: In the first system only the beginning and the end of the spoken word are determined, i.e. no further segmentation is performed. Dynamic interpolation is applied to the complete word pattern A-E resulting in a constant number of 38 loudness spectra for each word as shown in fig. 2c. Since the average number of syllables in the vocabulary is about 2, this system uses the same number of spectra as are applied in bisyllabic words in system II and III. The complete word pattern is brought to a constant average loudness (amplitude normalization) and is classified.

System II: In the second system a syllabic segmentation is achieved by evaluating the total loudness $N(t)$ and a modified loudness function $N_m(t)$ which is calculated from the weighted sum of all 22 channels and which is shown as a dotted line in fig. 2a. Local maxima of the smoothed modified loudness function $N_m(t)$ are proper candidates for syllable nuclei, as shown at B and D in fig. 2a. A crude discrimination between "vowel" and "non vowel" re-

Fig. 2
a) Loudness function of the German word "kommen" (/komən/)
b) 22 specific loudness functions
c) normalized word pattern without segmentation (system I)
d) normalized word pattern with syllabic segmentation (system II)
e) normalized demisyllable patterns (system III).

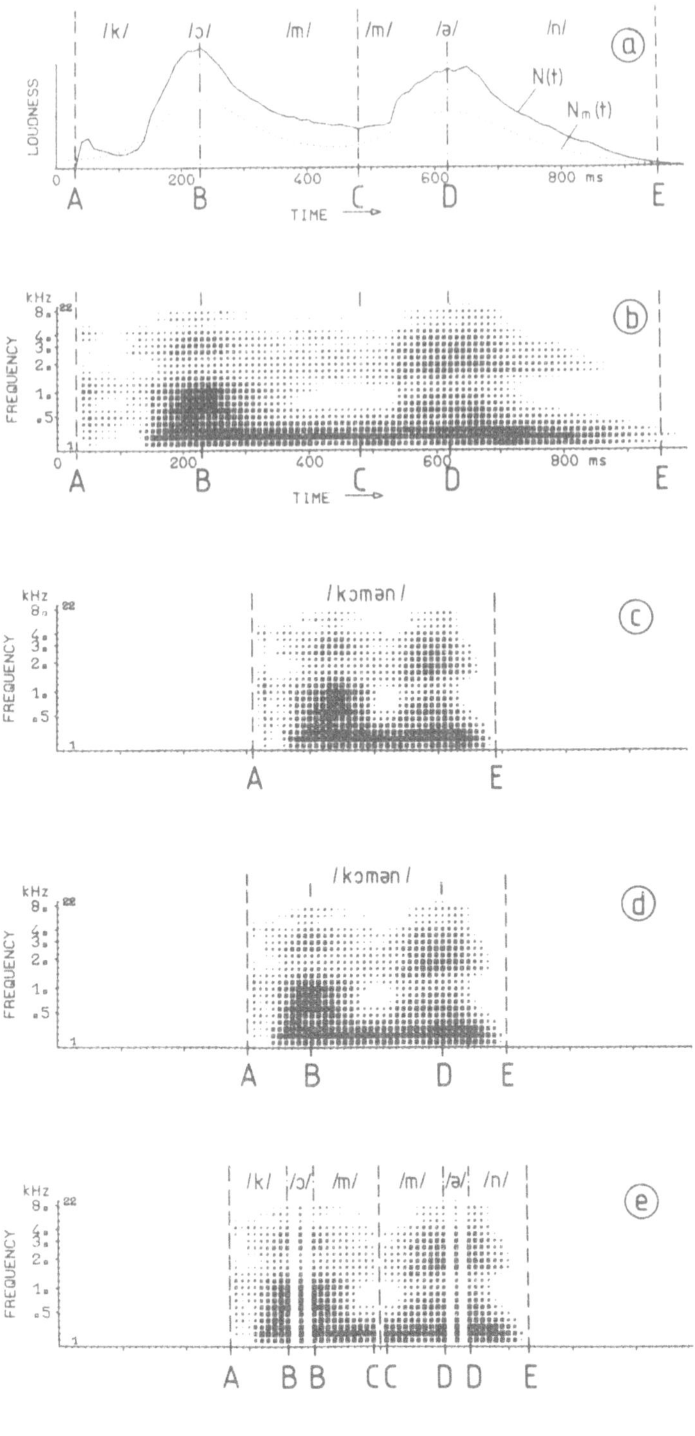
/k/ /ɔ/ /m/ /m/ /ə/ /n/
(a)
LOUDNESS
N(t)
Nm(t)
0 200 400 600 800 ms
A B C D E
TIME
(b)
kHz
FREQUENCY
8 4 3 2 1 .5
(c)
/kɔmən/
(d)
/kɔmən/
(e)
A B B C C D D E

jects maxima produced by fricatives and plosives. Then, the segment boundaries are established at the syllable nuclei, the word beginning, and the end, respectively; fig. 2d shows the resulting patterns. The first segment A-B and the last segment D-E were time normalized both by dynamic interpolation into 10 spectra. Since the medial segment B-D includes two demisyllables this segment was accordingly time normalized into 20 spectra. The complete word pattern consisting of all three segments is brought to a constant average loudness and is classified.

System III: In the third system a demisyllable segmentation is applied by which each syllable is divided into an initial and a final demisyllable. The demisyllable patterns A-B, B-C, C-D, and D-E are time normalized into 10 spectra each, using dynamic interpolation, see fig. 2e. Additionally, each loudness spectrum within a demisyllable is brought to a constant average loudness. The single vowel spectra at B and D in fig. 2e were constructed by averaging neighbouring spectra within the vowel region. Then, the demisyllables and vowel spectra are classified using initial consonant clusters, final consonant clusters and vowels as decision units (5). Consonant clusters are processed as a whole and are not further divided into single consonants. Medial consonant clusters are divided into one final and one initial consonant cluster. A suitable cutting point can be found at a minimum of the loudness function N(t) which can be seen at C in fig. 2a. Here, the medial consonant /m/ is cut in half resulting in a final /m/ and an initial /m/; this method was described in (4). The demisyllable recognition yields a phoneme string which is then compared with the lexical entries of the vocabulary. The similarity measurement with the lexical entries is based on statistics concerning the confusions between consonant clusters as well as between vowels and diphtongs.

3. SELECTION OF VOCABULARIES

The basic vocabulary for all experiments consists of the 1000 most frequent German words. Due to grammatical flexions this vocabulary contains many very similar words which are,for instance, identical in the root of the word and only differ in their endings. The total word list was spoken by one speaker three times giving a speech material of 3000 words. Since the aim of this paper is focussed on the influence of segmentation, the speech material of a single speaker was assumed to be sufficient for basic investigations. In the 1000-word experiments 2000 words were used as a training set and the remaining 1000 words as a test set. In system III the patterns for the consonant clusters and vowels were produced from 2000 words and tested using the remaining 1000 words. In order to be able to study the influence of the vocabulary size, subsets, selected at random, were generated from the basic 1000-word vocabulary giving subsets of 500, 250, and 125 words.

4. RESULTS

The segmentation errors and recognition errors are depicted in fig. 3. The syllabic systems (II and III) have a constant segmentation error rate of about 4%; since no attempt was made to correct a wrongly segmented word these errors result directly in recognition errors. Therefore, the total recognition error rate is the sum of segmentation and recognition errors shown in fig. 3. The dependence of the recognition scores on the vocabulary size is depicted in fig. 4. For a small vocabulary the word recognition without segmentation (system I) performs best; however, the larger the vocabulary the stronger the decay in the recognition rate. Due to the constant amount of segmentation errors the systems II and III show a lower performance for small vocabularies but the decay is not so strong. Demisyllable recognition (system III) performs best for vocabularies consisting of more than 500 words. As mentioned above system III is organized into two stages: demisyllable recognition and final correction using the lexicon. Altogether, 40 initial and 58 final consonant clusters as well as

Recognition system		Vocabulary size 1000	500	250	125
I. Without segmentation	recogn. error	18.7	14.4	8.4	6.4
II. Syllabic segmentation	segm. error	4.2	5.2	4.4	4.8
	recogn. error	13.0	8.2	6.8	3.2
III. Demisyllable recognition	segm. error	4.2	5.2	4.4	4.8
	recogn. error	10.0	7.7	4.4	3.5

Fig. 3 Segmentation and recognition errors of the systems I - III (in %).

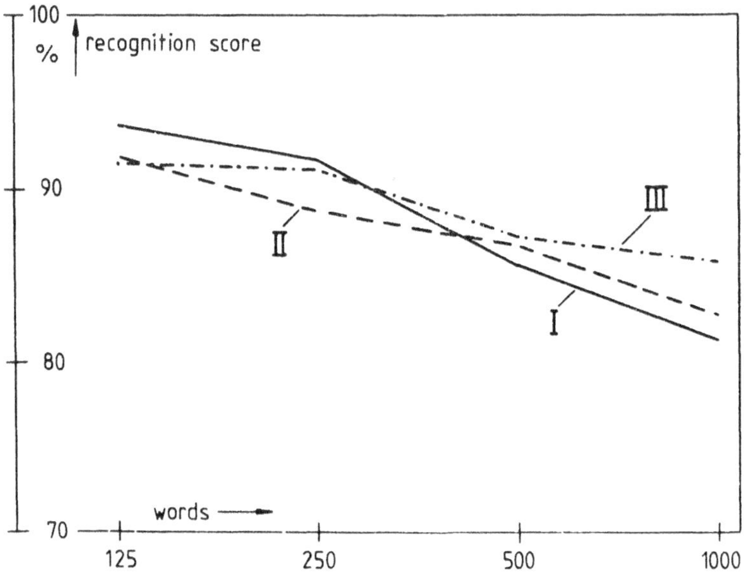

Fig. 4 Recognition scores as a function of vocabulary size; for system II and III segmentation and recognition errors were added.

19 vowels and diphtongs are needed to describe the 1000-word vocabulary. Some insights into the performance of demisyllable recognition (stage 1) can be obtained by confusion matrices of the consonant clusters. Fig. 5 shows the list of all confusions which occurred for initial consonant clusters standing in initial position of the word. Of course, the a-priori probabilities for the various clusters are very unequal and cannot be seen from the list in fig. 5 which displays percentage values. The average recognition rate is 91.3% for the initial consonant clusters in fig. 5, 94.7% for final consonant clusters standing in final position of the word, and 86.6% for those consonant clusters (initial and final) which stand in medial position between syllable nuclei. It is worth noticing that the phonetic output of demisyllable recognition contains the correct phonetic transcription, for the whole word, in about 47% of all cases. In the remaining cases the number of phonetic errors could be reduced by the lexical correction (stage 2). Lexical correction makes explicite use of the redundancy of the vocabulary as is also done implicitely in the word recognition systems I and II when measuring distances in the word pattern space. Correction was based on two different sets of confusion data: one set was adapted to this experiment and one was obtained from another vocabulary. The average error rates from

SPO-KEN	COR-RECT	CONFUSED WITH					
b	96	d 1.7	h .8	r .8	ʃp .8		
d	89	b 5.7	g 2.9	' 1	h 1	v .5	
g	94	d 4.5	' .9	k .5	h .5		
p	65	h 15	' 10	k 5	gr 5		
t	79	k 7.9	d 5.3	h 5.3	tr 2.6		
k	95	g 1.9	d .9	h .9	t .9		
'	95	h .24	d .6	g .6	j .6	k .2	f .2
						r .2	v .2
h	85	' 10.6	d 1.4	b .7	k .7	g .7	p .7
f	98	fr 1.6					
m	92	n 6.4	v 1.3				
n	88	m 11.2	v 1				
v	99	l .5	m .5				
j	98	' 1.6					
l	83	v 6.7	j 3.8	n 3.8	' 1	bl 1	m 1
bl	88	gl 5.9	l 5.9				
gl	67	bl 9.5	kl 9.5	g 4.8	h 4.8	pl 4.8	
pl	33	kl 50	gl 17				
kl	100						
fl	67	f 33.3					
pfl	33	f 66.6					
r	86	' 3.4	d 3.4	br 1.7	f 1.7	h 1.7	kr 1.7
br	41	r 29.4	b 17.6	dr 5.9	tr 5.9		
dr	50	b 33.3	d 16.7				
gr	89	k 2.8	d 2.8	fr 2.8	p 2.8		
pr	45	fr 11	gr 11	h 11	kr 11	ʃpr 11	
tr	76	t 9.5	d 4.8	f 4.8	fr 4.8		
kr	71	k 21.4	tr 7.1				
fr	74	f 20.7	' 1.7	b 1.7	pr 1.7		
ʃr	100						
ʃpr	67	f 8.3	gr 8.3	h 8.3	pr 8.3		
ʃl	89	ʃt 11					
ʃv	67	g 16.7	h 16.7				
ʃp	83	g 16.7					
ʃt	91	ʃ 5.3	t 3.5				
ʃtr	100						
ʃn	100						
ʃ	100						
z	96	l 1.2	f .6	v .6	ts .6	tsv .6	
ts	98	f 1.6					
tsv	88	ts 12					

Fig. 5 Confusion list of the 40 initial consonant clusters (in %).

both correction attempts are assumed to be fair estimates for the overall error rates of system III and are depicted in fig. 3.

5. DISCUSSION

All experiments were performed using the speech material spoken by a single male speaker. Since no speaker-specific information is evaluated during segmentation, it can be expected that the syllabic segmentation can be applied to other speakers in the same way. The word lists were spoken within one day resulting in relatively uniform word patterns. Therefore, Fig. 5 does not reflect long term intra-speaker variations, which could result in a decreasing recognition accuracy.

In the word recognition systems (I and II) the total word pattern was brought to a constant average loudness. That means that the loudness contour remains, giving additional information about the temporal structure of the word. This was advantageous in the experiments, since the three recordings of the word list were spoken relatively uniformly. For the same reason word recognition with syllabic segmentation (system II) could not exceed the performance of the simple system I to any great extent. We suppose that syllabic segmentation really only comes into its own when applied to nonuniformly spoken speech material. In system III the temporal loudness contour was not yet evaluated since each loudness spectrum was amplitude normalized; if the loudness contour were taken into account in a similar way the performance of this system could be increased further. Special problems are the segmentation errors in system II and III. The effect of these errors could be reduced if one or two alternative segmentations were provided in the lexicon for those words which cannot be segmented consistently. Or, one could try to get more than one segmentation from the signal, thus offering a lattice of segmentation results which are all processed further.

An important advantage of demisyllable recognition (system III) is the fact that the classification procedure only has to be trained on a limited number of decision units, even for very large vocabularies. Restriction to a small set of decision units is necessary, especially when the system is to be adaptive. Then the adaptation only has to be applied to the units and not to the complete vocabulary. Using smaller units offers a further advantage. Starting from a fixed training set (e.g. 2000 words) each demisyllable class can be represented by several (about 45) reference templates whereas each word class of the systems I and II is represented only by 2 templates. In our experiments system III performed best at a vocabulary size of more than about 250 words. In particular, the decay in the recognition performance of this system is less strong than in the other systems. Therefore demisyllable recognition seems to be advantageous for large vocabularies.

REFERENCES

(1) De Mori, R.: 1977, Syntactic recognition of speech patterns. In: Syntactic Pattern Rececognition, Applications (K.S.Fu, ed.). Springer, pp. 65-94.

(2) Gresser, J.Y., and Mercier, G.: 1975, Automatic segmentation of speech into syllabic and phonemic units. In: G.Fant and M.A.A.Tatham (eds.), Audit. Anal. and Perc. of Speech, Academic Press, pp. 359-382.

(3) Mermelstein, P.: 1975, Automatic segmentation of speech into syllabic units. JASA 58, pp. 880-883.

(4) Ruske, G., and Schotola, T.: 1978, An approach to speech recognition using syllabic decision units. Proc. IEEE Int. Conf. on Acoustics, Speech and Signal Processing, Tulsa, pp. 722-725.

(5) Schotola, T.: 1974, Automatic recognition of consonant clusters. 8th Int. Congr. Acoust., London, p. 253.

(6) Zwicker, E., Terhardt, E., and Paulus, E.: 1979, Automatic speech recognition using psychoacoustic models. JASA 65, pp. 487--498.

INTERACTION BETWEEN AUDITORY, SYLLABIC AND LEXICAL KNOWLEDGE IN A SPEECH UNDERSTANDING SYSTEM

P. Demichelis[-], R. De Mori[+], P. Laface[-]

- CENS - Politecnico di Torino - C.so Duca degli Abruzzi 24
+ ISI - C.so Massimo d'Azeglio 42 - 10125 TORINO (Italy)

Abstract

The paper describes how does the syllabic knowledge interact with the lexicon and the auditory knowledge in a Speech Understanding System.

1. Introduction

Research on Speech Understanding Systems (SUS) offers a good opportunity for conceiving and evaluating complex parallel processing systems capable of modeling human perception. This topic is relevant to Computer Science. Perception, Cognitive Sciences, Pattern Recognition and Artificial Intelligence. The conception of Speech Understanding Systems as perceptual models may also be useful for designing machines capable of performing more ambitious tasks than the ones actually achieved by commercial machines and laboratory prototypes offering the challenge of experimenting new parallel algorithms and non-conventional system architectures.

Central to the organization of a Speech Understanding System (SUS) are the representation of knowledge structured on several levels of abstraction, and the control strategy that has to use the knowledge efficiently.

This note deals with the problem of accessing a large lexicon in continuous speech.

In spite of the fact that lexical access is constrained by semantic, syntactic and pragmatic predictions, the number of word candidates whose first syllable agree with the acoustic cues in a given time interval of the signal may still be large.
Lexical hypotheses are progressively reduced by parallel al-

J.-P. Haton (ed.), Automatic Speech Analysis and Recognition, 165–178.

gorithms which focus the attention on the most evident interpretations.

At the beginning, in parallel with the generation of the first set of lexical hypotheses, other more detailed acoustic cues are extracted. These detailed acoustic cues and the active set of lexical hypotheses act as constraints for syllable hypothesization.

Acoustic cues are extracted independently of the lexicon and top-down constraints which do not modify lower level acoustic analysis. This characterizes a "parsimonious" control strategy in agreement with the one that is supposed to be adopted by the human brain.

2. Generation of lexical hypotheses

The generation of lexical hypotheses is seen as a complex problem to be solved which is decomposed into simpler sub-problems. The structure of this decomposition is represented by a generalized AND/OR graph called Problem-Reduction-Representation (PRR). Fig. 1 shows, as an example, the PRR of the Italian word /tono/ ('tone' in English).

In order to generate the word hypothesis /tono/, the prosodic cues and the phonetic features that are sufficient for the generation of this word hypothesis have to be detected. This means that both the corresponding sub-problems have to be solved in the same time-interval. The word problem /tono/ is thus represented by an AND node whose sub-problems are reached by arrows connected by a solid arc-segment. Often, in practice, these two sub-problems may be solved only to some extent. In other words, features can be hypothesized from the speech spectra with degrees of evidence that determine a degree of solution for the sub-problems.

Interesting work using other approaches are those by Baker (1), De Mori (2), Jelinek (3), Vintsjuk (4), Mc Keown (5). The main characteristic of the present approach is the way the lexical knowledge is structured. Here a word is represented by a network of sub-problems rather than by a network of phone symbols obtained by matching the input data with prototype spectra. The sub-problems of the network contain relations and rules for evaluating the degrees of solution. Some sub-problems concern the hypothesization of syllabic segments represented by strings of characters. Each character is associated a set of phonetic features. This association can be context-dependent.

Furthermore, unlike the approaches based on stochastic automata, the solution of sub-problems like those of Fig. 1, may contain rules for combining structural primitives, for dealing with numerical attributes and also procedural rules for moving the focus of attention in order to generate lexical hypotheses.

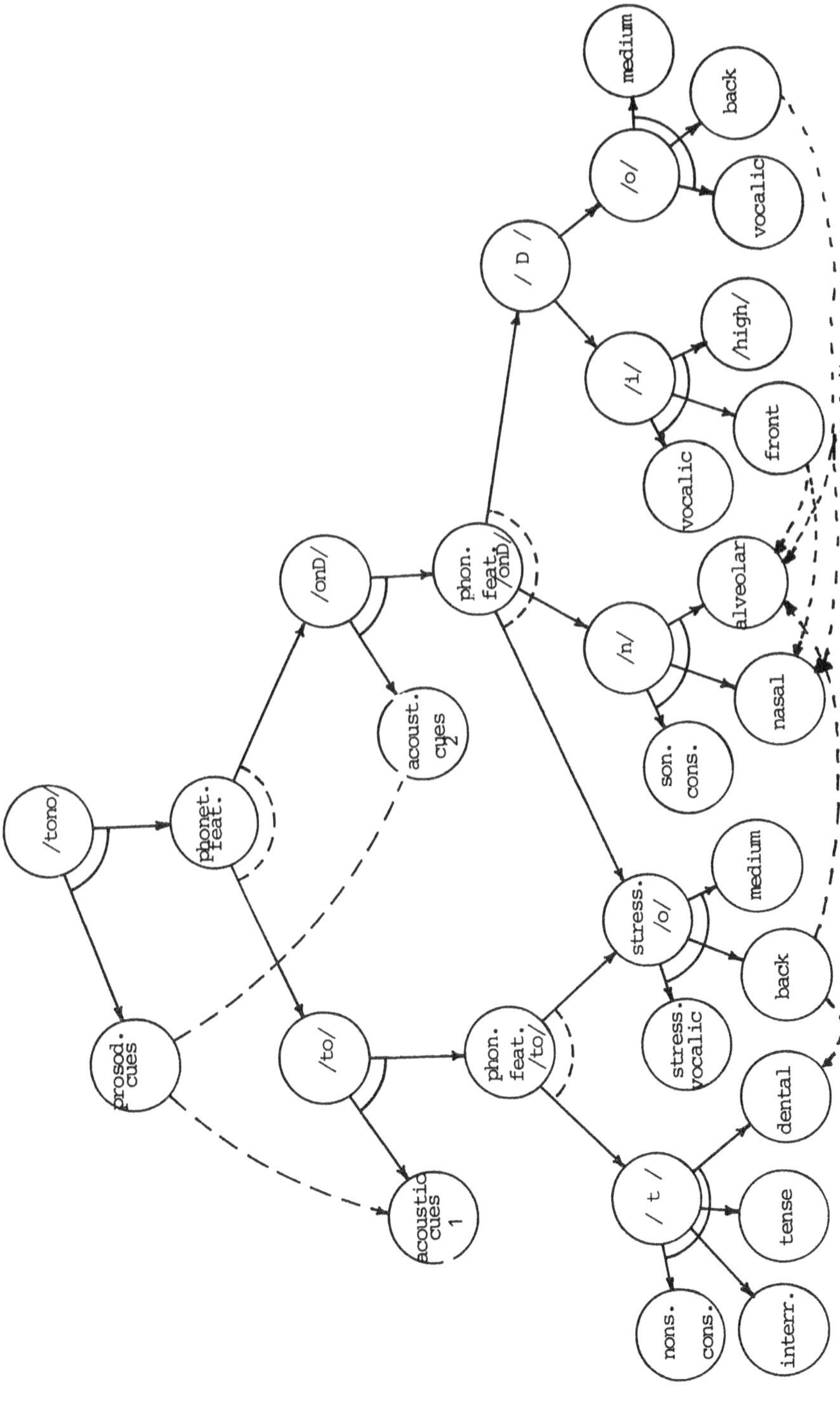

Fig. 1 - Graphical representation of a lexical item.

Movements of the focus of attention are not only determined by probabilities or purely numerical computations. They may depend on qualitative consideration also.

The process of matching a word hypothesis with acoustic cues may be stopped when enough syllables have been detected for distinguishing a word among its competitors. This is in accordance with perceptual findings (Marslen-Wilson (6)).

At the beginning of a sentence, words are allowed to start almost everywhere but, as soon as hypotheses of high evidence are found, an inhibition mechanism can stop the proliferation of competing hypotheses. Procedural rules may implement this mechanism of "focus by inhibition" which inhibits the generation of word hypotheses in a time interval in which already existing word hypotheses consistent with syntactic and semantic constrints proceed to completion with a high degree of evidence. Marslen-Wilson (6) has argued that a mechanism of this type is also present in human perception.

Continuing the description of Fig. 1, the sub-problem consisting in the detection of phonetic features has two sub-problems pointed by arrows joined by a dashed arc. The dashed arc represents a temporal relation for which the sub-problems have to be solved in successive and contiguous temporal intervals and the first problem has to be solved in the first interval, the second problem in the second interval and so on. Phonetic features are hypothesized by two syllabic sub-problems indicated as /to/ and /onD/. The syllabic sub-problem /to/ has two sub-problems. The sub-problem indicated as 'acoustic cues 1' contains a precondition for focusing the attention of the lexicon on the word /tono/.

The second sub-problem of /to/ concerns the hypothesization of the set of phonetic features characterizing this pseudo-syllabic segment.

Pseudo-Syllabic-Segments (PSS) are considered here rather than true syllables. PSSs may overlap in time and delimit time intervals in which context-dependencies due to coarticulation effects can be described by a proper set of rules.
A formal definition of PSSs in terms of a generative grammar has been proposed by De Mori et al. (7)

The sub-problems of /to/ are those of the phonetic features of /t/ and of the stressed vowel /o/. Each sub-problem corresponds to the detection of a phonetic feature in a segment of the speech pattern. The set of phonetic features characterizing a symbol depends on the context and is determined by phonological rules. Each phonetic feature is related to a set of acoustic cues.

Some of these relations are context-independent, some others are context-dependent. Context-dependencies are represented in Fig. 1 by dashed arrows joining sub-problems labelled by names of phonetic features. Every arrow represents the fact that the solution of the sub-problems pointed by the arrow is influenced by data produced for the solution of the sub-problem which the arrow come out from. For example, the evidence of the phonetic feature /back/ for the vowel /o/ appears in the relation between the phonetic feature /dental/ and its corresponding set of acoustic cues.

The relations between phonetic features and acoustic cues are not represented in Fig. 1 because they belong to the syllabic knowledge.

The second PSS /onD/ shares the first /o/ with /to/ and has an OR node labelled D for representing all the possible terminations of the word (the word is a noun which can be singular or plural). The possible terminations are the unstressed /o/ for the singular and the unstressed /i/ for the plural. The structure of D is common to many nouns in Italian and is defined once for all of them. The possibility of using the same structure for representing the termination of many words is very useful for languages like French, Spanish and Italian. This possibility is not available when the lexicon is represented by a tree like in Woods et al. (8), Klatt (9), Smith and Erman (10).

A formal definition of lexical structures and their relations with acoustic cues can be done in the framework of context-sensitive grammars and will be given in Appendix A.

Generating lexical hypotheses is a problem solving activity with some important differences with respect to the classical way in which problem solving is seen in Artificial Intelligence. The lexical task, in fact, is not that of finding the best solution; rather, all the solutions consistent with top-down expectations and with the evidence of a-acoustic data have to be kept. Furthermore, some solution have to be accepted even if all the corresponding sub-problems have not been completely solved.

In other words, the task here is not accomplished by using an admissible search algorithms for finding the best path through the space of all the possible solutions. Rather, the goal is that of keeping as small as possible the probability of missing correct hypotheses. False alarms can be tolerated although they also must be kept as rare as possible.

The efficiency of a Speech Understanding System does not depend only on the search strategies implemented in it. Efficient systems can be designed by introducing rich knowledge representations and wise procedures for their use. Knowledge representation, particularly for what concerns the relations between auditory, syllabic and lexical knowledge is one of the peculiar aspects of the research described in

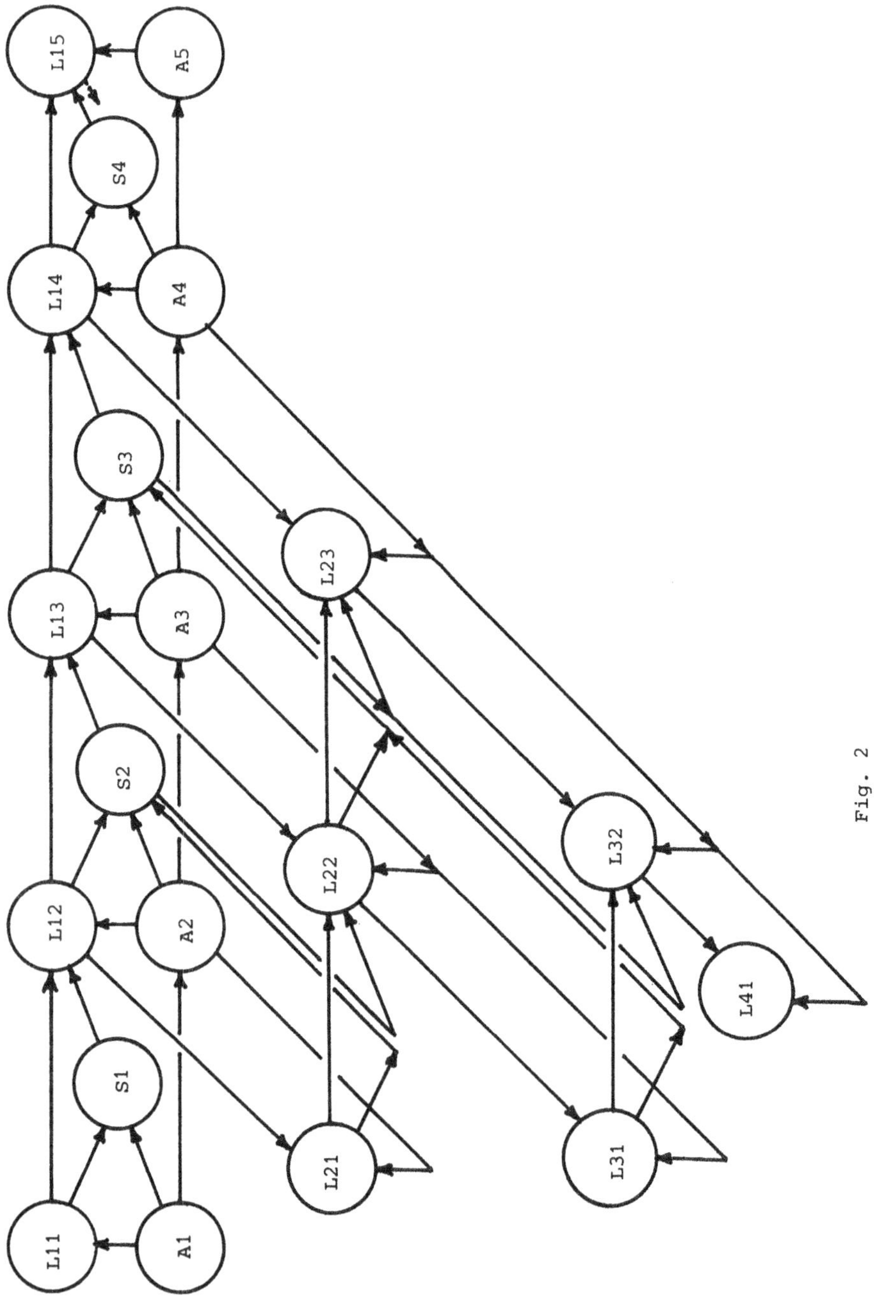
L15
A5
S4
L14
A4
S3
L23
L13
A3
S2
L22
L32
L12
A2
L41
S1
L21
L31
L11
A1

Fig. 2

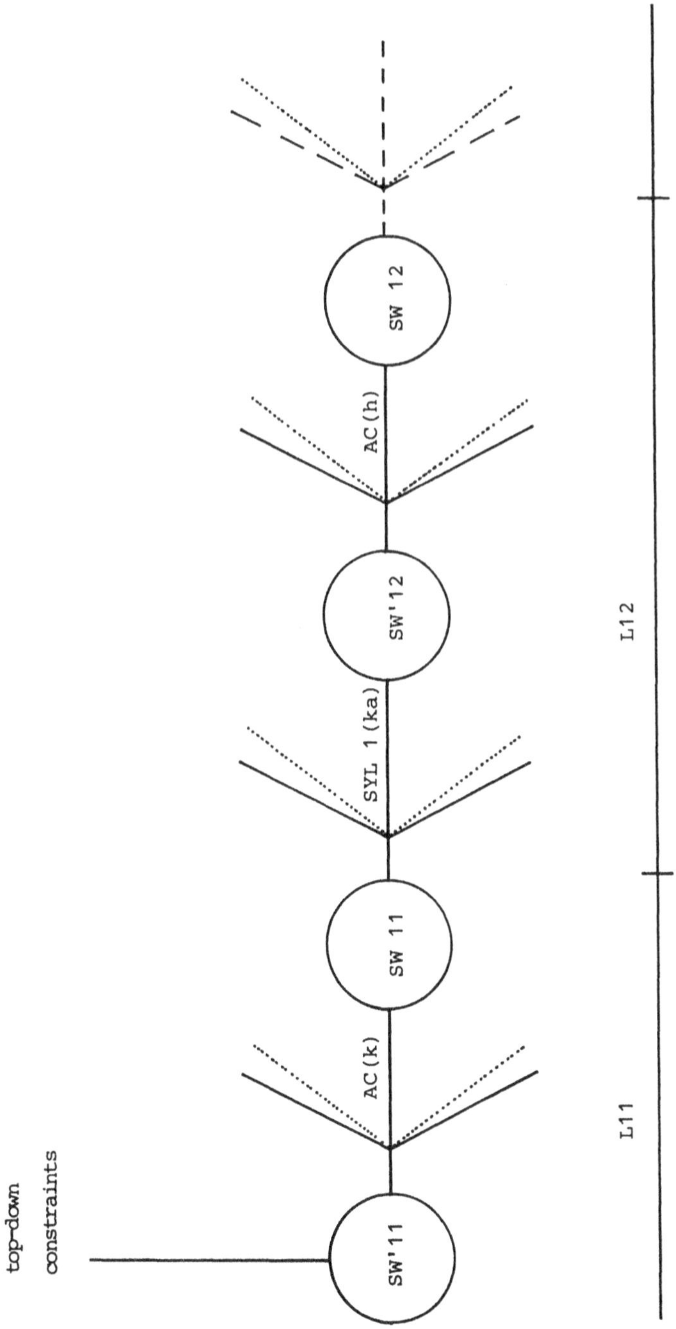

Fig. 3 - Tree structure for lexical access

this paper.
An efficient knowledge use is made possible by a task decomposition represented by an iterative array of computational activities. This is a novelty introduced in this paper which extends to complex tasks, such as perceptual models the advantages of using iterative computational structures.

3. A network model for lexical access.

Figure 2 shows an example of a three-dimensional iterative network of computational activities. Each computational activity is performed by a set of concurrent or cooperating processes. The horizontal dimension of the network corresponds to time, while the vertical dimension corresponds to levels of abstraction; along the third dimension, there are circles representing processes for the generation of sets of lexical hypotheses starting at different time instants. These time instants can only correspond to the beginning of PSS hypotheses.

The activities A1,A2,....,An,..AN (1<=n<=N) are auditory activities. A collection of computational activities called "experts" transforms the input signal into a description of acoustic cues. When the acoustic cues of a pseudo-syllabic segment are found, messages containing descriptions of acoustic cues are sent to upper perceptual levels. The first activity A1 is accomplished after a set of acoustic cues is obtained and transmitted to the lexical level and another set of detailed acoustic cues is sent to the syllabic level. When a message from the auditory level is received by the lexicon, a lexical activity L11 is created. Let T11 be the time interval in which the acoustic cues, sent to the lexixon, have been detected.

The second auditory acitvity A2 attempts to perform the same task as A1 but in a time interval T12 immediately following T11.
Some of the operations performed by A2 can be concurrent with some others performed by A1. These concurrent operations can be the execution of different program segments using disjoint sets of data in the same interval of the signal or the execution of the same program on different intervals of the signal.

L11 uses a lexical-access tree structure of pointers to the graphs representing lexical items.

An example of a lexical-access tree structure is shown in Fig. 3. All the words of the lexicon can be accessed from the root. The branches coming out from the root of the tree are labelled by acoustic preconditions AC(k), (1<= k <=K), K is the number of branches coming out from the root.
The acoustic cues generated by A1 are such that they can-

not match more than one acoustic precondition AC(k). Acoustic preconditions roughly correspond to pseudo-syllabic segments.

An acoustic precondition AC(k) is a sequence of acoustic cues (ac(k1), ac(k2),...,ac(ki),...,ac(kI(k))), each cue may have associated conditions on its parameters. I(k) is the number of acoustic cues in AC(k).
The acoustic cues of AC(k) have to be detected in the indicated sequence even if they are allowed to be not contiguous in time. The duration of a gap between the detection of two successive acoustic cues in the signal cannot exceed a threshold tg(k). Values of the threshold gaps tg(k) ($1 <= k < = K$) are determined by a small set of word-independent rules.

Let AC(k) be the acoustic precondition which matches with the string of acoustic cues A1AC generated by A1, AC(k) points to all the words of the lexicon for which AC(k) solves the sub-problem of the type indicated as "acoustic cues 1" in the example of Fig. 1. Matching an acoustic precondition is what Erman et al. (11) call "stimulus". The main difference is that here a stimulus does not come from a blackboard of hypotheses, but it is a message sent by an auditory activity to a lexical computational activity. Furthermore, this message describes, unambiguously, a segment of speech data.

Acoustic preconditions may correspond to phonetic or prosodic features.
The dashed arrows in Fig. 1 coming out from the prosodic cues sub-problem represent the influence of the prosodic cues on the solution of the acoustic precondition sub-problem. In Fig. 3, each branch labelled by an acoustic precondition AC(k) ends in a set of word candidates. Let SW11(k) be the set following the branch labelled AC(k) and let:

$$SP1 = (SW11(k) \;/\; 1 <= k <= K).$$

Let S1(t) be a function of time ($0 < = t <= TPH$), TPH being the duration of the sentence to be interpreted) taking values on SP1.
Let S1(ta) be the value taken by S1(t) at time ta. S1 (ta) indicates a set of word hypotheses generated by the matching of a sequence of acoustic cues detected in the signal and a precondition AC(k). Only the most plausible of these word hypotheses (hopefully no more than one) will be retained after more descriptions of acoustic cues are received and processed by lexical activities.

The selection of lexical hypotheses may be also constrained by top-down syntactic and semantic predictions. All the active word hypotheses at a given time generate predictions for the following pseudo-syllabic segment.

Generation of syllabic hypotheses is constrained by lexical expectations and the detection of acoustic cues. Syllabic hypotheses are generated by a "syllabic expert"

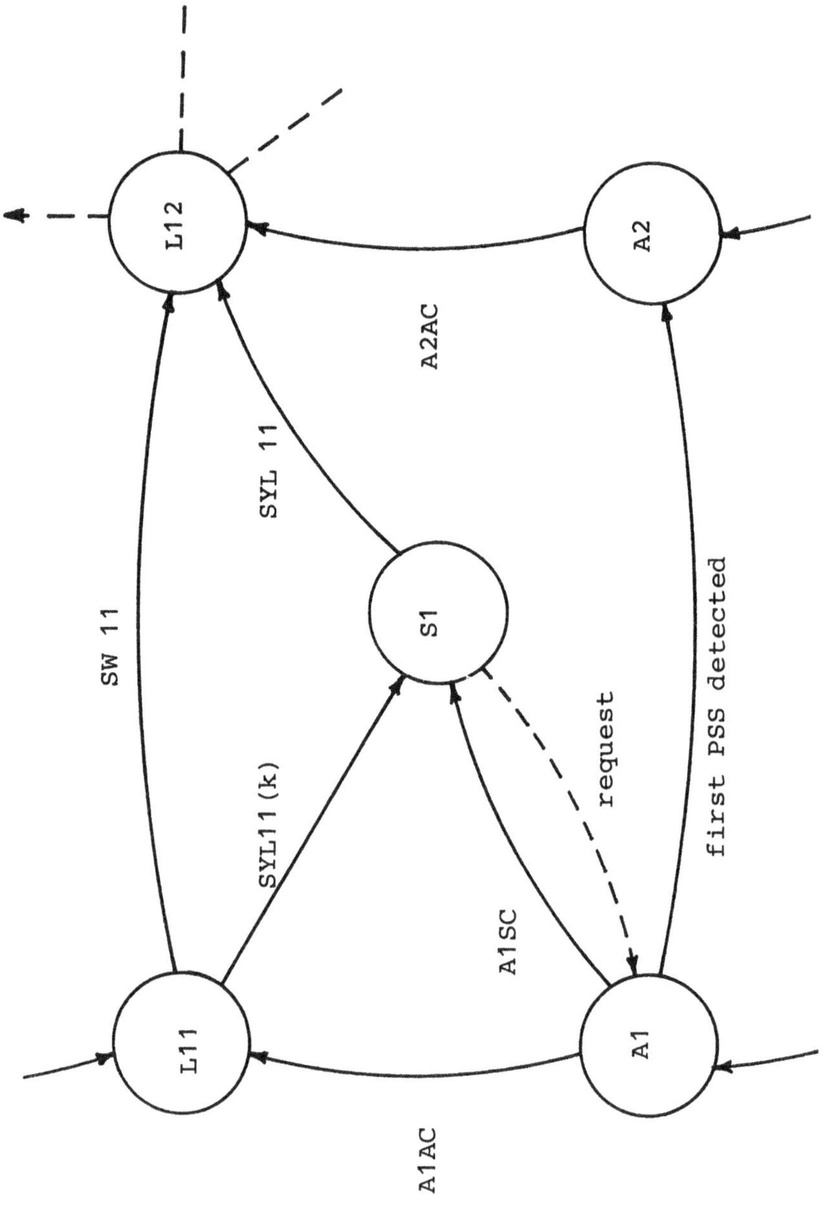

Fig. 4 - Messages exchanged between computational activities.

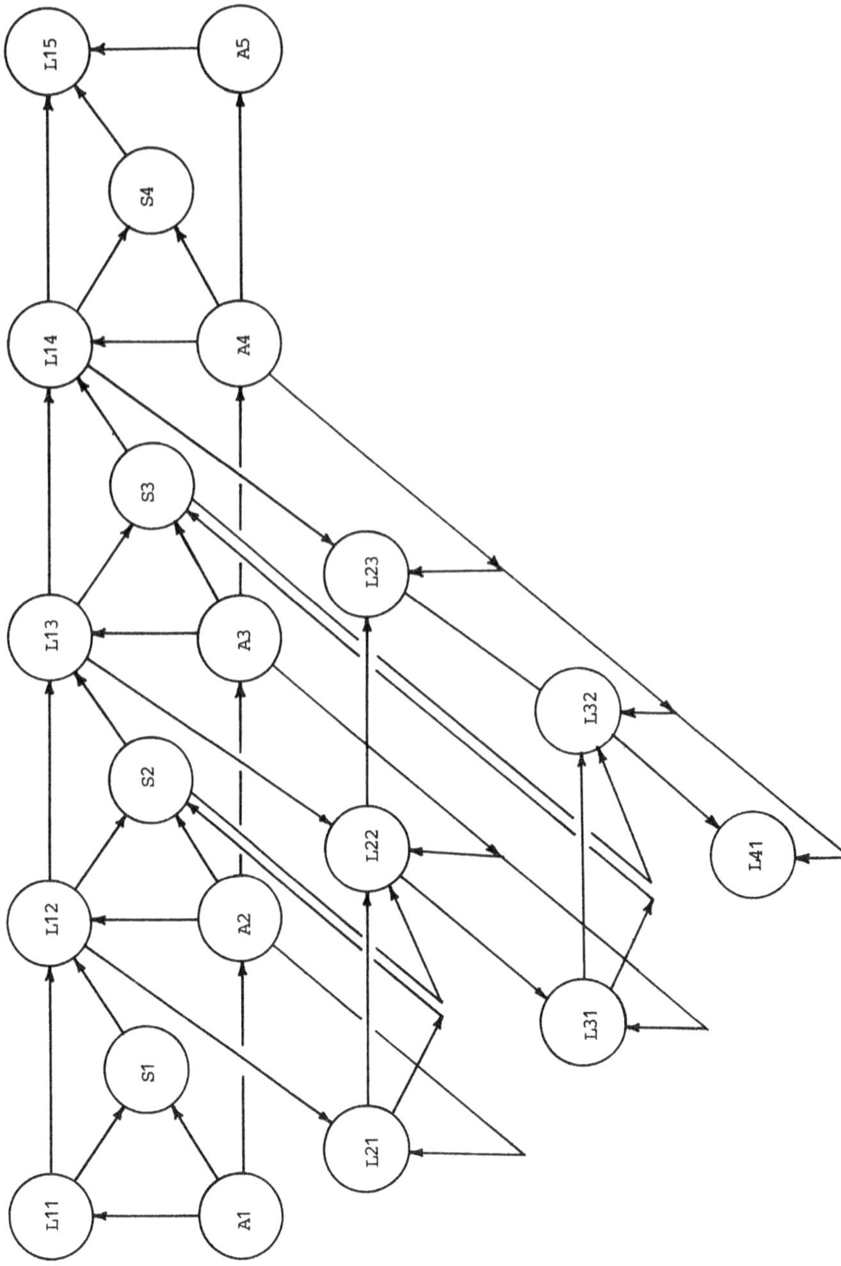

Fig. 5 - Iterative network of computational activities for lexical access in continuous speech.

whose composition in terms of cooperating and concurrent processes is described in another paper (7).

Exceptionally, generation of hypotheses inside a syllabic segment (PSS) may require the extraction of further acoustic cues which won't be in competition with the ones already available but will be just added to them. Model-driven acoustic cues extraction would make the control strategy less parsimonious and should be avoided.

Fig. 4 shows a detail of the interaction between L1, A1 and S1.

The acoustic cues extracted by A1 are sent as a message A1AC to L11. Let assume that A1AC matches AC(k), then L11 generates a set of word hypotheses SW11 compatible with A1AC and top-down constraints SW11 $\subseteq$ SW11 (k). Based on SW11, L11 sends a request to S1 for generating hypotheses in the set SYL11(k) made of the first syllabic segments of the words in SW11. S1 receives a set of acoustic cues AISC from A1 and generates a set of syllabic hypotheses SYL11 compatible with the acoustic cues:

$$SYL11 \subseteq SYL11(k).$$

The set SYL11 is sent to L12.

The request for the extraction of more acoustic cues is indicated by a dashed arrow coming out of S1. Interactions with levels superior to the lexicon are also indicated.

After having received the hypotheses about possible word initial PSSs, L12 evaluates the evidences of partial word hypotheses and decides which of them can be retained.

Generating syllabic hypotheses corresponds to solving some of the sub-problems appearing in the representation of a lexical item. The order of solution of these sub-problems is not determined a priori as in the problem solving strategy proposed by Stockman (13) and Kanal (14), by a fixed ordering of the sons of each node in a Problem Reduction Representation. Rather it is dinamically determined by a procedural knowledge directly invoked by the detection of the acoustic cues.

The structure of this knowledge will be described later in this paper. Again, computational activities are determined by messages intentionally exchanged by processes and not just by the appearance of hypotheses in a blackboard. Furthermore, in the project proposed here, procedural knowledge is integrated into the structural one.

L12 keeps active a set of word hypotheses SW'12 such that:

$$SW'12 \subseteq SW11.$$

The basic scheme described so far is repeated iteratively. A2 generates acoustic cues based on which SW'12 can be further reduced to SW12, L12 will then send a message to S2 for the verification of the second syllables of the active word hypotheses. These successive refinements of lexical hypotheses can be reduced to a simple consistency evaluation when a single evident word hypothesis remains that univoquely allows to predict its continuation.

A new lexical activity L21 is created by A2 in parallel with L12. L21 accounts for words which may start with a syllable in the time interval T12. Inhibition messages, represented by arrows in the third dimension in Fig. 2, may inhibit the generation of a new set of lexical hypotheses or the continuation of the verification of the hypotheses in already generated sets. Notice that a single auditory and a single syllabic activity are performed for many lexical activities in the same interval even if lexical activities constrain the generation of syllable hypotheses. This is another peculiar characteristic of the model proposed in this paper.

Inhibition may be resumed and generation of lexical hypotheses may restart at the end of evident hypotheses. The inhibition mechanism is controlled by rules that must be tailored in such a way that semantically significant words are never missed because of faults of the inhibition mechanism. Inhibitions are messages sent from one lexical activity to another and their effect is such that already existing hypotheses influence the generation of other hypotheses on successive portions of the signal. This is a novel conception of focus of attention rules in lexical access suggested by speech perception theories (Marslen-Wilson (6)).

Acknowledgments

This work has been supported by the Italy-France cooperation program.

Bibliography

(1) J.K. Baker (1975) - The DRAGON System: An Overview- IEEE Trans. on Acoustics Speech and Signal Processing vol. ASSP-23, pp. 24-29

(2) R. De Mori - Syntactic Recognition of Speech Patterns, Applications - Ed. K.S. Fu, Springer Verlag, Cap. 4

(3) F. Jelinek - Continuous Speech Recognition by Statistical Methods - Proceedings IEEE, vol.64, pp. 532-556

(4) T.K. Vintsjuk (1976) - Generative Grammars and Dynamic Programming in Speech Recognition with Learning - Proc. ICASSP (Philadelphia, PA), pp. 446-449

(5) D.M. Mc Keown (1977) - Word Verification in the HEARSAY-II Speech Understanding System - Proc. ICASSP (Hartford, CN), pp. 795-798

(6) W.D. Marslen-Wilson (1975) - Sentence Perception as an Interactive Parallel Processing - Science, vol. 189, pp. 487-501

(7) R. De Mori, P. Laface and E. Piccolo (1976) Automatic Detection and Description of Syllabic Features in Continuous Speech - IEEE Trans. on Acoustics Speech and Signal Processing, vol. ASSP-24, pp. 365-379.

(8) W.A. Woods, M. Bates, G. Brown, B. Bruce, C.Cook J. Klovstad, J. Makhoul, B. Nash-Webber, R.Schwartz, J. Wolf and V. Zue (1976) - Speech Understanding Systems - Final Technical Progress Report. Volumes I-V. Report n. 3438, Bolt Beranek and Newman, Cambridge,MA

(9) D.H. Klatt (1979) - Speech Perception: A Model of Acoustic Phonetic Analysis and Lexical Access. Journ. of Phonetics, vol. 7, pp. 279-312

(10) L.D. Erman and A.R. Smith (1981) - NOAH - A Bottom--Up Word Hypothesizer for Large Vocabulary Speech Understanding Systems - IEEE Trans. on Pattern Analysis and Machine Intelligence, vol. PAMI-3, pp. 41-51

(11) L.D. Erman, F. Hayes-Roth, V.R. Lesser and D.R.Reddy (1980) - The HEARSAY-II Speech Understanding System. Integrating Knowledge to Resolve Uncertainty. - Computing Surveys, vol. 12, pp. 213-253.

(12) R. De Mori - Computer Models of Speech using Fuzzy Algorithms - Plenum Press, N.Y., 1982

(13) G.C. Stockman (1977) - A Problem Reduction Approach to the Linguistic Analysis of Waveforms. - Ph.D. Thesis, University of Maryland, College Park, MD.

(14) L.N. Kanal (1979) - Problem-Solving Models and Search Strategies for Pattern Recognition - Trans.on Pattern Analysis and Machine Intelligence, vol.PAMI-1, pp. 193-201.

SPEAKER INDEPENDENT CONNECTED WORD RECOGNITION

S. E. Levinson

Acoustic Research Department
Bell Laboratories, Murray Hill, New Jersey 07974

Abstract

A method for speaker independent connected word recognition is described. Speaker independence is achieved by clustering isolated word utterances of a 100 speaker population. Connected word recognition is based on a syntax-directed dynamic programming algorithm which matches the isolated word templates to sentence length utterances. The method has been tested on a task oriented English-like language based on a 127 word vocabulary. Four subjects, two men and two women, spoke a total of 209 sentences comprising 1750 words. At an average speaking rate of 171 words per minute over dialed-up telephone lines, a correct word recognition rate of 97% was observed.

I. INTRODUCTION

Between the relatively well understood task of speaker trained isolated word recognition and the more problematic goal of unrestricted continuous speech recognition, there is a type of automatic speech recognition with which we have been experimenting for approximately three years. We have chosen to call it connected word recognition by which we mean that we are attempting to recognize strings of words articulated carefully but without pauses between them. In our experiments, we have been using reference templates derived from isolated word utterances and have employed dynamic programming algorithms to match the reference templates to the connected speech stream.

J.-P. Haton (ed.), Automatic Analysis and Recognition of Speech, 179–190.

At the same time we have been experimenting with applying statistical clustering techniques to large collections of isolated word templates as a means of achieving some degree of speaker independence. We have also been employing some simple methods of grammatical analysis both as an error correction device and as an aid in semantic analysis.

Recently we have attempted a synthesis of these diverse investigations and have met with some success in speaker independent connected word recognition. As a test vehicle for our method we have used an airline information and reservation task. We have designed a simple formal language of high relative redundancy (.75) and moderate vocabulary size (127 words) for the purpose. Speaker independent templates for each of these vocabulary words were fabricated by clustering isolated word utterances from 100 speakers.

Four subjects, two male and two female, provided test data for an experimental evaluation of our method. They spoke a total of 209 sentences comprising 1750 words at an average speaking rate of 171 words per minute. For these utterances an average correct word recognition rate of 97% was observed resulting in a correct sentence recognition rate of 87%. When used in conjunction with a semantic processor this accuracy is sufficient to provide robust human/machine communication on the "airline" task.

The remainder of this paper is organized as follows. In Section II a brief overview is given of the development of our clustering methodology and we discuss in detail the preparation of the speaker independent reference templates for the "airline" vocabulary. Section III is devoted to a discussion of the development of our dynamic programming algorithms for connected word recognition with special emphasis on the syntax-directed approach on which this study is based. In Section IV we give the details of the experimental evaluation of the syntax-directed system. Finally in Section V we examine the implications of our experimental results with respect to practical technology and further research.

II. CONSTRUCTION OF SPEAKER INDEPENDENT TEMPLATES

In many isolated word recognition experiments including some of the very earliest, it has been observed that performance degrades, in some cases only slightly and in others disasterously, when used by a speaker other than the one for whom the machine was trained. The cause of this phenomenon is that the features chosen for recognition exhibit a high degree of variability across speakers.

There are several ways in which an isolated word recognizer might deal with variability due to differences among speakers. One possibility is to find or construct a set of features which is invariant across a speaker population. Alternatively the recognizer could be made to adapt to a particular speaker by making appropriate transformations on a set of "universal" templates during the recognition process.

Although both of these approaches are very attractive, there is, at present, not enough detailed knowledge to implement them and obtain robust speech recognition. A third approach exists in which one copes with speaker variability by characterizing it statistically. We have studied this method and in particular have chosen to perform the statistical analysis by means of clustering techniques.

Initially, Rabiner [1] proposed an ad hoc clustering technique based on thresholding the Itakura distance [2] between pairs of templates. The method was applied to a recognition task using a 54 word vocabulary of computer terms. After clustering word templates of four speakers, recognition tests on eight new speakers were performed. The 85% accuracy observed in these tests was not as good as desired but it was substantially better than the 77% rate obtained without clustering.

On the strength of Rabiner's results, a more intensive study, the results of which are reported by Levinson et. al. [3], was undertaken. The goal of this investigation was to employ clustering techniques which were known to be effective. In addition to the primary objective of locating significant clusters, we sought to identify outliers and measure cluster overlap. To this end we selected the ISODATA technique of Ball and Hall [4] augmented by some simpler and less sophisticated procedures whose intended purpose it was to reveal the coarse structure of the data. In actuality, these auxiliary procedures were of little use but the ISODATA procedure supervised by a human operator proved to by highly effective.

Like many other clustering methods, ISODATA is a heuristic for finding an assignment of samples to classes so that the value of an objective function is optimized. ISODATA is unique, however, in that the number of clusters can be altered by splitting and merging procedures invoked when preset thresholds on intercluster and intracluster dispersion measures are exceded. In our implementation we allowed a human operator to alter the splitting and merging criteria while the clustering operation was in progress.

As in the earlier study we retained the Itakura metric as

the fundamental distance measure. We defined cluster center to mean that token whose maximum distance to all other members of the cluster is least. Based on this definition, inter-cluster distance was defined as the average pairwise distance between cluster centers. Similarly the intracluster distance was defined to be the average of the distances from all members of the cluster to the cluster center. Finally our figure of merit was chosen to be the ratio of intercluster distance to average (over all clusters) intracluster distance. In the case of two spherically symmetric clusters, a value of 2.0 for this ratio signifies no overlap.

The ISODATA procedure we used was the following. Initially a guess is made at the number of clusters and tokens are chosen at random to be the cluster centers. Then:

i.) Samples are assigned to a cluster according to the nearest neighbor rule.
ii.) New cluster centers are computed according to the minimax definition.
iii.) The dispersion measures are computed according to the definitions given above.
iv.) Clusters are split and merged as indicated by the dispersion measures.

The above steps are repeated until the figure of merit is sufficiently large or a stable configuration, indicated by two successive identical iterations, is reached.

This procedure was applied to a 39 word "alpha-digit" vocabulary comprising the ten digits, 26 letters of the alphabet and three additional words, STOP, ERROR, and REPEAT. We collected utterances of the vocabulary from 100 speakers, 50 male and 50 female. All were native speakers of American English. These data were hand edited to remove artifacts near the endpoints and clustered one word at a time. We found clusters ranging in number from 6 for the word F to 19 for REPEAT. The largest clusters contained a minimum of 12 and a maximum of 36 tokens. There were as few as 3 and as many as 16 outliers for a single word. The value of the figure of merit was least at 2.4 for the word ONE and greatest at 3.9 for E.

By using the cluster centers as speaker independent templates, Rabiner et. al. [5] obtained a respectable 80% correct recognition rate for 20 new speakers.

The principal disadvantage to the supervised ISODATA procedure is that a substantial amount of human participation is required. For large data sets, this cost is prohibitive. Based upon experience gained in the ISODATA experiments,

Rabiner and Wilpon [6] devised a fully automatic procedure using the earlier Rabiner [1] study as a starting point. Essentially, their method is the following agglomerative procedure in which the entire data pool is taken as the initial configuration. Then:

i.) Find the minimax center of the current set of samples.
ii.) Locate all samples whose distance to the center is less than some preset threshold value.
iii.) Label these samples as a cluster and remove them from the pool.

The above procedure is repeated either until all samples are classified or until no more clusters can be found. Cluster centers may then be determined by methods described below.

This technique was applied to the 39 word "alpha-digit" vocabulary and performed only slightly worse than the interactive procedure. In a subsequent study on the 54 word computer instruction vocabulary, Rabiner and Wilpon [7] achieved a correct recognition rate of greater than 95%. In both of these studies the speaker independent recognition rates were at least as good as those obtained in a speaker trained mode. In addition the clustering procedure was shown to give a significant increase in performance over use of an equal number of randomly selected tokens as templates. In this study, cluster centers derived by averaging the time-registered autocorrelation coeficients were shown to be superior to the minimax centers.

Finally, Wilpon et. al. [8] have applied the automatic clustering procedure to the 127 word "airline" vocabulary described by Levison [9]. Again 100 speakers, 50 men and 50 women, spoke the entire vocabulary. This time, however, the robust training procedure of Rabiner and Wilpon [10] was used instead of the manual editing procedure to eliminate artifacts. The centers of the 12 largest clusters were used as templates in an isolated word recognition experiment on 20 new speakers. A correct word recognition rate of 89% was obtained. This is slightly better than the speaker dependent result reported for the same vocabulary by Levinson et. al. [11]. The speaker independent templates are the ones used in the connected speech recognition experiment we shall describe in Section IV.

III. USING ISOLATED WORD TEMPLATES FOR CONNECTED WORD RECOGNITION

Since whole word templates have proven to be adequate for

robust recognition of isolated words it is reasonable to try to use them for connected speech. The apparent advantages of this idea are that whole word templates give good representations of within-word coarticulation, since the natural articulation of the word is preserved in the template, and the complexity of such a system is reduced by eliminating a hierarchy of acoustic-phonetic templates. Equally obvious is the principal disadvantage that concatenating isolated word templates does not take into account coarticulartory effects which occur at word junctures in fluent speech. To a limited but perhaps useful extent, the undesirable effects of word juncture phenomena can be avoided by a temporal registration technique based on well known dynamic programming principles.

The use of techniques based on dynamic programming to compensate for non-linear variation in speaking rates seems to have begun over a decade ago with independent studies by Velichko and Zagoruiko [12] and Sakoe and Chiba [13]. Although the intended use was not for connected word recognition, the extension to that problem is quite obvious. Explicit use of the technique for connected word recognition was first reported by Sakoe and Chiba [14], Sakoe [15] and Bridle and Brown [16]. Our experience with the method comes from studies by Itakura [2] and Rabiner et. al. [17] on isolated word recognition. In the later, we defined the unconstrained endpoint-local minimum procedure (UELM) on which subsequent work in connected word recognition by Rabiner and Schmidt [18], and Myers and Rabiner [19] was based. Myers and Rabiner [20] devised the level building algorithm which they showed to be equivalent to, but substantially more efficient than, the Sakoe and Chiba two level procedure. In an effort to use the UELM procedure for larger vocabularies and longer utterances Levinson and Rosenberg [21] introduced syntactic constraints in a form used previously [22].

The syntax-directed procedure of Levinson and Rosenberg used a finite state parser in conjunction with the UELM algorithm in a hypothesize and test paradigm. The parser gave the UELM algorithm a word candidate and a nominal starting point in the speech stream. The UELM procedure returned the score of the match and its nominal end point which would then serve as the initial point for a transition out of the appropriate state. Thus the two programs could be operated in a loop to recognize a sentence in left-to-right order.

Though not without merit, the system was plagued by its inability to preserve more than one path per state of the parser. This introduced a failure mode in which a word classification error might, due to extreme differences in template lengths, also cause a segmentation error from which recovery was unlikely.

It seemed that the only solution to the failure mode was to parse in the much larger state space formed by taking the cartesian product of the original set of states with the discrete time axis. This appeared to make the execution time exponential in the input length. In fact, however, Myers and Levinson [23] showed how to use the level building algorithm to overcome the seeemingly unavoidable combinatorial explosion.

The efficiency is a direct consequence of the local constraints inherent in the UELM recursion which cause only that portion of the cartesion product which is actually needed to be formed. To make the level building procedure syntax-directed three modifications are made:

i.) The levels are uniquely identified with states rather than word position so that the ith level need not be temporally contiguous to the 1+1st level.

ii.) The vocabulary is partitioned so that templates for only those words leaving the ith state are matched at the ith level.

iii.) An additional set of pointers is required to link the levels in temporal order.

These modifications are most conveniently implented if the states of the finite state parser are topologically sorted, that is numbered so that a transition from state i to state j forces i<j. Otherwise the algorithm is exactly as stated in [20].

IV. EXPERIMENTAL RESULTS

The syntax-directed level building procedure described above was subjected to an experimental performance evaluation with an airline information and reservation task language [9], a high relative redundancy (0.75) subset of English [24] containing more than 6×10^9 sentences whose finite state transition diagram has 144 states, 450 transitions and a 127 word vocabulary.

A set of 50 sentences using every vocabulary word and every state transition was constructed. These sentences were then spoken over dialed-up telephone lines by four speakers, two male and two female, at an average rate of 171 words per minute. The test sentences ranged in length from 4 to 17 words. Some containing telephone numbers were subject to few syntactic constraints while others requesting or giving information were quite stylized.

TABLE I
Experimental Results

SPEAKER	TOTAL SENTENCES	TOTAL WORDS	SPEAKER TRAINED		SPEAKER INDEPENDENT	
			CORRECT SENTENCES	CORRECT WORDS	CORRECT SENTENCES	CORRECT WORDS
SEL	59	478	56	470	55	470
AER	50	424	46	417	42	413
KLS	50	424	42	393	43	413
CES	50	424	35	349	41	408
TOTALS	209	1750	179	1629	181	1704
			86%	93%	87%	97%

TABLE II
Error Probabilities and Speaking Rates

SPEAKER	TEST RATE	TRAINED			INDEPENDENT		
		REF. RATE	RATIO	Pe	REF. RATE	RATIO	Pe
SEL	165	136	.83	.02	113	.68	.02
AER	188	129	.69	.02	113	.60	.03
KLS	159	101	.64	.07	113	.71	.03
CES	172	100	.58	.18	113	.65	.04

The test corpus was used in two experiments, one with speaker dependent templates and the other using the speaker independent templates described in Section II. The results of these experiments are summarized in Table I. Note that performance across speakers is highly variable in the speaker trained mode whereas speaker independent mode performance is more uniform and slightly better in the average. A possible explanation for this phenomenon may be found in Table II in which speaking and error rates are compared. Notice that speakers who performed poorly in the speaker trained mode were those whose ratio of reference speech rate to test speech rate is lowest. The figures are low enough to suggest that the UELM local constraints may have been violated. It is very difficult to estimate the reference speech rate for the speaker independent case since the templates for a given word vary greatly in length. Based on a crude estimate, it appears that the ratio of speaking rates is increased for the poor subjects in the speaker independent case and a concomitant decrease in error rate is obtained.

V. CONCLUSION

We have described the construction of speaker independent word reference templates and a syntax-directed dynamic programming algorithm which together permit fairly robust connected word recognition. In fact the accuracy is sufficient to allow reasonable performance when used as the "front end" of an existing conversational mode system [25]. We expect that this system will be further enhanced when the "front end" computation is carried out in special purpose hardware the prototype for which is now operating [26].

Unfortunately, the method cannot be offered as a general solution to the problem of fluent speech recognition. We believe in fact that the "airline" task is close to the limit of complexity which the system can comfortably support.

On the other hand, the experiments are worthwhile in two ways. First, there are some applications for which the simplicity and performance of the method is appropriate. Second, the general principles can be applied to units other than words, demisyllables [27] for example, and may thus be useful in more difficult speech recognition problems.

REFERENCES

1. Rabiner, L. R., "On Creating Reference Templates for Speaker Independent Recognition of Isolated Words," IEEE Trans. ASSP, ASSP-26(3), pp. 34-42, 1978.

2. Itakura, F., "Minimum Prediction Residual Principle Applied to Speech Recognition," IEEE Trans. ASSP, ASSP-23(1), pp. 67-72, 1975.

3. Levinson, S. E., Rabiner, L. R., Rosenberg, A. E. and Wilpon, J. G., "Interactive Clustering Techniques for Selecting Speaker Independent Reference Templates for Isolated Word Recognition," IEEE Trans. ASSP, ASSP-27(2), pp. 134-141, 1979.

4. Ball, G. H. and Hall, D. J., "Isodata, an Interactive Method of Multivariate Analysis for Pattern Classification," Proc. IFIPS Congress, 1965.

5. Rabiner, L. R., Levinson, S. E., Rosenberg, A. E. and Wilpon, J. G., "Speaker Independent Recognition of Isolated Words Using Clustering Techniques," IEEE Trans. ASSP, ASSP-27(4), pp. 336-349, 1979.

6. Rabiner, L. R. and Wilpon, J. G., "Considerations in Applying Clustering Techniques to Speaker Independent Word Recognition," J. Acoust. Soc. Am., 66(3), pp. 663-673, 1979.

7. Rabiner, L. R. and Wilpon, J. G., "Speaker Independent Isolated Word Recognition for a Moderate Size (54 Word) Vocabulary," IEEE Trans. ASSP, ASSP-27(6), pp. 583-587, 1979.

8. Wilpon, J. G., Rabiner, L. R. and Bergh, A., work in progress.

9. Levinson, S. E., "The Effects of Syntactic Analysis on Word Recognition Accuracy," Bell. Sys. Tech. J., 57(5), pp. 1627-1645, 1978.

10. Rabiner, L. R. and Wilpon, J. G., "A Simplified Robust Training Procedure for Speaker Trained Isolated Word Recognition Systems," J. Acoust. Soc. Am., 68(5), pp. 1271-1276, 1980.

11. Levinson, S. E., Rosenberg, A. E. and Flanagan, J. L., "Evaluation of a Word Recognition System Using Syntax Analysis," Bell. Sys. Tech. J., 57(5), pp. 1619-1626, 1978.

12. Velichko, V. M. and Zagoruiko, N. G., "Automatic Recognition of 200 Words," Int. J. Man-Machine Studies, 2, p. 223, 1970. There seems to be work by Russian investigators which predates this study. Several obscure references are given by Zagoruiko in a bibliographical paper entitled "Automatic Recognition of Acoustic Patterns the Problem State Review" which was to have appeared in the Procedings of the 5th International Conference on Pattern Recognition but arrived too late for inclusion in that volume. Copies were distributed at the conference and reprints may be available from the International Association for Pattern Recognition or the IEEE Computer Society.

13. Sakoe, H. and Chiba, S., "A Similarity Evaluation of Speech Patterns by Dynamic Programming," Digest of the 1970 National Meeting of IECEJ, 1970.

14. Sakoe, H. and Chiba, S. "A Dynamic Programming Approach to Continuous Speech Recognition," Proc. 7th ICA, 1971.

15. Sakoe, H. "Two-level DP Matching - A Dynamic Programming Based Pattern Matching Algorithm for Connected Word Recognition," IEEE Trans. ASSP, ASSP-27(6), pp. 588-595, 1979.

16. Bridle, J. S. and Brown, M. D., "Connected Word Recognition Using Whole Word Templates" Proc. Inst. Acoust. Autumn Conf., 1979.

17. Rabiner, L. R., Rosenberg, A. E. and Levinson, S. E., "Considerations in Dynamic Time Warping for Discrete Word Recognition," IEEE Trans. ASSP, ASSP-26(6), pp. 575-582, 1978.

18. Rabiner, L. R. and Schmidt, C. E., "Application of Dynamic Time Warping to Connected Digit Recognition," IEEE Trans. ASSP, ASSP-28(4), pp. 377-388, 1980.

19. Myers, C. S. and Rabiner, L. R., "Connected Digit Recognition Using a Level Building DTW Algorithm" IEEE Trans. ASSP, in press.

20. Myers, C. S. and Rabiner, L. R., "A Level Building Dynamic Time Warping Algorithm for Connected Word Recognition," IEEE Trans. ASSP, ASSP-29(2), pp. 284-296, 1981.

21. Levinson, S. E. and Rosenberg, A. E., "A New System for Continuous Speech Recognition - Preliminary Results," Proc. ICASSP-79, pp. 239-243, 1979.

22. Levinson, S. E. and Rosenberg, A. E., "Some Experiments with a Syntax-Directed Speech Recognition System," Proc. ICASSP-78, pp. 409-412, 1978.

23. Myers, C. S. and Levinson, S. E., "Connected Word Recognition Using a Syntax-Directed Dynamic Programming Temporal Alignment Procedure," Proc. ICASSP-81, pp.956-959, 1981.

24. Sondhi, M. M. and Levinson, S. E., "Computing Relative Redundancy to Measure Grammatical Constraint in Speech Recognition Tasks," Proc. ICASSP-78, pp. 409-412, 1978.

25. Levinson, S. E. and Shipley, K. L., "A Conversational Mode Airline Information and Reservation System Using Speech Input and Output," Bell Sys. Tech. J., 59(1), pp. 119-137, 1980.

26. Ackenhusen, J. G. and Rabiner, L. R., "Microprocessor Implementation of an LPC Based Word Recognizer," Proc. ICASSP-81, pp. 746-749, 1981.

27. Rosenberg, A. E., Rabiner, L. R., Levinson, S. E. and Wilpon, J. G. "A Preliminary Study on the Use of Demisyllables in Automatic Speech Recognition," Proc. ICASSP-81, pp. 967-970, 1981.

AN ALGORITHM FOR CONNECTED WORD RECOGNITION

J.S. Bridle, M.D. Brown and R.M. Chamberlain

Joint Speech Research Unit, Cheltenham

Abstract

The principles of an efficient one-pass Dynamic Programming whole-word pattern matching algorithm for the recognition of spoken sequences of connected words are described. The results of preliminary experiments are included, using just one utterance to derive the reference pattern for each of the vocabulary words. Some extensions and limitations of the technique are discussed.

1.0 INTRODUCTION

It has been known for some time that it is possible for a machine to recognise isolated words from a small vocabulary for a known speaker, using the method of whole-word template matching (References 1-5). The machine's information about each vocabulary word is limited to one or more whole-word patterns ("templates") derived by acoustic analysis of utterances of the word, previously spoken by the particular person using the machine. The pattern derived from an unknown word is compared with all the stored templates for that user, and the most similar template determines the machine's decision about the identity of the word spoken. All commercially available automatic speech recognition machines use some form of whole-word template matching.

J.-P. Haton (ed.), Automatic Speech Analysis and Recognition, 191–204.

At first sight whole-word template matching techniques would seem to be limited to isolated word recognition, because we would not expect to be able to segment a connected utterance into words. However, it is possible to extend the whole-word pattern matching idea to deal in a natural way with connected words by temporarily ignoring the segmentation problem. We define the best word sequence for a given input utterance as the one for which the corresponding templates, when joined together end to end, make a "composite template" which matches the input pattern best (better than any other template sequence). Connected-word recognition using whole-word matching techniques will then be possible if we can devise an efficient algorithm for finding the best word sequence and also if the best word sequence corresponds to the actual words spoken often enough to be useful.

Sakoe (Reference 6) describes a two-stage Dynamic Programming algorithm solution in which the first stage computes and stores the score for every template matched against every candidate portion of the input. The second stage computes the best sequence of templates using this stored information. Machines based on Sakoe's algorithm have been constructed, but are still unacceptable for many applications because of the cost to implement the large amount of computation required. However, these machines have demonstrated that the above approach to connected word recognition can successfully recognise speech.

Research at JSRU has produced an alternative to Sakoe's algorithm, which, by definition, should produce the same results given the same data and parameters, but uses far less computation. The JSRU method (described briefly in reference 7) uses an efficient one-pass Dynamic Programming algorithm to find the sequence of templates which best matches the spoken sequence of unknown words, allowing for timescale distortion. The recognition can be guided and improved by applying simple syntax rules. The main features of the algorithm have been implemented for some years in a non-real-time computer program.

The following sections describe the JSRU algorithm and give the results of some preliminary experiments using a small amount of test data. A real-time equipment based on this algorithm has recently been designed and constructed under contract by Logica Ltd., of London. This equipment, and further models, will be used to explore both the template matching technique itself and the applications of speech recognition in many military and civil areas.

2.0 THE ALGORITHM

To describe the algorithm, it is assumed that the input and the template patterns are represented as sequences of units which are called "frames" (from vocoder terminology). In our experiments reported in a later section, each frame is a 19-point spectrum cross-section corresponding to 20ms of speech, but the main requirement for the matching algorithm is that any frame of the input can be compared with any frame of the template to give a measure of "distance" or "dissimilarity" between the two frames. Using these distances, a score, which is a sum of distances between corresponding frames, is calculated for an explanation of the input as a sequence of templates. This score, which is calculated by a Dynamic Programming (DP) algorithm, is the score for the sequence of templates and set of correspondences between frames that minimises the sum of distances between frames, allowing for distortions in the timescales.

To set the scene and introduce some terminology, we shall first consider the problem of isolated word recognition, then lead on to our connected word algorithm.

2.1 Dynamic Programming for Isolated Word Recognition

Assume that an input pattern of length M frames has been detected and extracted, and that this input pattern is to be compared with all the templates. Since all the templates are considered independently we shall concentrate on just one, the t′th template. The problem to be addressed is that of allowing for differences in the timescales of the input and the template.

A "time registration path" maps the timescale of the input to that of a template (Figure 1). For each input frame there is one corresponding template frame, but not necessarily vice versa.

The Score for a given template is defined as the sum of frame distances for the best time registration path (i.e. the smallest possible sum).

The possible sequences of template frames need to be restricted, otherwise spurious matches may be found. A suitable restriction that will be used to illustrate the principle is that any template frame may be repeated or single template frames may be skipped. Even with this restriction there is an enormous number of potential time registration paths. For example, if template and input are both 25 frames long, there are more than 10 to the power 10 possibilities!

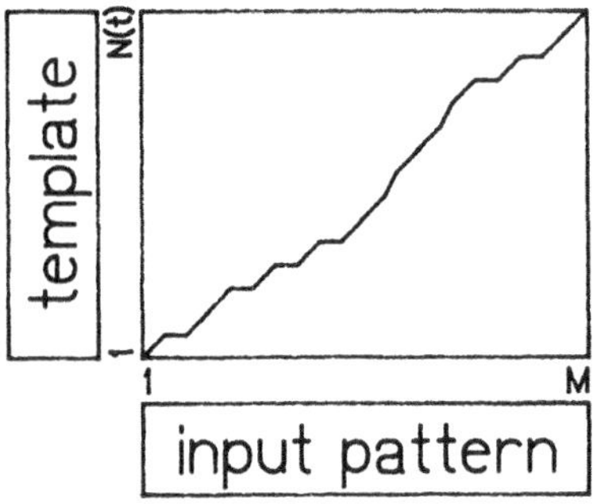

Fig.1 An isolated word recognition time registration path

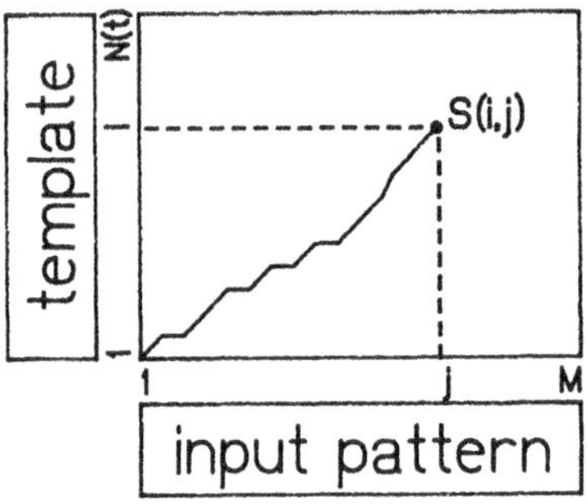

Fig.2 The isolated word recognition score

To apply Dynamic Programming to this problem we start by generalising the idea of a score for a whole template. The score S(i,j) will be the sum of the frame distances for the best way of aligning the first j frames of the input against the t'th template up to and including the i'th frame of the template (Figure 2). This score for any point on the time-time plane can be computed quite easily if it has already been computed for the previous input frame.

$$S(i,j) = \underset{a=0,1,2}{\text{Minimum}} \left(S(i-a,j-1) + \text{dist}(i,j) \right) \quad \ldots\ldots\ldots(1)$$

The score for the t'th template, of length N(t) frames, matched with the complete input pattern of length M is S(N(t),M). It is usual in isolated word recognition to consider the templates in turn and select the one with the best score. If the best score is better than some preset threshold, then the appropriate recognition response may be made.

There are several related DP algorithms in use for isolated word recognition. (See reference 8 for a survey.) The method above was chosen to fit in well with our connected word recognition algorithm. In fact, penalties P(a) are also included so that the score can include a penalty for local timescale distortion (i.e. when a=0 or 2).

2.2 Extension to Connected-Word Recognition

In connected word recognition a "time-registration path" maps the time-scale of the input to the timescale of a sequence of templates. This can be illustrated for any particular sequence of templates as in Figure 3. This diagram can be re-drawn, as in Figure 4, so that any template sequence can be accommodated. In Figure 4 a valid path starts at one of the points A,B,C (the start of the input and the start of one of the templates) and ends at one of the points X,Y,Z (the end of the input and the end of one of the templates).

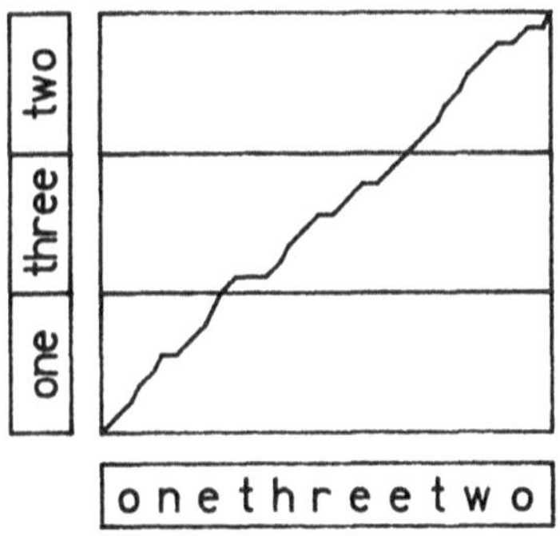

Fig.3 Time alignment of a sequence of templates to a connected word input

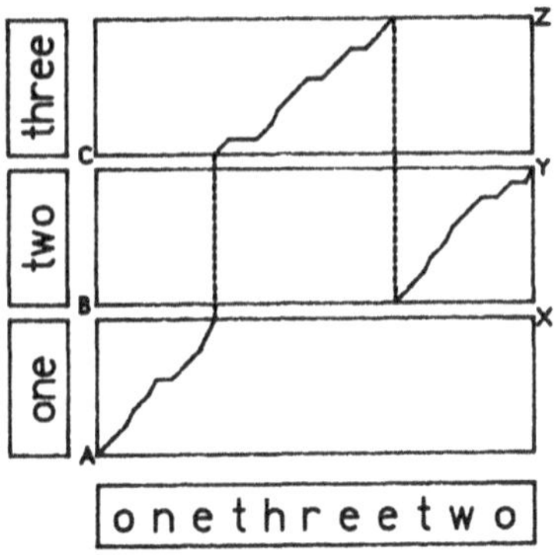

Fig.4 An alternative representation of the time alignment

Within templates the usual DP steps (e.g. a=0, 1 or 2) can be allowed, but transitions between templates are always from the end of one template to the beginning of any other.

Scoring an arbitrary connected-template path will be very similar to the isolated-word case: we simply sum all the between-frame distances and time-distortion penalties along the path. The best explanation of the input corresponds to the best-scoring path, but it should be noted that in the case of connected words there is less interest in the score itself than in the sequence of templates through which the best path passes.

The score, C(t,i,j), is computed for the best way of matching the first j input frames with any permissible sequence of templates followed by the first i frames of the t′th template (Figure 5).

Within each template, the same basic operation is performed as in isolated word recognition:

$$C(t,i,j) = \underset{a=0,1,2}{\text{Minimum}} \left(C(t,i-a,j-1) + P(a) + dist(t,i,j) \right) \quad . \ . (2)$$

where P(a) is a timescale distortion penalty.

At the start of each template, the ends of the preceding permissible templates must be examined and the best score selected:

$$C(t,1,j) = \underset{r \in R(t)}{\text{Minimum}} \left(C(r,N(r),j-1) + dist(t,1,j) \right) \quad . \ . \ . (3)$$

where R(t) is the set of templates that can precede the t'th template, according to the syntax.

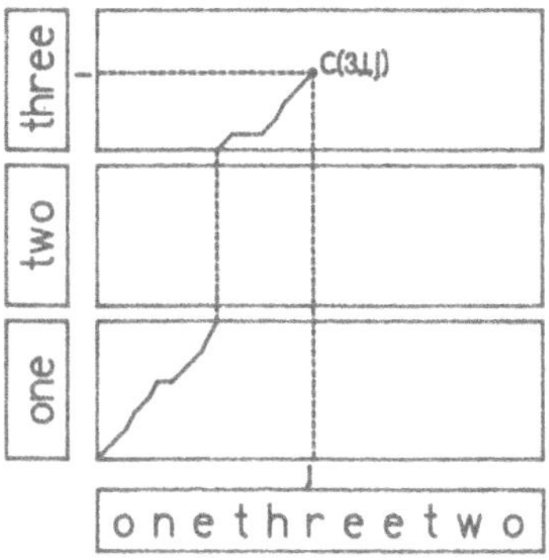

Fig.5 Connected word recognition score

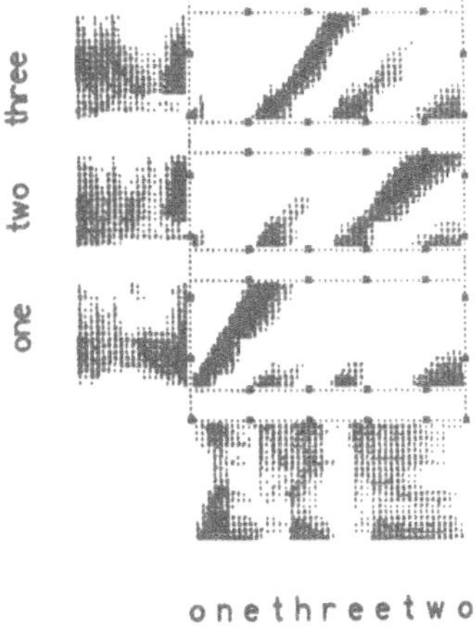

Fig.6 Connected word recognition scores using typical speech patterns

Computation proceeds for all templates in parallel in one forward pass through the input pattern. Only one array (of "cells") is needed to store the scores, with one value per template frame. At the end of the input, the score for the best interpretation can be found by examining the scores at the ends of all the templates that can end a phrase. However, we are far more interested in the actual sequence of templates which produced this best score, so during the main pass over the input we must keep track of the word decisions along all the current time-registration paths, then trace them back at the end of the phrase. Methods of doing this are described in subsequent sections.

Figure 6 shows some scores computed for templates and input derived from speech using a 19-channel filter bank analyser. The best score for each input frame is displayed as black, and the other scores for the same input frame are displayed relative to the best.

2.3 Primitive Syntax - Vintsyuk's Algorithm

Consider first the simple case in which any of the R templates can follow any other. Vintsyuk (Reference 9) suggests the following method. At each input frame, the best scoring template, if any, ending at that frame is:

$$T(j) = \underset{1 \leqslant r \leqslant R}{\mathrm{ArgMin}} \left(C(r,N(r),j-1) \right) \quad \ldots\ldots\ldots \quad (4)$$

(ArgMin means the value of the index which minimises the expression.) Two arrays are required to hold the number of the best scoring template, T, for each input frame and the frame number, F, of the start of that template. F(j) records the starting frame of the template which ends the best-scoring sequence at input frame j-1. The values held in the T and F arrays are sufficient to recover the best word sequence.

In order to fill in the array F, we need another array, L, holding one entry per template frame, which propagates through the cells with the scores. L(t,i,j) is the input frame at which the best time-registration path up to (t,i,j) entered the current template. This information passes through the template, propagated with the scores, so that:

$$L(t,1,j) = j \quad \ldots\ldots\ldots\ldots\ldots\ldots \quad (5)$$

$$L(t,i,j) = L(\ t,i-a,j-1\ ) \quad \ldots\ldots\ldots\ldots \quad (6)$$

where a is the index chosen in equation (2).

The array F can now be filled with the L value from the end of the best ending template at each input frame:

$$F(j) = L(\ T(j),N(T(j)),j\ ) \quad \ldots\ldots\ldots\ldots \quad (7)$$

At the end of an input of length M, the best sequence of templates is recovered as follows. The final template in the best sequence is T(M), the last but one is T(F(M)), the next is T(F(F(M))), and so on until we come to the template corresponding to the beginning of the input.

2.4 Extension to Arbitrary Finite State Syntax

To deal with an arbitrary finite state syntax, we first introduce the concept of "nodes". A node is a place in the syntax where a particular set of templates is preceded by some other defined set of templates. (The "primitive syntax" dealt with above has only one node.) We extend Vintsyuk's algorithm by applying the approach described in the last section at every node. Several entries in the arrays F and T may be needed for each input frame, so we abandon the indexing of the T and F arrays by input frame, and let the arrays fill up as entries are recorded. (This technique saves a considerable amount of storage space in the case of complex syntaxes when beam clipping (see section 3.4) is employed.) The array T is used to store the number of the best scoring template at each node at every frame. The array F now contains indices into T (and into itself) which link elements of T lying on a time registration path. Thus if, at a node, an entry is made in, say, the k'th element of the array T, then the "word links" for the first frame of each template following that node satisfy:

$$L(t,1,j) = k \quad \ldots\ldots\ldots\ldots\ldots\ldots \quad (8)$$

and, as before, this propagates through the template according to equation 6. Thus the values of k are propagated through the templates and, when an entry is recorded in T, the value of the corresponding word link, L(t,N(t),j), which is stored in F, is the value of k which indexes the record of the start of the template whose end is now being recorded.

The sequence of templates for the best explanation of the input can now be recovered as before by chaining down the array F and using the values F(M), F(F(M)), F(F(F(M)))... as indices into the array T. These indices no longer correspond to frame numbers so if we wish to recover the input frame numbers corresponding to the first frame of each template, another array must be added in parallel with F and T to hold the input frame at which the entry was made. A set of values in the arrays F and T (and any others in parallel with them), which share the same array index, is referred to as a "word-link record".

2.5 Summary

The above algorithm will by definition find the best "explanation" of the input pattern in terms of a sequence of template patterns but it may seem rather expensive in computation and storage. However, compared with an isolated word recogniser with the same size vocabulary, using the same methods of pattern representation and Dynamic Programming matching algorithm, it takes about the same amount of computation per input frame although it needs a greater amount of working storage. Because all words must be considered in parallel, the working storage for the scores is increased by a factor of the vocabulary size. The word links L and the arrays F and T are also extra storage, but the total amount of working storage will probably be less than that required to hold the template patterns themselves.

3.0 POTENTIAL PROBLEMS AND EXTENSIONS

3.1 Word Boundaries

In our method there is no need for any segmentation of the input speech to find the boundaries of the words. After the pattern matching process, when the input speech has been recognised, the best time registration path can be traced back (using the word-link information) to find, as a by-product, the points in the input which correspond to the junctions between the templates.

3.2 Coarticulation

A major problem in connected-word recognition is that the concatenation of standard patterns is not a good model of connected speech. In our method it is reasonable to hope that most of the recognition will be determined by the main part of the word; the coarticulation will then just have the effect of making the scores worse, but the correct sequence will still normally match better than any incorrect sequence. However, the "wildcard" template (see section 3.6) may be of assistance here by matching a few poor frames around word boundaries.

3.3 Phrase Endpoints

One of the most difficult problems in isolated-word recognition is "endpoint detection" - determining where the word starts and finishes - because this usually precedes pattern matching. The technique we are using avoids deciding the positions of the boundaries between words before deciding the identity of the words, and the same method can also be used to deal with the phrase endpoint problem. We use phrase syntaxes which start and end with "silence templates". It is then only necessary to find the start point very roughly, move back far enough to include some silence and start recognition using a silence template. A silence template can similarly be used to deal with the end of the phrase.

This technique can also be applied to isolated word recognition. The choice of endpoints is important to the recognition and the method avoids the need to decide about the endpoints prior to pattern matching.

3.4 Beam Clipping and Score Scaling

The amount of computation can be reduced significantly by pruning the DP scores (Lowerre's "Beam Search", Reference 11). For each input frame, all scores which are more than some "beam width factor" away from the best score for that input frame are removed from further consideration. This avoids considering relatively unlikely interpretations of the input pattern so far, but keeps the options open if there seems to be some ambiguity. Our present computer program also reduces the range of numbers needed to represent the scores by setting the best score for each input frame to zero, by subtraction.

3.5 Continuous Operation

The algorithm described above is suitable for recognising isolated phrases of limited length, but it can be extended to deal with continuous speech. This removes the need to even detect the presence of a phrase before pattern matching, and it means that some words may be recognised before the speaker has finished. In the isolated phrase version above, the word-link records are followed at the end of the phrase to retrieve the recognised word sequence. Usually, all the word-link records that are referenced by "active cells" (i.e. cells currently being processed) can be traced back through F to a word-link record before which there is an unambiguous path. The recogniser can operate continuously by outputting this unambiguous path and then freeing the storage in the word-link record list, both for the unambiguous path records and for word-link records that cannot be accessed from active cells. Spohrer et. al. (Reference 10) describe a related continuous operation algorithm.

3.6 Non-Vocabulary Words

It is possible to make provision for the speaker uttering words that are not in the vocabulary. In isolated word recognisers, there is usually provision for rejecting an utterance for which all the templates score worse than some preset threshold, but for connected-word recognition we wish to be able to reject a portion of an utterance while accepting the rest. Our method is to have a very elastic "pseudo-template" which always produces the same "distance" when matched with any input frame. By incorporating this "wildcard" template into the syntax, spurious inputs, such as breath noises and illegal words, can be permitted at selected points in the syntax. The "distance" for the "wildcard" template has to be chosen carefully to make it unlikely that the "wildcard" template will be recognised when a word in the vocabulary is spoken. When the wildcard is chosen in preference to the correct vocabulary word, it implies that the word has been spoken very differently from its template version.

3.7 Syntax Constraints

The use of a syntax allows the speaker to follow only certain sequences of words. If the speaker deviates from the syntax, then the recognition process will not be successful because the sequence of words spoken will not be considered. However, the use of syntax does have two important advantages. Firstly, it aids the recognition by reducing the choice of words at particular places in the syntax. This can prevent the recogniser from following paths that correspond to nonsensical phrases in the particular language. Secondly, it can reduce substantially the amount of computation needed to achieve recognition. We note that nothing is lost by providing a syntax; if nothing is known about the syntax of the utterances then the recognition syntax may be defined as allowing any word to follow any word.

In our method, the form of the syntax is restricted to finite state, including loops. This places a restriction on the syntaxes that can be specified, but the limitation is not at all severe for the type of application envisaged. Baker (Reference 12) presents a possible method of extending to a more general type of syntax.

4.0 PRELIMINARY RESULTS

Our algorithm has been available for some time on a mini computer, using a 19-channel vocoder filter bank for acoustic analysis because it was conveniently available (see reference 13). We have tried to recognise recorded groups of three and five digits, using digit templates each derived from one utterance of an isolated digit spoken in the same session by the same speaker. The algorithm would accept any number of digits in a group (the word sequence rules were particularly simple). On a total of 350 digits the digit error rate was about 2.5%.

This is neither a good test nor a particularly good result. In practice much depends on the speaker and the manner of speaking. The NEC DP-100 machine, which implements Sakoe's algorithm, made about as many errors on the same recorded material. To put the performance of the automatic speech recognisers in perspective, it is interesting that the speaker himself (JSB) made about 1% digit errors in merely reading out the strings.

5.0 CONCLUSIONS

The above technique should be useful in many applications of simple connected-word voice input to machines and conceivably for very low-bit-rate voice communications for specialised applications.

However, the experiments which we have carried out so far using whole-word template matching have been limited. In particular only small vocabularies have been used in these tests. The full power and the limitations of these methods has yet to be ascertained and the hardware equipment that has recently been built for JSRU will help to explore the potential of these methods. This equipment has storage for more than 100 templates, accepts syntax definitions and operates continuously. It also incorporates several refinements in acoustic analysis and frame distance computation which are described elsewhere.

6.0 ACKNOWLEDGEMENTS

The authors wish to thank Alan Borrett for his help with the running of the experiments reported here, and for his assistance with the preparation of the figures.

7.0 REFERENCES

1. P.DENES and M.V.MATTHEWS, "Spoken digit recognition using time-frequency pattern matching", J. Acoust. Soc. Am. 32(11), pp.1450-1455, 1960.

2. T.K.VINTSYUK, "Speech recognition by Dynamic Programming methods", Kibernetika (Cybernetics), No.1, 1968.

3. Z.M.VELICHKO and N.G.ZAGOROYKO, "Automatic recognition of 200 words", Int. J. Man-Machine Studies, 2(3), pp.222-234, 1970.

4. L.C.W.POLS, "Real time recognition of spoken words", IEEE Trans. Comput., Vol.C20, pp.972-978, September 1971.

5. F.ITAKURA, "Minimum residual principle applied to speech recognition", IEEE Trans. Acoustics, Speech and Signal Processing, Vol.ASSP-23, pp.67-72, February 1975.

6. H.SAKOE, "Two-level DP matching - A Dynamic Programming-based pattern matching algorithm for connected word recognition", IEEE Trans. Acoustics, Speech and Signal Processing, Vol.ASSP-27, No.6, pp.588-595, December 1979.

7. J.S.BRIDLE and M.D.BROWN, "Connected word recognition using whole word templates", Proc. Institute of Acoustics, Autumn Conference, pp.25-28, November 1979.

8. H.SAKOE and S.CHIBA, "Dynamic Programming algorithm optimisation for spoken word recognition", IEEE Trans. Acoustics, Speech and Signal Processing, Vol.ASSP-26, pp.43-49, February 1978.

9. T.K.VINTSYUK, "Element-wise recognition of continuous speech

consisting of words of a given vocabulary", Kibernetika (Cybernetics), No.2, 1971.

10. J.C.SPOHRER, P.F.BROWN, P.H.HOCHSCHILD and J.K.BAKER, "Partial traceback in continuous speech recognition", Proc. Int. Conf. Cybernetics and Society, Boston, MA, USA, October 1980.

11. B.T.LOWERRE, "The HARPY Speech Recognition System", Carnegie-Mellon University, Dept. Computer Science (Dissertation), April 1976.

12. J.K.BAKER, "Trainable grammars for speech recognition", in "Speech communication papers presented at the 97th Meeting of the Acoustical Society of America", pp.547-550, 1979.

13. J.N.HOLMES, "The JSRU Channel Vocoder", IEE Proc., Vol.127, Part F, No.1, pp.53-60, February 1980.

SERAPHINE : a connected word recognition system

C. GAGNOULET, M. COUVRAT, D. JOUVET

Centre National d'Etudes des Télécommunications
LAA/TSS/ATP B.P. 40 22301 LANNION FRANCE

Abstract

This paper is intended to describe the studies conducted at C.N.E.T. in Lannion, in the field of speech recognition by words. These studies led us to develop a prototype system for the recognition of short sentences (up to ten words), from a vocabulary of about one hundred words. We give a short description of the various modules constituting this system, together with the results we obtained in some specific applications. Finally, we give an idea of the future possible development of this system.

I. General description

A speech recognition system based on isolated words is not very suitable for an easy communication, though offering a wide application field. This drawback can be overcome by using a syntax with a sequence of isolated words, which make it possible to use a more extensive vocabulary. The constraint represented by a word-to-word pronunciation is still a major problem, though. In addition, it is often natural to pronounce several words together (in numbers, for instance). Only continuous speech can express this natural tendency.

Difficulty arises from the non-stationarity aspect of the voice signal. It appears when determining the boundaries between linguistic units. Actually, the acoustical features of sounds (phonemes) have several contextual variants (coarticulation).

Two different approches have been envisaged for the recognition of sentences : first, an analytical approch, which consists in splitting the signal into phonemic constituants, and

J.-P. Haton (ed.), Automatic Speech Analysis and Recognition, 205–215.

then identifying the lexical units (words). Secondly, a global approach, consisting in detecting the word sequence which most ressembles the sentence to be recognized according to a time-warping measure. These words are memorized in a reference dictionary, constituted during a learning period. When no syntactic knowledge is used, we speak of a connected words recognition.

We shall now describe this approach. Figure 1 shows the structure of the system. Two phases can be made out :

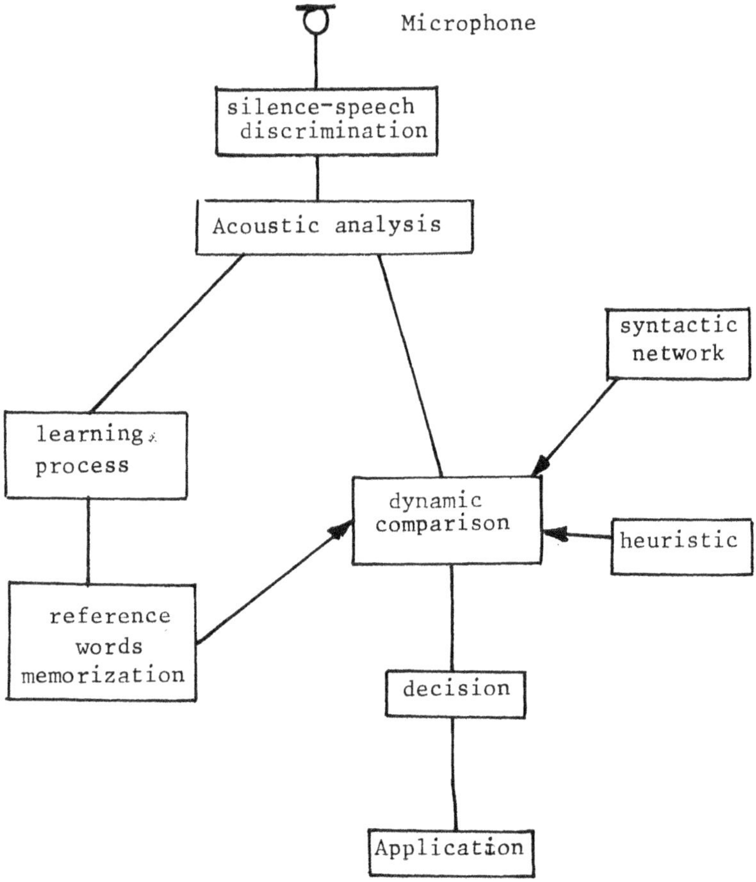

Fig. 1 Organization of a connected word recognition system

1. The learning phase consists in memorizing the reference units. The extraction of relevant parameters is performed by the acoustic analyser, which reduces the necessary memory space.

These reference units generally consist of isolated words. Additional informations must be introduced during this stage, depending on the application envisaged, and on the recognition system adopted : syntax, segmentation rules, co-articulation rules...

2. The recognition phase. The recognizer computes distances between the references and speech segments in the sentence, then identifies the unknown sentence. It may use information given by a syntactic module, or by learning process.

The aim of such a system is to recognize few words sentences. Two possible applications are the recognition of numbers (automatic dialing) and the recognition of figure sequences.

We now intend to study the system more accurately. First, we shall study the acoustic analyser, which is not specific of our approach, then the part devoted to recognition.

II. The acoustic analyzer

An acoustic analyzer is designed for two purposes : it enables to reduce the voice signal redundancy on the one hand, and to perform the extraction of relevant parameters on the other hand. In addition the memory size required directly depends on the type of analyzer.

Different studies which were conducted enabled to make a comparison between different types of analyzers [1], [2]. The most recent studies proved the interest of performing a cepstrum analysis for speech recognition [3], instead of the usual spectral analysis or the linear prediction analysis.

The complex cepstrum of a sequence x(n) is defined as a sequence $\hat{x}(n)$ whose the Fourier transform is the complex logarithm of the Fourier transform of x(n) :

$$X(k) = \sum_{n=0}^{N-1} x(n) \exp\left(-j \frac{2\pi}{N} k n\right)$$

$$\hat{X}(k) = \operatorname{Log}(X(k))$$

$$\hat{x}(n) = \frac{1}{N} \sum_{k=0}^{N-1} \hat{X}(k) \exp\left(j \frac{2\pi}{N} k n\right)$$

When minimal phase sequences (poles and zeros inside the unit circle), the complex cepstrum is related to the real cepstrum [4] (defined as the inverse Fourier transform of the logarithm of the Fourier transform module). This caracteristic makes it possible to avoid using a complex logarithm.

Davis and Mermelstein [3] recently proposed a different analysis procedure. This transform does not involve a linear frequency range but consists of a sequence of values, which are the outputs of a filter bank, whose distribution matches with the "Mel" frequency range. This distribution lies on perceptive criteria [5] ; it can be approximated by performing a linear distribution up to 1000 hz, and a logarithmic distribution above.

In the 0-6400 Hz range, we used a bank of 24 filters, distributed according to the Mel scale. The cepstrum coefficients are then computed by taking the cosine transform of the filter outputs.

The experiments conducted in C.N.E.T. consisted in comparing three different analyzers [6]

- a spectrum analysis using a channel vocoder
- a cepstrum analysis using L.P.C. analysis (LPCC)
- a Mel-frequency cepstrum analysis. (MFCC)

Table I shows the results obtained for an isolated word recognition experiment, with difficult vocabularies. It gives evidence of the advantage of using Mel-frequency cepstrum coefficients for a recognition process. Though in a different context, it confirms previous results obtained by Davis and Mermelstein [3].

Table 1

Vocabulary	Size	channel vocoder	(12) LPCC (18)	MFCC (6)	(8)	(10)
poids, toi, loi quoi, foi, roi...	17	84	87	93	93	92
harpe, arme, arc harde, art	9	91	96	96	96	97
p, t, d, les, ré, gué, gai...	21	91	96	93	96	97

The results of this study proved :

- the interest of using cepstrum coefficients to obtain an economical and accurate representation of the voice signal.
- the importance of a spectral analysis based on perceptive criteria
- a minor importance of the following parameters :
    - shape of the filters (rectangular, triangular, trapezoidal filters)
    - lenght of the analysis frame

The acoustic processing in the "SERAPHINE" system is thus the following :

The analog signal received in the microphone is filtered, then sampled (12.8 kHz), and coded with 12 bits. During the recording phase , a silence-speech discrimination (based on energy criteria) is performed every 10 ms, in order to extract the effective part of the signal.

Figure 2 sums up the processings carried out for each frame of lenght Tf. These processings can be described as follows :

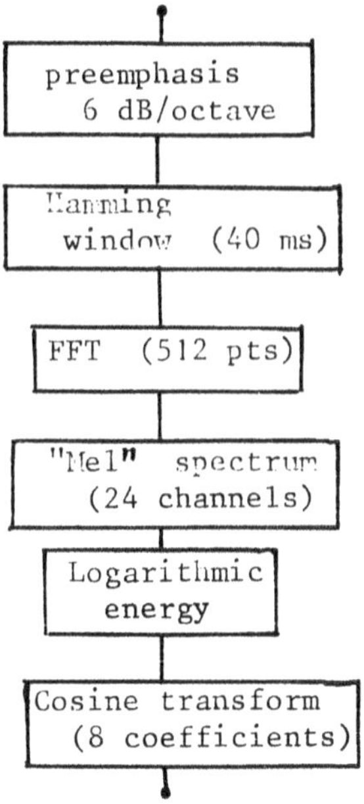

Fig. 2 A frame analysis in SERAPHINE

- a preemphasis filter $P(z)= 1-b.z^{-1}$ (b=.96) which is intended to compensate for the distorsions due to the shape of the glottal pulse. It should be noted that the operator P performs only an approximation of the derivation (6dB/octave).
- a Hamming windowing
- a Fast Fourier transform (512 points)
- a channel filtering
- a cosine transform to compute the cepstrum coefficients

All these computations are performed every Ta ms. The 12.8 sampling frequency was associated to the following values of Tf and Ta : Ta=20ms and Tf=40ms (50% overlapping for each analysis frame).

We used this numerical procedure for experiments. However, it would be possible to obtain the Mel frequency cepstrum coefficients by usinga channel vocoder, the channels of which would be distributed according to the Mel frequency scale, and then performing a cosine transform on the outputs of the channels.

We now describe the recognition part of the system.

III. Dynamic comparison of words

We shall introduce the comparison of words in the context of isolated words recognition. The same methods will be used for the connected word recognition, using distances between reference words and speech segments extracted from the sentence to identify.

Two standardization processes are involved in the determination of a similarity index between the word to be recognized and the reference word : an energy standardization and a time adjustment. Actually, two different acoustic templates of the same word generally present distinct temporal structures, and so, a time warping adjustment is necessary.

Consider two words :

$$R = (R(n), n=1..N)$$
$$T = (T(n), n=1..M)$$

where

$$R(n) = (r_1(n), r_2(n) \dots r_p(n))$$
$$T(n) = (t_1(n), t_2(n) \dots t_p(n))$$

(p representing the number of parameters for each frame)

Figure 3 shows the time warping principle (for simplicity, R(n) and T(n) are represented as scalar functions) ; it consists in eliminating the time distorsions and finding a coincidence function F to represent them. The time axis of a word is transformed so as to obtain the best coincidence with the time axis of the other word.

$$F = (F(1), F(2), \dots F(K)) \quad \text{for K values,}$$

$$\text{with } F(k) = (n(k), m(k))$$

For the isolated word recognition, it is assumed that :

$$F(1) = (1, 1)$$
$$F(K) = (N, M) \qquad \text{(boundary conditions)}$$

To determine the coincidence function F, it is possible to calculate the sum S(F) of distances d(n(k),m(k)), weighted by a function w(k), on the path associated with F :

$$S(F) = \sum_{1}^{K} d(n(k), m(k))\, w(k)$$

This gives a logical definition of a distance (normalized with respect to w), between the words R and T

$$D(R,T) = \min_{K,(n,m)} \frac{\sum_{1}^{K} d(n(k),m(k))\,w(k)}{\sum_{1}^{K} w(k)}$$

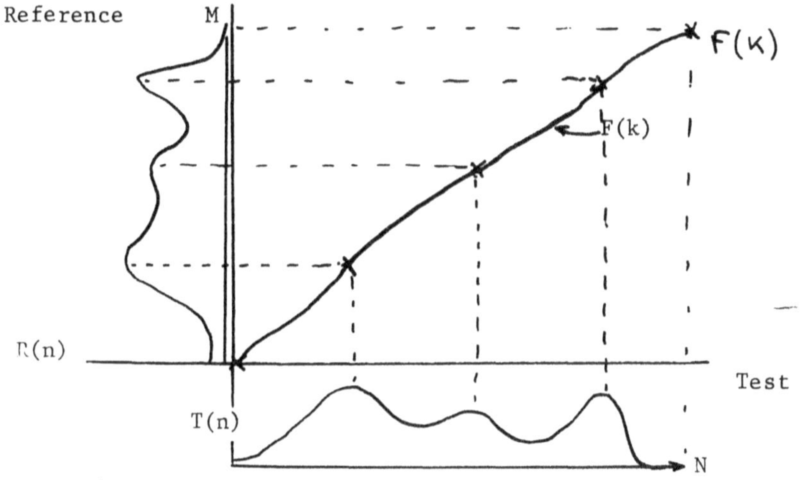

Fig. 3 Dynamic time warping

The characteristics of this distance depend upon the constraints on F and w.

Actually, the coincidence function must be in agreement with the main characteristics of the evolution of the speech signal :

- monotony $n(k-1) \leq n(k)$
  $m(k-1) \leq m(k)$

- continuity $n(k)-n(k-1) \leq 1$
  $m(k)-m(k-1) \leq 1$

These conditions enable to state the following relation :

$$F(k-1) = \begin{cases} (n(k),m(k)-1) \\ (n(k)-1,m(k)-1) \\ (n(k)-1,m(k)) \end{cases} \quad \text{with } F(k) = (n(k),m(k))$$

- slope condition : it consists in allowing a discrete set of derivative values for F, introducing restrictions in the possible sequences of values for F.

Figure 4 shows the two types of constraints we used, These types were cho sen in view of their performance [7],[8], and of the ease of use. It should be noted that these constraints led us to define a validity domain for the global path, since they limit the compression and expansion rates of the word time axis (in the limits of .5 and 2 for both types).

 (Sakoe type) 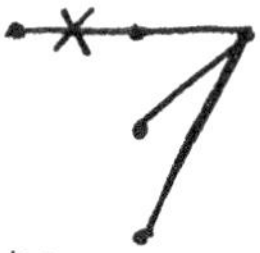 (Itakura type)

Fig. 4 Local constraints

Two different weighting functions can be defined [7]

- the symetrical form : $w(k) = n(k)-n(k-1)+m(k)-m(k-1)$
- the asymetrical form : $w(k) = n(k)-n(k-1)$ (with respect to R)

The symetrical form gives better results for isolated word recognition, but an asymetrical form is easier to use for a connected word recognition [9]

A dynamic programming procedure can be used to solve the optimization problem. Considering the second type of constraints, associated with an asymetrical form, we obtain the following iterative equations :

$$D(n,m) = \mathrm{Min} \begin{cases} D(n-1, m-2) + d(n,m) \\ D(n-1, m-1) + d(n,m) \\ D(n-1, m) + g(k)\, d(n,m) \end{cases}$$

$$\text{with } g(k) = \begin{cases} 1 & \text{if } m(k-1) \neq m(k-2) \\ \infty & \text{else} \end{cases}$$

where $D(n,m)$ is the total distance on the optimal path from the initial state (1,1) to the state (n,m)

IV. Recognition of sentences

Various problems arise when applying to sentences the above described method :

- the minimization must be performed on the whole sentence instead of on words
- the coarticulation between words must be taken into account.
- the syntactic knowledge must be used to eliminate erroneous sequences.

Several solutions were proposed to solve this problem :

- the 2-level DP matching [9]
- the sampling approach [10]
- the level building procedure [11]

These three procedures were simulated, in association with the cepstrum analysis described above. The first procedure appeared to be rather awkward and little adapted to the implementation of syntax and coarticulation rules. The results of the second procedure were unsatisfactory. Hence, we chose the level-building procedure [11].

Let us shortly recall the principle : the searching for optimal sub-sequences is performed in various steps. One step corresponds to the transition from a sequence of lenght L to a sequence of lenght L+1. At each step, the most likely partial sequences are simultaneously developped and factorized using a dynamic programming procedure, as described above.

An asymetrical form, combined with the second type of local constraints were used.

A few interesting features were added to this algorithm :

- a heuristic helps suppressing at each step the least credible solutions. The recognition process then consists of a beam search, which works out several simultaneous solutions at a given stage.

- a filtering of the vocabulary is performed at each stage, using a set of transition tables obtained during the learning phase from the syntactic knowledge.

- the coarticulation can be processed, thanks to a modification of local constraints at the beginning and at the end of the reference words. The assumption that some samples of the reference could be omitted is allowed.

These coarticulation rules depend on the reference word. The adjustment is still performed manually. We are presently studying an automatic adjustment procedure.

The system previously described was first tested with 3 different applications :

- numbers from 0 to 100 (30 words vocabulary)
- numbers from 0 to 1000 (40 words vocabulary)

- a pocket computer simulation (35 words vocabulary)

Table 2 gives the results we obtained with 3 speakers (2 male and 1 female). The preliminary learning consisted in a one pass recording of the isolated words.

Table 2

Application	Loc 1 (trained)	Loc 2 (untrained)	Loc 3 (untrained)
nb. 0 .. 100	99.0	98.6	92.3
nb. 0 .. 1000	95.8	97.0	82
pocket computer	99.4	95	87

Recognition rates for three applications

V. Conclusion

The recognition rate of the system we studied is satisfactory enough to consider its industrialization

The originality of this system results from the simultaneous implementation of a cepstrum analysis and of a time-warping adjustment for sentences.

Analysing the operating and the errors of the system shows how important the learning phase is. So, we are presently conducting studies to optimize this learning phase, through an automatic computing of the optimum overlapping of words, and an automatic detection of erroneous reference words.

The running of programs (presently written in FORTRAN IV language on a SEMS M 115 mini-computer) takes today about 30 times the real time for a 40 words vocabulary. Therefore, we are presently developping a real-time version of the system on a bit-slice microcomputer system.

Bibliography

[1] WHITE, NEELY Speech recognition experiments with linear predic tion,bandpass filtering, and dynamic programming IEEE - ASSP 24 n° 2 (1976)

[2] MICLET, NEHME Expériences en reconnaissance de la parole par prédiction linéaire 2ème congrès AFCET-IRIA reconnaissance des formes et intelligence artificielle Toulouse Septembre 1979

[3] DAVIS, MERMELSTEIN Comparison of parametric representations for monosyllabic word recognition in continuously spoken sentences IEEE - ASSP 28 n° 4, 1980

[4] OPPENHEIM, SCHAFER Homomorphic analysis of speech IEEE - Trans. on audio 16 n° 2, 1968

[5] FANT Speech sounds and features MIT Press, 1973

[6] COUVRAT, GAGNOULET Optimisation de l'analyse en vue de la reconnaissance globale CNET ref. 286/LAA/TSS/ATP , 1981

[7] SAKOE, CHIBA Dynamic programming algorithm optimization for spoken word recognition IEEE - ASSP 26 n° 1, 1978

[8] MYERS, RABINER, ROSENBERG Performance tradeoffs in dynamic time warping algorithms IEEE - ASSP 28 n° 6, 1980

[9] SAKOE Two level DP matching IEEE - ASSP 27 n° 6 , 1979

[10] RABINER, SCHMIDT Application of dynamic time warping to connected digit recognition IEEE - ASSP 28 n° 4, 1980

[11] MYERS, RABINER. A level building dynamic time warping algorithm for connected word recognition IEEE - ASSP 29 n° 2, 1981

IMPLEMENTATION OF MODELS FOR SPEECH RECOGNITION WITH VERY LARGE SCALE INTEGRATED CIRCUIT TECHNOLOGY

Jonathan Allen

Research Laboratory of Electronics and
Dept. of Electrical Engineering and Computer Science
Massachusetts Institute of Technology
Cambridge, Massachusetts 02139 USA

ABSTRACT

System design requires matching of several different structures, including the algorithm, the architecture, and the technology. In this paper, we examine how these factors influence the design of practical speech recognition implementations. In some parts of the system, custom integrated circuits can be useful to provide high-speed low-cost functionality, while in other parts conventional processors are used. Analog and digital technologies are examined, and predictions for future trends are made.

1. INTRODUCTION

The design of a successful computational system to perform a particular task is influenced by many different factors. The major attributes to be considered include:

1. Technology
2. Algorithm
3. Data structure
4. Programming language
5. Architectural units.

J.-P. Haton (ed.), Automatic Speech Analysis and Recognition, 217–229.

A principle of design to which we adhere is that the best systems result when these five attributes interact in a mutually complementary way. Such a relationship is referred to as "structural match," connoting the cohesive way in which each of the five structural subsystems contributes to overall system performance. There is not usually any uniquely optimal way in which these structures should be related, but any system designer should strive to pick these elements in a way consistent with system performance goals.

Speech recognition systems are a particularly interesting class of systems from the implementation point of view since they require several kinds of computation (e.g. symbolic and numeric) inter-related in potentially highly parallel networks. Although all five factors mentioned above are important influences to a system design, we will focus on the influence of technology, algorithm, and architectural units on the design of speech recognition systems. It will be convenient to consider two classes of speech recognition algorithms, and to examine the corresponding architectural and technological considerations for each of these classes.

2. GENERAL ARCHITECTURAL AND TECHNOLOGICAL CONSIDERATION

Computer architecture may be described as the study of computer performance, which is often quantified in terms of size, speed, and power. A given technology (e.g. TTL or NMOS) sets a base level of performance utilized by an architecture, which utilizes fundamental space/time tradeoffs to achieve performance goals. Thus, for example, a parallel combinational multiplier requires many logic circuits but is very fast, whereas a serial multiplier time-multiplexes a smaller number of circuits and hence is slower, relative to any given technology. When parallelism is implicit in a task, as it certainly is in speech recognition, it can be exploited at the expense of additional control while requiring less basic technological speed than a less parallel implementation. Frequently highly repetitive tasks (e.g. the repeated use of the "butterfly" calculation in a FFT) permit extensive parallelism with relatively modest control complexity, and hence are attractive design alternatives. Nevertheless, some designers prefer to exploit the maximum technological speed available, hence minimizing the need for parallelism.

Currently, the most attractive technology for large scale integration is the N-channel Metal-Oxide-Semiconductor process, often called NMOS. This technology, in its most advanced forms, can provide circuits with approximately 500,000 transistors, and clock speeds of 20 MHz and beyond, with chip power dissipations of several watts. We may consider this to be the base technology to be utilized in speech recognition system architectures. It makes sense to take full advantage of this technological capability, from both the memory and processing points of view, thereby minimizing the need for complicated control implied by extensive parallelism. Furthermore, it must be emphasized that technological improvement is taking place at a very rapid pace, much faster than innovations in speech recognition algorithms or computer architecture. For this reason, the reader should be aware that emerging aggressive technologies can render practical algorithms that could not previously be realistically contemplated. In the same way, highly parallel architectures involving (perhaps) thousands of processors with distributed memory should not be ruled out a priori. Investigators should bear in mind then, that it is not unreasonable to consider algorithms and architectures that imply vast computational resources. Indeed, if such techniques are otherwise satisfactory, the technology to support them will almost surely be available within a few years.

3. WAVEFORM-BASED SPEECH RECOGNITION OF ALGORITHMS

As the first class of speech recognition algorithm to be considered, we describe those systems based on the implicit assumption that all of the information needed to recognize a given input word is contained in the input audio waveform plus templates derived from previously input utterance signals. For many years, this assumption was so universally held that most researchers felt no need to explicitly recognize it, although there is now much evidence to suggest that many other language-related constraints facilitate the recognition of utterances by human listeners. These systems may be thought of as a cascade of two transformations as shown in Figure 1. In the first transformation, the input speech waveform is converted to a reduced parametric description. There are two reasons for this procedure. First, since there is much redundant information in the audio speech signal, it is desirable to reduce the number of bits needed to represent the signal to as little as a few hundred bits per second, in order to reduce the computational load of further processing. Secondly, this conversion should suppress irrelevant detail while displaying those parameters that are robust for recognition purposes. In most cases, the transformation consists of a spectral analysis, since the spectrum changes relatively slowly, can be characterized by a small number of parameters, and forms the basis for much of the knowledge of human speech perception.

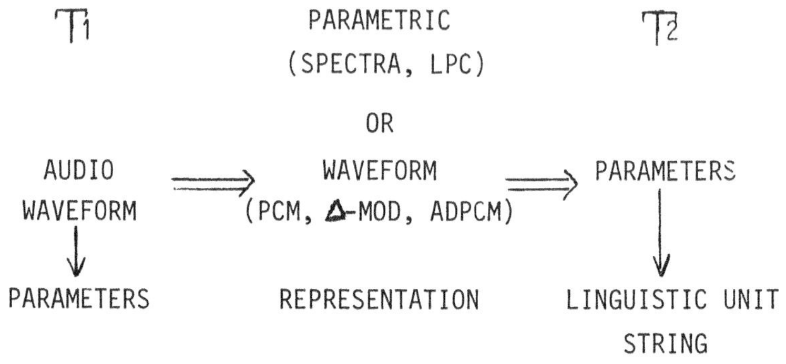

SPEECH RECOGNITION TRANSFORMATIONS
Figure 1

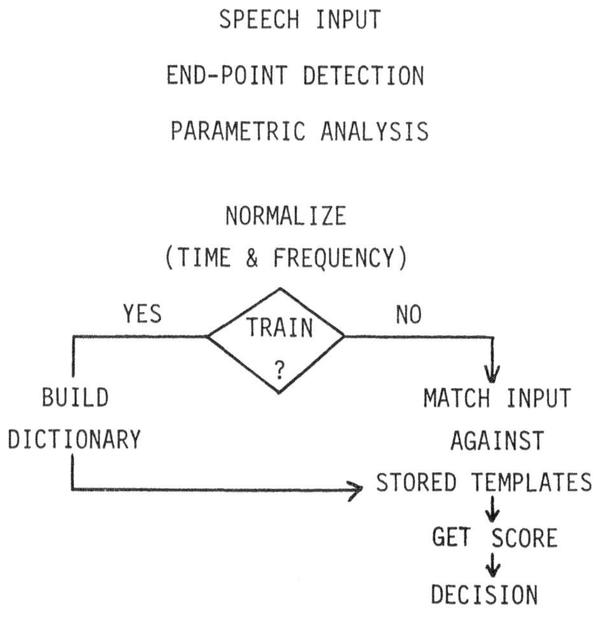

WAVEFORM-BASED SPEECH RECOGNITION
Figure 2

The second transformation seeks to convert the previously computed parametric representation to the linguistic text string that constitutes the name of the utterance. This is usually a search and match process that compares the reduced parametric form derived from the input waveform with similarly computed templates for all of the candidate utterances. This comparison must utilize some sort of distance metric that characterizes the degree of closeness between the input and reference template.

From this general view of speech recognition as a cascade of two transformations, we can show a more concrete process chart of these recognition schemes in Figure 2. Here we comment on the several computational tasks shown in the Figure.

(a) The input signal, after digitization, is examined for end points of the utterance. In some systems, end point determination is done after the parametric (usually spectral) analysis.

(b) Next, a parametric analysis is performed. Several representations, including the spectrum, formants, linear predictive coefficients, or zero crossings can be obtained.

(c) Normalization in time and frequency is often performed in order to compensate for differing vocal tract length and rate of speech.

(d) Assuming that a set of parametric templates has already been assembled to form a dictionary of candidate utterances, a match (or lexical lookup) is performed of the input against each stored template. The precise nature of the matching procedure, including the weighting of selected attributes of the utterance, is of great importance to the success of the recognition scheme.

(e) Finally, based on scores that reflect the distance between the input and the dictionary templates, a decision is made in favor of the lexical item closest to the input.

4. TECHNOLOGY FOR WAVEFORM-BASED SPEECH RECOGNITION ALGORITHMS

Given the procedures outlined above, we can now comment on the appropriate technology for waveform-based systems.

Several algorithms for determination of end points have been developed, and there is need for continuing research in this area. The main problems to be dealt with include weak fricatives at the beginning or end of an utterance, weak plosive bursts, final nasals, devoiced fricatives at the end of words, and the loss of voicing in some final vowels. While some schemes require predictor or autocorrelation coefficients, one procedure uses zero crossing rate and short term energy (based on an average of absolute waveform values over a small [10 msec] window (1). Such an algorithm can be readily computed with contemporary microprocessor technology, and there is little motivation for special circuits. If, however, the problem of endpoint detection is considered to be part of the larger problem of voiced/unvoiced/silence detection, then much more complicated algorithms may be used that not only require extensive signal analysis, but also sophisticated decision procedures (2). Such techniques, while promising, are not yet appropriate for practical implementations.

The next major computational task is parametric analysis. Almost all systems utilize some form of spectral analysis, which can be computed in a variety of ways. Perhaps the simplest scheme is to utilize a filter bank such as those used in the analysis portion of a channel vocoder. There are four technologies available for this purpose. First, the filter bank can be constructed by utilization of solid-state operational amplifiers with appropriate RC input and feedback networks to achieve the desired filtering function, which is usually some approximation of a pole pair. The entire set (about twenty) of filters needed to cover the speech band can be placed on one chip in conventional technology. The major problem with these "active-RC" schemes is the dependence of RC products on processing variations as well as drift due to temperature and voltage changes, which leads to a lack of stability in filter performance. To meet these objections, several new analog schemes have been developed, the most important being CCD and switched-capacitor filters (3,4). Charge-coupled devices (CCD's) are descendents of bucket-brigade devices that shift charge packets through MOS devices. It is possible to sample the voltage associated with these packets and add the voltages in a weighted summation to provide a sampled-data transversal filter. Such filters can be realized in a small area, but experience has revealed several difficulties with their use. First of all, in several designs, the input-driving and output-sensing circuits have been realized off-chip, thus diminishing the area advantage. Also, since the input signal is sampled in time, a continuous-time anti-aliasing filter must be provided, which is relatively complex for the low sampling rates required for CCDs. Finally, since CCD filters are nonrecursive, transfer functions with a narrow passband (and hence long impulse response) require a very long CCD filter which utilizes a large silicon area. For all of these reasons, CCD filters are no longer seen as a viable technology for speech filter banks.

The CCD filters do, however, exploit charge-storing capability in MOS devices, and the sample-data control is attractive if it can be accurately exploited. Furthermore, while we can avoid transversal filters by designing analog recursive filters using operational amplifiers, these designs are limited by their dependence on ratioed passive components to determine filter properties. What is desired is a technique that takes advantage of MOS charge-storing, but avoids the need for stable passive components with values that can vary over a wide range. One solution to this problem has been the "switched capacitor" filter, in which resistances are simulated by switching of a capacitor. Accurate capacitor values can be obtained over a wide range by modularly adding incremental areas of capacitance. By switching one side of a capacitor (the other side is grounded) back and forth between voltages V_1 and V_2, an average current between these voltages of $C(V_2-V_1)f_c$ (where f_c is the switching frequency) can be made to flow. Thus the size of the equivalent resistor that would give the same average current between V_2 and V_1 is $R = 1/cf_c$. In practice, f_c can be made very high so that the sampling effect does not affect the signal band of interest. It is interesting to note that as C gets smaller, the effective R gets bigger, so that the switched-capacitor technique can be used to realize very large resistances. For example, a 10 megohm resistor may be needed together with a monolithic capacitor of a reasonable size, say 10 pf. This value of resistance is easily achieved by switching a 1-pf capacitor at a 100 KHz rate, requiring an area of only 0.01 mm^2. Direct use of polysilicon or diffusion layers in the integrated circuit for this resistance would require an area at least 100 times larger.

It is possible to build highly selective bandpass filters using the switched-capacitor techniques which are stable and accurate due to their dependence on f_c and capacitor area ratios for passive component values. These filters also consume little chip area, and do not require specialized processing, so that this is an inexpensive technology for filter banks made in the early 1980s. In the author's opinion, however, switched-capacitor technology will give way to all-digital techniques by the mid 80s due to rapid improvements in MOS digital circuits, and the effects of process scaling. Digital filters, which are not only discrete (sampled) in time but also in sample values, provide repeatability, arbitrary accuracy, flexible control, infinitely long memory, ease of processing, lack of dependence on environmental variables, and straightforward design techniques. In the digital world, a pole-pair can be described as a second-order difference equation, which in turn can be implemented as a second-order "section." This structure is a simple interconnection of multipliers, adders, and delays, and forms the backbone of digital recursive implementation. It is easy to build the required adders and delays, but a great deal of effort has been expended on the design of fast, small digital multipliers (5). Both parallel and serial multiplier designs have been developed, however, that provide high-speed operation in a small area. For example, in NMOS technology a 16 bit by 16 bit combinational multiplier can operate in less than 300 nsec, and be implemented in less than 10 mm^2 of area. It is this level of performance that has made digital techniques so useful for speech applications, and continued improvement of circuit speeds and densities will only increase this attractiveness.

Using digital second order sections, a filter bank can be easily made. Usually about 20 filters are provided, each having a bandwidth that increases with center frequency. This variation in bandwidth corresponds to known properties of the inner ear transduction mechanism. Thus the response time is faster for high frequencies than for low tones. There is no agreement on the best kind of filters to use, some designs being borrowed from vocoder practice and some being derived from 1/3 octave characteristics. All of these designs, however, are easily realized as digital circuits. Other forms of spectral analysis have also been utilized. The Fourier spectrum can be computed when adequate computational power is available. Fast multiply-accumulate modules are needed as well as flexible address arithmetic. When the Fast Fourier Transform is used to compute the spectrum, a basic module is the "butterfly" which includes one complex multiply, one complex add, and one complex subtraction. The butterfly module is just now appearing in bipolar integrated circuit technology, and is frequently used in custom MOS digital designs. Similar computational capability is needed to generate LPC predictor coefficients, the most time-consuming element being the calculation of autocorrelation coefficients. Here again, the ability to rapidly compute a sum of products is required, and this basic function can be utilized in integrated circuit designs for both recursive and nonrecursive filters, autocorrelation, and Fourier transforms. It is apparent then, that a few basic digital computational elements can be combined to compute spectra in a variety of different ways, and this contributes to the popularity of the digital approach. As process scaling continues, MOS technology can be expected to comfortably perform even complicated spectral analysis routines at audio sample rates, with all of the advantages of the digital approach. The author believes that digital techniques will continue to take over where analog techniques are presently used, even when cost is a major consideration.

The next major task to be considered is dynamic time warping, which uses a non-linear time warp to find the best match of an input spectrum with a reference spectrum template. This time-warping uses a form of dynamic programming (DP) which is also known as the Viterbi algorithm. An exposition of the DP techniques is beyond the scope of this paper, but is readily available in the literature (6). Here, it suffices to say that a set of recursive equations can be solved to locally propagate the warping along the utterance. This calculation can make use of a basic computational element at each temporal grid point defined by the sampled input utterance and reference template. Architectures for such computations have been presented (7,8) but no complete chip has been demonstrated as yet. Since DP can improve template-oriented recognition by large amounts, the appearance of these chips will be very useful. It is estimated that DP comparisons of an input spectrum with as many as 1000 reference templates can be performed in real time. It is appropriate, however, to give a word of caution about DP techniques. The brute-force application of DP does not take into account the fact that some parts of the utterance (in both space and time) are more important to perception than others. That is, DP is a general technique applicable to any waveform, and should not be thought of as inherently related to speech recognition. There is now a trend away from DP to more specialized techniques that seek to weight the spectrum in some perceptually relevant way, an example being the use of the energy envelope as a weighting function (9), so that more intensive portions of the utterance contribute relatively more to the DP matching process. This trend can be expected to continue.

In current technology, only the input filtering and DP matching tasks have given rise to special-purpose chips. In fact many of the tasks can be accommodated by conventional general purpose parts. These range from digital signal processing chips through numerical functional units to general purpose microprocessors. The digital signal processing chips have been designed mainly for recursive filtering tasks, and are capable of computing approximately 50 second order sections with 16 bit accuracy at a 10 KHz sample rate, in NMOS technology. This is clearly sufficient for input filtering as well as for speech synthesis. Difficulties have been encountered when there was insufficient on-chip memory, and also with the lack of appropriate addressing modes required for Fast Fourier Transforms, so that there will continue to be a role for custom designs that provide all of the needed capability on one chip. Several chips have appeared that provide special functional units for rapid computation of transcendental functions as well as floating point capability. These are not currently critical to template-oriented speech recognition schemes, but may greatly increase performance in more sophisticated future systems. Lastly, we should mention general purpose microprocessors, which are becoming increasingly powerful. These processors can be used for overall control and general purpose calculation, and can be readily combined in multi-processor configurations. 32-bit microprocessors with large address space capability and high speed will shortly be available for use in high-performance speech recognition schemes. We emphasize again that the predominant technology for all of these devices is NMOS, which provides high speed, high circuit density, and acceptable power levels. When low power consumption is important, at some expense in area, then CMOS will find increasing use.

5. MULTIPLE KNOWLEDGE SOURCE SYSTEMS

It has been apparent for some time that humans do not recognize speech by relying solely on the input speech waveform. While the techniques described in the previous sections may be acceptable in applications requiring a vocabulary of about 100 words spoken in isolation with little or no syntactic constraints, such means fail to perform well for large (over 1000 words) vocabularies used with normal syntax in continuous speech. There is by no means any consensus on the best way to solve this larger problem, and the reader is referred to the literature for a view of the several approaches (10). It does appear, however, that speech perception can be modeled as an integrative process in which constraints expressed in the input waveform, phonological structure, lexicon, syntax, semantics, and pragmatics are combined (in ways not yet understood) to form the percept. From this viewpoint, the information in the waveform is but one of several factors contributing to the percept, and signal properties, including even any invariant properties derived from the signal alone, cannot be solely relied on for computation of the percept. Instead, these constraint systems are utilized through a set of cooperating knowledge sources of diverse type.

While innovation in speech recognition strategies can be expected to continue for some time, there are a number of design issues that must be confronted. First, there is the question of knowledge-source representation. A major problem is posed by the diverse kinds of information that must be combined to make a decision, ranging from numerical attributes of the speech signal through symbol semantic relations. As research continues in the area of knowledge representation and belief structures, we ca expect new architectural schemes to be suggested, but it must be emphasized that this is a very young field. Another area of research is the group of topics centered around search strategy, matching, and scoring. Many of the schemes utilized now use simple Euclidean distance metrics and search methods (coupled with scoring algorithms) that have little demonstrable relation to human speech processing. There is a vigorous debate in the field as to the need or utility in "copying" human processing, but it is the author's bias that the space of possible systems is so large that we can profitably take guidance from psycholinguistic models for human speech perception. The last major design issue is the selection of a control strategy. Some systems utilize a bottom-up search starting from the phonetic details, whereas others employ a top-down procedure, only checking those structures that can confirm or deny the predictions made by the control. The author believes that there is little evidence for either of these strategies in human performance, ad that a much more complex interaction is needed to properly integrate all of the constraining information needed for the search.

The preceding discussion indicates that there are many degrees of freedom in the design of systems for large vocabulary, natural syntax, continuous speech recognition systems. It should not be surprising, then, that there have not been any special purpose chips designed for these systems. Instead, large general purpose computers have been used that allow the system designer considerable flexibility in the implementation of algorithms.

There is one system, however, that utilizes a very specialized knowledge source implementation. This is the HARPY system, based on DP search of a transition network. In the HARPY system, all knowledge source constraints are collapsed onto one network of chained events, each of which is assigned an a priori probability through training. The network represents all anticipated sequences of phonetic elements, as constrained by the given lexicon and syntax. A modified DP search (called a beam search) is then used to find the optimal path through the network. This network is clearly a prime candidate for special architecture and custom chips. It is a well-defined structure where the details are well-understood. As yet no custom architectures for this system have appeared, but they may be reasonably expected in the next year or two, due to the demonstrated success of HARPY and its clarity of design.

Other schemes for large vocabulary continuous speech recognition utilize loosely coupled knowledge sources that are currently represented in very general software structures on general purpose machines. There is need for much conceptual understanding of these schemes before it makes sense to devote any substantial design effort to custom chips. It is, however, almost certain that multi-processor architectures will be needed, and that extensive parallelism will be exploited. The study of parallel computational structures and their control is still in its infancy, and for the present, it is important to note that no one parallel architecture can be demonstrated to be inherently superior apart from considerations of the application domain. The choice of system structure, including interconnect networks (rings, trees, matrices) must wait until a much higher level of algorithmic understanding is present. As we have noted before, it is almost certain that the technological capability for these systems will be appropriately available when the system design understanding requires its utilization.

6. FUTURE TRENDS

In this section, we mention trends in three areas. First, we consider technology. It is important to emphasize that technology is changing very rapidly, and that researchers should not be discouraged by the computational demands of proposed algorithms. Memories of 1 million bits on a chip, and one-chip processors with main-frame capability will soon be available. Furthermore, the prospect of wafer-scale integration raises a view of integrated systems on a wafer with multiprocessor and large memory capabilities. Thus the future of technology is very bright, and is not a limitation to improved practical speech recognition systems.

Second, trends in computer architecture need to be considered. Special function units, such as those for DP, will continue to appear, which will greatly improve performance. Also, continued research in communication and the control of parallelism should be of benefit to the designers of speech recognition systems. It is felt, however, that progress in parallel architectures is most likely through careful consideration of specific cases rather than general theory.

Lastly, there is the question of trends in speech recognition algorithms. Problems of knowledge representation, integration of cues, determination of distance metrics, matching strategies, and search techniques require much research. These are exceedingly difficult questions, requiring long and intensive investigations. What can be said with some optimism, however, is that when highly capable algorithms are devised, the technological means to support them will be available.

References

(1) Rabiner, L.R. and Sambur, M.R. "An Algorithm for Determining Endpoints of Isolated Utterances," B.S.T.J., Feb. 1975.

(2) Atal, B.S. and Rabiner, L.R. "A Pattern Recognition Approach to Voiced-Unvoiced-Silence Classification with Applications to Speech Recognition," IEEE Trans. ASSP, ASSP-24, June 1976.

(3) Hewes, C.R., Brodersen, R.W., and Buss, D.D. "Applications of CCD and Switched Capacitor Filter Technology," Proc. IEEE, 67, Oct. 1979.

(4) Brodersen, R.W., Gray, P.R., and Hodges, D.A. "MOS Switched-Capacitor Filters," Proc. IEEE, 63, Apr. 1975.

(5) Allen, J. "Computer Architecture for Signal Processing," Proc. IEEE, 63, Apr. 1975.

(6) Myers, C., Rabiner, L.R., and Rosenberg, A.E. "Performance Tradeoffs in Dynamic Time Warping Algorithms for Isolated Word Recognition," IEEE Trans. on ASSP, Dec. 1980.

(7) Burr, D.J., Ackland, B., and Weste, N. "A High Speed Array computer for Dynamic Time Warping" ICASSP 81 Proceedings.

(8) Murveit, H., Lowy, M., and Brodersen, R.W. "An Architecture of an MOS-LSI Speech Recognition System Using Dynamic Programming" (A), J. Acoust. Soc. Amer., 69, Spring 1981, Supp. 1.

(9) Shikano, K. "Acoustic Processing in the Conversational Speech Recognition System," ICASSP 81 Proceedings.

(10) Lea, W.A. Trends in Speech Recognition, Prentice-Hall, 1980.

Acknowledgements

This work was supported in part by United States Air Force Contract AFOSR F49620-81-C-0054.

SELF-ORGANIZED CONTINUOUS SPEECH RECOGNITION

Frederick JELINEK

Continuous Speech Recognition Group
Computer Sciences Department
IBM T.J. Watson Research Center
Yorktown Heights, N.Y. 10598

Abstract

Current efforts to recognize continuous (or "connected") speech are aimed at constructing a voice-excited "typewriter" that automatically transcribes natural speech into ordinary (e.g. English) written form. So far, however, only very restricted speech has been recognized. The sentences that are spoken must either be prescribed a priori by an artificial grammar which the experimenter has designed, or else limited by a vocabulary and a restricted area of discourse such as that used in business letters, book reviews, or airline reservation systems. These latter so-called natural tasks are generally much more difficult than the artificial ones (given a fixed vocabulary).

At IBM-Research we are now working on the recognition of continuous natural speech, without the aid of artificial pauses between words, or unnaturally small vocabularies or artificial constraints on syntax. As of this writing ours is the only place in the United States (and, as far as we know, the world) with this goal. The particular utterances to be recognized are chosen from the Laser Patent Text, consisting of patent disclosures involving laser devices. Previously (1978) we succeeded in recognizing perfectly sentences generated by an artificial grammar known as New Raleigh.

OVERVIEW OF THE IBM-RESEARCH APPROACH

The distinctive feature of our approach is that we formulate the recognition problem in terms of communication theory. As a consequence, we base our system on careful statistical modeling of all speech processes involved : sentence production, speaker's pronunciation, and the recognizer's speech signal proces-

J.-P. Haton (ed.), Automatic Speech Analysis and Recognition, 231–238.

sing.

Our models are derived from as much actual speech data as we can obtain and handle computationally. We have devised methods of automatic model computation, thus minimizing or completely eliminating human intervention. Our strategies are not based on rules devel oped from trying to intuit how people recognize sentences (as is prevalent elsewhere), although the basic structure of our models is, of course, man-made. This approach is both more accurate and more flexible ; as the speaker or the components of the system change, our self-organizing programs remain valid, and computer time is all that is required to adjust to a new configuration.

More detailed descriptions of our methods can be found in reference (1) and (2).

All our experiments have been carried out with a system implemented on a general-purpose IBM 370/168 and tuned automatically to a designated speaker who was recorded (with one exception) in a quiet room with high-fidelity equipment.

THE IBM-RESEARCH RECOGNIZER

The typical recognition experiment employs a speech recognizer consisting of an acoustic processor (AP) followed by a linguistic decoder (LD). Traditionally, the AP has been viewed as an artificial phonetican which attempts to transcribe the spoken waveform into a string of phonetic symbols. The LD then translates that string into written form.

At IBM we formulate the recognition process as a communication problem. This is shown in Figure 1, where the speaker and the AP are conceptually viewed as combined in an acoustic channel which provides the LD with information y from which the LD can deduce the channel's input word sequence $\mathbf{w}$. Actually, the LD tries to find that word string $\mathbf{w}$ which maximizes the probability $P(\mathbf{w}, \mathbf{y})$ of the joint observation of the input-output pair $\mathbf{w}, \mathbf{y}$ at the terminals of the channel. As is well known, $P(\mathbf{w}, \mathbf{y})$ can be rewritten as

$$P(\mathbf{w},\mathbf{y}) = P(\mathbf{w})\ P(\mathbf{y}|\mathbf{w})$$

Here $P(\mathbf{w})$ denotes the probability that $\mathbf{w}$ was generated by the sentence generator, and $P(\mathbf{y}|\mathbf{w})$ the probability that as a result of the speaker's reading w the AP put out the string $\mathbf{y}$.

In order for the LD to carry out its search, it thus needs statistical models of text generation and of acoustic channel performance to supply the probabilities $P(\mathbf{w})$ and $P(\mathbf{y}|\mathbf{w})$. I shall discuss these models in the next section.

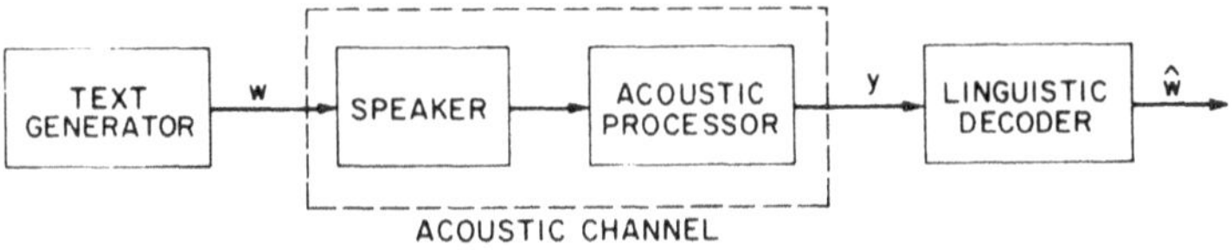

Figure 1
The communication theory view of speech recognition

Because we have found it more fruitful to view the AP as a data compressor than as an artificial phonetician, we do not attempt to identify phonetic segments* in the continuous speech. Our AP consists of a signal processor followed by a pattern recognizer. The signal processor first digitizes the speech at a 20 KHz sampling rate. Then every centisecond a discrete Fourier transform (DFT) is computed from a 2-centisecond segment of the sampled waveform. In this way an 80-component vector, called a spectral time sample (STS), is produced. Each component is the average magnitude of the DFT over some frequency band.

The pattern recognizer has stored in it a set of 45 prototype vectors of the same dimensionality as the STS. As an STS is generated, the pattern recognizer measures its similarity to each of its prototypes and puts out the identifier y of the most similar one. Thus the AP output y is a string of prototype identifiers, one for each centisecond of speech.

The prototypes correspond to the steady state portions of various phones of English. They are selected automatically from a relatively large sample of the speaker's speech and are thus tailored exclusively to him. Since English phonemes have varying derivation, a varying number of STS's results from a spoken phone. Neither the AP nor any other component of the recognizer determines the phone-- STS correspondence explicity.

MODEL OF SPEECH PROCESSES

The LD requires models of text generation (the language model, LM) and of acoustic channel performance. If the text is generated by an artificial grammar, then the latter serves as the LM. An example is the New Raleigh grammar shown in Figure 2. It has the form of a Markov source (2) consisting of states (indicated by circles in Fig. 2) interconnected by transitions with which are associated sets of outputs (the boxed word lists of Fig. 2). The source is placed in the start state (the leftmost state

*Phones are the basic units used by linguits to transcribe the pronunciation of words. Reference (3) contains a good basic discussion of phonology and phonetics.

of Fig. 2), a transition leading to the next state is chosen at random, and so is an output from the set associated with the transition. The process of choosing a transition and an ouput is repeated until the end state is reached (far right of Fig. 2). The chosen words are strung together into the sentence to be read by the speaker. Once the transition and output list choice probabilities are specified, it is possible to compute the probability $P(\mathbf{w})$ that the grammar genarates a given word string $\mathbf{w}$.

Figure 2
Grammar of the Raleigh language

In the future the design of a LM for a naturally generated text will probably involve considerations of syntax, semantics, and discourse pragmatics. So far no one has accomplished this. At present, the LM for our natural Laser Patent Text computes the probability of a word string $\mathbf{w}$ from the local probabilities $P(w_i|w_{i-1},w_{i-2})$ that any word w_i of $\mathbf{w}$ will follow its two preceeding words w_{i-1},w_{i-2}. Thus the LM predicts the next word by knowing the identity of the last two. The local probabilities are estimated from a large amount of Laser Text set aside for this purpose.

The acoustic channel performance model giving the probability $P(\mathbf{y}|\mathbf{w})$ is also based on the Markov source concept. Our approach is to model the channel performance for all the words in in the vocabulary and then obtain the model for the word string $\mathbf{w}$ by connecting together all the models of its individual words.

To obtain a performance model for any given word we first construct a phonetic source whose possible output strings are the transcriptions of all the different phonetic pronunciations of the word. Figure 3 depicts such a source for the word two whose pronunciations expressed in the international phonetic alphabet are t^hU^u, t^hu, tU^u and tu. Next, we replace every transition in the phonetic source by another type of Markov source-- an AP performance source corresponding to the phonetic symbol associated with that transition. Figure 4 illustrates this replacement.

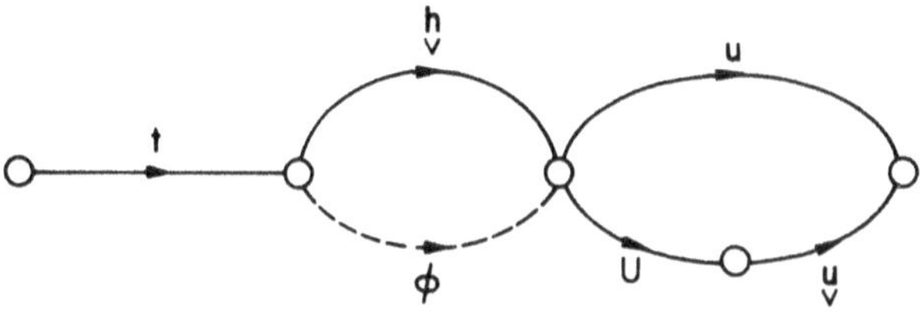

Figure 3
A phonetic Markov subsource (word "two")

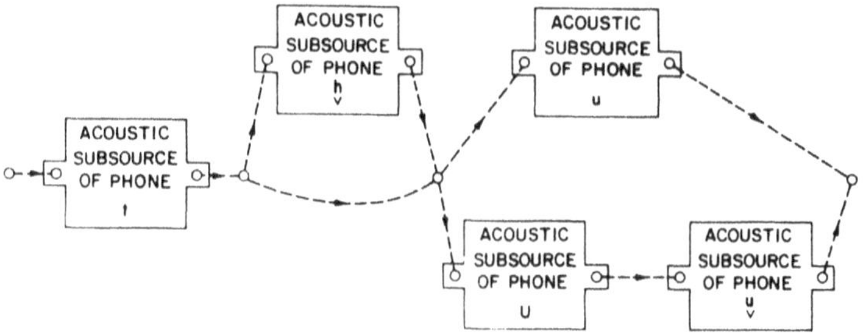

Figure 4
A phone-based Markov source based on the phonetic subsource of Figure 3

Space limitations prevent giving any details about AP performance sources beyond stating that they all have the same transition structure, and that a different source generating substrings **y** of AP output symbols exists for each phonetic symbol. The structures for the phonetic sources are obtained automatically by rule from basic pronunciations of the corresponding words, such as can be found in standard dictionaries. To obtain the probabilities determining the different sources underlying the acoustic channel model we let the speaker read a known script into the AP and observe the corresponding output string **y**. An iterative algorithm then makes optimal probability adjustments starting from some initial guess.

THE HYPOTHESIS SEARCH AND TASK COMPLEXITY

In attempting to find that word string **w** which maximizes the probability product P(**w**) P(**y**|**w**), the LD finds that sentence which best fits the observed AP output y, taking into account the likelihood of uttering the sentences in the first place. As there are obviously too many possible sentences for an exhaustive search, we use a search strategy that only looks at the most promising word sequences. At any given moment there is a stack which contains as entries all of the word sequences (partial sentences) examined so far. The stack is ordered according to the LD's estimate of the likelihood (relative to the observed AP string **y**) that the entry in question is the beginning of the actual sentence spoken. The likelihood of all the sequences resulting from one-word extensions of the sequence are reinserted into the stack so as to preserve its order. The search ends when a complete sentence is found at the top of the stack.

However, since very many subsequences (forming the beginning of a natural sentence) can be followed by most of the words in the vocabulary, the above search must be further abbreviated. We do this by considering only those words which have been predetermined to "sound like" the beginning of the unused part of the AP output string **y** which has not been caused by the word subsequence being extended. This short cut causes occasional unrecoverable errors when the true word uttered is not in the "sound-alike" set. This usually results in a whole string of further errors because the set of likely words following a wrong word would ordinarily not contain the correct word.

RESULTS

We have tested the IBM approach on both artificial and natural tasks. In a laboratory setting we have achieved perfect recognition in the first, New Raleigh language case (Figure 2). Our latest results indicate that we can correctly identify 91% of words contained in sentences of the Laser Patent Subtext experiment.

The vocabulary size of Raleigh is 250 while that of Laser is 1000. But vocabulary size is a poor measure of recognition task complexity. Perfect recognition could be achieved trivially for a language consisting os a single 1000-word-sentence. The difficulty of a task can be estimated from its entropy H, (or perplexity 2^H) because Information Theory shows that a recognizer can ignore all but the 2^{nH} most probable word strings of length n (for large n) and yet achieve any prescribed recognition error tolerance. The entropy of the New Raleigh task is 2.86* while that of Laser is 4.6**.

An interpretation of perplexity is that it corresponds to the appropriately computed average number (relative to the probability structure of the language) of different word branches leaving each state. The canonical language of perplexity 2^H would

---

*This is relatively high as artificial tasks go. The entropy of the CMU AIX-05 language on which the Carnegie-Mellon recognizer Harpy runs is 2.18 (4).

** In a famous experiment, Shannon has estimated the entropy of unconstrained English to be about 5.5.

have exactly 2^H_n equipropable branches leaving each state, and would then produce exactly 2^{nH} equipropable different word strings of length n. Actually, natural tasks are more difficult than artificial ones for reasons beyond their larger entropy. Observe in Figure 2 that only very small parts of the vocabulary can be used at any given point of a Raleigh sentence, while a Laser sentence can branch in many more ways, although most of the branches at any point are unlikely (see the last paragraph of the preceding section). Also, natural sentences are usually much longer than those generated by artificial tasks, so that the chance that unrecoverable errors will take place is increased. As already stated, most of our results pertain to a single speaker recorded in a sound treated room.

Except for the recording and analog-to-digital conversion of the signal, the only hardware involved is an IBM 370/168 computer system. Depending on the recognition task, processing time is between 10 and 300 times real time.

When we achieved the perfect recognition results on the Raleigh task, we tried to make it harder in several ways. First, we deleted some of the syntax constraints from the decoder's language model, raising its perplexity (average fan-out) from 7.27 to 13.89. We still got perfect recognition. Next, we removed all constraints, thus making the decoder "believe" that any of the 250 words could follow any other with equal probability. This time we correctly decoded only 91.2% of the words spoken. 47% of the sentences were decoded completely without error.

We then decided to loosen also the grammar that generated the spoken sentences. We asked colleagues to construct any English sentences they wished out of the 250 word vocabulary. A sample of these is shown in Table I. We attempted recognition using the last mentioned unconstrained language model, and correctly identified 81.6% of the words spoken. Since the decoder frequently inserted short words, the ratio of correct words to words recognized was 0.73.

Finally, we carried out the original Raleigh experiment on speech recorded by a close speaking, noise cancelling microphone in an office environment. We did not observe any deterioration in our correct recognition rate.

Some additional remarks about our Laser experiments are in order. The original Laser text as obtained from the U.S. Patent Office has a vocabulary of about 12.000 words. We thought this to be too large an increase from the 250 word vocabulary we handled in the Raleigh experiment. We found the 1.000 most frequent words, and conducted recognition on test sentences from the text consisting of those words only. A sample of these sentences is presented in Table II. Tables III and IV show typical alignments of the decoded sentence with that actually spoken. In the example of Table III, although the was changed into all of the. In the example of Table IV, adapted to it active and stable output to stata below point.

A lot of work remains before continuous speech recognizers will become an economical commercial product. However, we believe that our laboratory results show that this goal will be achieved in the not too distant future.

ACKNOWLEDGMENTS

The research reported in this article is the joint work of the members of the IBM Continuous Speech Recognition Group.

REFERENCES

(1). F. JELINEK, "Continuous Speech Recognition by Statistical Methods", Proceedings of the IEEE, Vol. 64, N° 4, April 1976.

(2). F. JELINEK , L.R. BAHL and R.L. MERCER "Continuous Speech Recognition : Statistical Methods", to appear in Handbook of Statistics, Vol. 2, P.R. Krishnaiah ed., North-Holland, Amsterdam, 1980.

(3). J. LYONS "Introduction to Theoretical Linguistics", Cambridge, England, Cambridge University Press, 1969.

(4). D.R. REDDY et al. "Speech Understanding Systems Final Report", Computer Science Department, Carnegie-Mellon University, 1977.

THE ERLANGEN SPEECH UNDERSTANDING PROJECT

Hans-Werner Hein

Lehrstuhl fuer Informatik 5 (Mustererkennung)
Universitaet Erlangen-Nuernberg
Martensstr. 3, 8520 Erlangen, FRG

ABSTRACT

A system for the automatic understanding of continuous German speech is introduced. Main principles of its design, the blackboard model structure and the parallel processing aspect are motivated and the essential subsystems described:
The relational database system to realize the blackboard, the parallel computer simulation and the experimentation frame for interactive structure definition and detailed performance analysis of various speech understanding configurations. The specifications for the first generation of knowledge modules, currently under development are also given.

1. RESEARCH OBJECTIVES

Since 1979 a project about the automatic understanding of normally spoken German sentences is running at Erlangen [1-4]. The evolved speech understanding system should be able to speaker-independently process signals of telephone bandwidth quality. Furthermore it should not be a one-purpose solution but an experimental tool to aid the development of optimized and specialized future speech understanders for many applications.

It follows that the system as under construction allows modular and easy exchange of its vocabulary, grammar, and task specific knowledge.

The interfaces are designed as if the system is part of a Conversation System (Fig. 1). Expected input is the acoustical data of a speech signal, the output will be the (hopefully) correct semantical deep-structure description of the utterance.

J.-P. Haton (ed.), Automatic Speech Analysis and Recognition, 239–251.

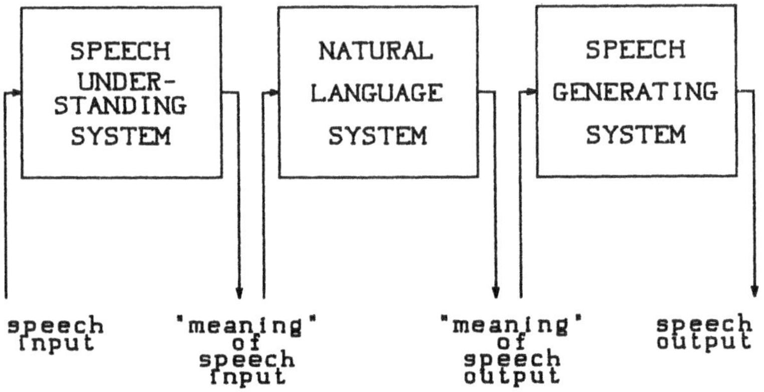

Fig. 1: Conversation System Structure

Given this, a connected Natural Language System may then produce a response at the same description level and a Speech Generating System may transform this into a surface-structure sentence which is output acoustically.

2. SYSTEM DESIGN PHILOSOPHY

Under the given objectives we choose as a principle system structure a database-oriented system (blackboard model) similar to HEARSAY-II [5]. Therefrom we get advantages compared to other structural principles in three main points: Experimental flexibility, conversation system software-economy, and parallel processing.

The advantage in experimental flexibility is easy to realize. Various systems assembled from a library of existing knowledge modules may be measured against the requirements of a certain application. Furthermore it is possible, and we already use this feature, to let a person simulate a non-existing knowledge module interactively, either to find out proper specifications for new or alternative knowledge modules, or to experiment with so far incomplete understanding systems.

The software-economy aspect concerns the following considerations: In a conversation system we have a speech understanding part, a natural language part, and one for speech generation. To do the speech understanding task one needs very probably modules also being inevitable in the natural language task (e.g. a grammar and a parser for top-down word hypothetisation and syntax analysis respectively) or in the speech generation task (e.g. a kind of synthesizer for word verification and final speech output respectively).

Those modules then only have to be developed once but will be used in connection with more than one part of the conversation system.

The database-oriented system is also the design key to parallel processing. Future speech understanding systems will have to operate nearly in realtime [6]. This requires either a very fast monoprocessor which then will be too expensive for practical applications, or a special and parallel processing hardware out of cheap components.

We follow the second alternative and develop the knowledge modules with simple and uniform interfaces to make them useable on a parallel processing hardware. Therefore we already run our speech understanding software on a parallel computer simulation.

After we get a pilot application working satisfactory at software level we plan to realize it on a parallel computing system. This will be either a saleable general purpose device or a special construction. A specialized architecture will have the long-term gain that for some of the system components the most time-consuming algorithms may be implemented directly in hardware, too.

3. SYSTEM STRUCTURE

A general view of the structure of a speech understanding system according to our approach is given in Fig. 2.

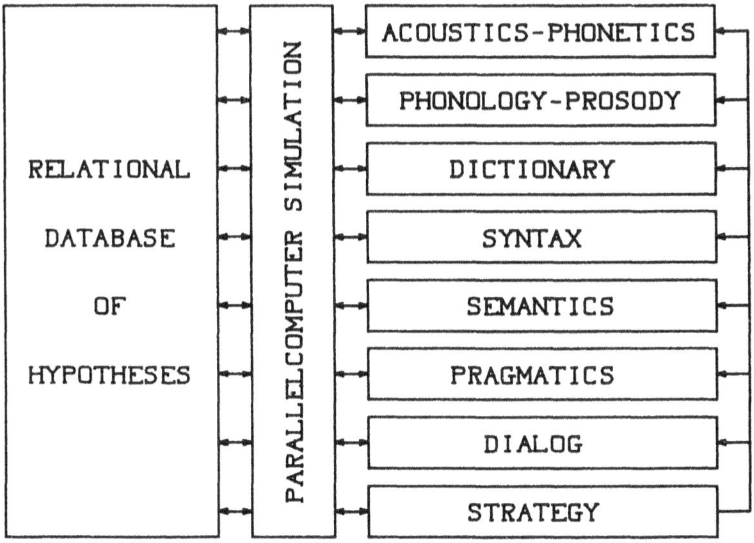

Fig. 2: Speech Understanding System Structure

3.1. Results Database

One main part of the system is the blackboard, provided with the means of a relational database system. It contains all intermediate analysis results being of importance to more than one knowledge module.

The intermediate results, called hypotheses, are stored under a basic data structure scheme which is uniform for all knowledge modules. A hypothesis owns among others attributes for its actual statement, temporal and logical predecessors, competitors, and successors, for single scores given by certain knowledge modules, and a total score.

The relational database of hypotheses is modularized into parallel accessible abstraction levels. Main levels are those of phone, word, partial sentence structure, and partial sentence meaning hypotheses but other non-intuitive levels easily will be integrated if required.

Because each knowledge module corresponding with its special nature works only on some of the database abstraction levels, access conflicts are minimized and successively appearing typical database subtasks, e.g. sorting and searching, are split up into parallel executable portions. From this results an evident reduction in database response time.

3.2. Knowledge Modules

The knowledge modules currently developed for the speech understanding system are mainly oriented towards linguistic viewpoints relating to the German language. An exception is the strategy module which only contains information about the actual speech understanding system, especially its structure and resources.

Generally a knowledge module has four main components and two standardized interfaces. Example knowledge module structures are shown later in Fig. 5 and Fig. 6.

The main components of a knowledge module are:

* Each time the knowledge module is activated a MONITOR decides from its internal state and the actual content of the hypotheses database, what to do next, and then initiates execution. The monitor so far is the local and unique task-oriented strategy of a knowledge module.

* The monitor does its decisions under the constraints of BEHAVIOUR PARAMETERS which may be looked at as global variables. They are updated periodically by the strategy module. To a certain extent the strategy module thus is

able to tune the local strategy of any other knowledge module. The behaviour parameters are defined by each knowledge module designer himself according to the knowledge-specific circumstances.

* Each knowledge module usually needs for the handling of hypotheses and knowledge data some well adapted ALGORITHMS.

* Together with the privat accessible KNOWLEDGE DATA the algorithms constitute the strictly speaking knowledge source. It is intended that especially for higher abstraction levels knowledge data is altered and extended by the knowledge module itself during the analysis of a speech signal. This feature may be very useful when analysing a sentence within the context of a longer dialog.

The interfaces of a knowledge module are the strategy module interface and the hypotheses database interface. The strategy module interface enables the strategy module to access the respective behaviour parameters.

Through the hypotheses database interface a knowledge module reads hypotheses from all attached abstraction levels. It also is thus enabled to verify existing and create new hypotheses dependent on its special knowledge and strategy.

3.3. Parallel Computer Simulation

For experimentation purposes we developed a parallel computer simulation software. It enables to define configurations consisting of i uniform processors with private memory of optional size, and j parallel accessible common memories of optional size. In our application each common memory may contain one abstraction level of the database of hypotheses.

In every simulation cycle each of the processors is enabled to access each common memory in read-mode, but it is not possible for more than one processor to write to a common memory module at the same time. If an access conflict occurs all but one (choosen at random) of the processors are delayed until the next simulation cycle.

The parallel computer simulation includes tracing options of various refinement. So the bottlenecks within a certain speech understanding configuration may be detected by automatic evaluation of the hardware-trace files after a speech signal analysis. The possibility of trace-data output in graphic manner for visual inspection is also provided.

When simulating speech understanding configurations with more knowledge modules than processors available, special commands for knowledge module scheduling will be required.

Therefore we introduced to the simulation a queuing mechanism which may be used by the strategy module according to the respective state of speech signal analysis. E.g. during the first stage of the analysis it may be useful to allocate less computation time to knowledge modules of higher abstraction levels than to lower level knowledge modules; and vice versa during the last stage of the analysis.

The scheduling feature of the parallel computer simulation is designed so universal that it even enables the strategy module to disactivate and activate itself at arbitrary simulation cycles.

3.4. Experimentation Frame

Due to the high complexity of the speech understanding system as a whole, we put the Relational Database System, the Parallel Computer Simulation, and the Modules under an Experimentation Frame.

It allows interactive definition of a speech understanding configuration, selection of knowledge and speech data, supervision of the speech signal understanding process, and participation

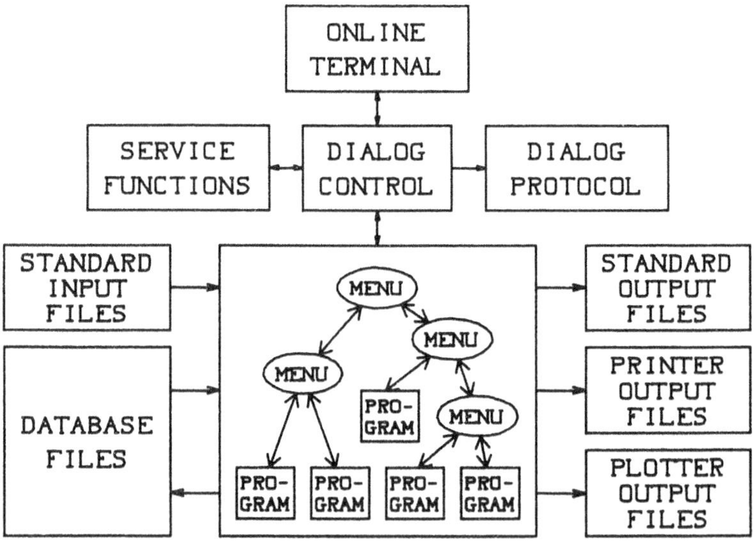

Fig. 3: Experimentation Frame Structure

in this process as an additional knowledge module. Furthermore it provides an acquisition of protocols and statistics in printed or graphic manner at every state of speech signal analysis.

All program modules belonging to the speech understanding software are arranged in a tree structure. Through menus a user initiates the execution of program modules in arbitrary order according to its intended experiment.

The menu-tree feature is very helpful for modul specialists developing and testing their programs within an otherwise unfamiliar software surrounding. It also supports an easy addition and exchange of program modules while it always guarantees the availability of an operable system version for all project participants.

Another aspect of the experimentation frame is that it provides standard input/output interfaces of various types together with lots of service functions for diverse purposes at a higher level than usual operating systems. Software modules independently developed by different specialists under the experimentation frame thus always keep compatible which facilitates their integration into the speech understanding system.

Fig. 3 gives an impression of the experimentation frame structure together with a small menu-tree.

4. MODULES

The following chapter introduces some knowledge modules currently under development for our system. According to the purpose of this paper only the top-level specifications will be given. More details may be found in the cited literatur.

4.1. Acoustics-Phonetics Module

This module uses the data produced by an offline running acoustic-phonetic transcription system for fluently spoken German speech [7,8]. The transcription system was programmed on a PDP11 with speech input devices. The acoustics-phonetics module makes the transcription system virtually present in the speech understanding system.

The advantages are a general repeatability of experiments using the same "speech signal" and the possibility of testing the system with synthetic data at phone level.

The structure of the transcription system is given in Fig. 4.

After bandpass filtering (100 to 3400 Hz) the signal is sampled at 10 kHz with 12 bit and then normalized.

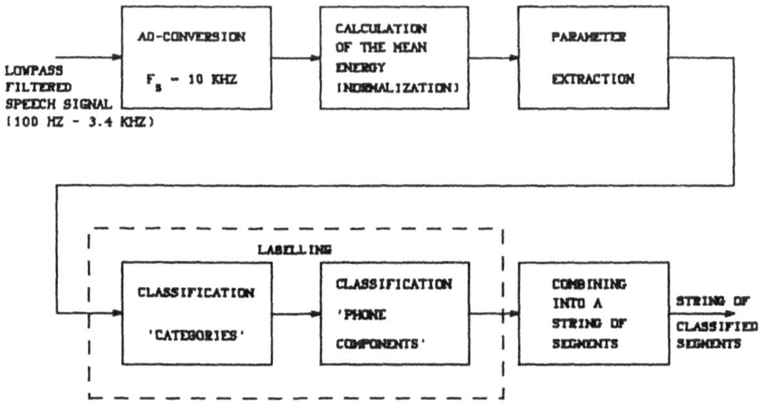

Fig. 4: Acoustic-Phonetic Transcription System

For overlapping Hamming-windows various parameters are extracted and a first classification into the four categories "silence", "voiceless", "voiced fricative", and "voiced non-fricative" is done.

A second classification of the windows follows using different sets of parameters for each category. The parameters are also different to those used for the category classification step. After that several heuristics are introduced to combine the labelled windows into phone segments.

Final output is a segmentation of the input speech signal with some alternative phone labels (together with probability scores), pitch frequency, and loudness information for each segment.

4.2. Dictionary Knowledge Module

Two types of knowledge data are contained in the dictionary knowledge module (Fig. 5):

* A dictionary with phonetic, morphemic, syntactic, and semantic information for each wordform. This dictionary is extracted from a large linguistic database and prestructured in the means of the used Viterbi algorithm to speech up e.g. searches for best fitting words in a given area of the speech signal.
* Statistic data (phone similarity weights) about relative grades of identity between phones used for phonetic word representation in the dictionary and phones equivalently recognized from the acoustics-phonetics module.

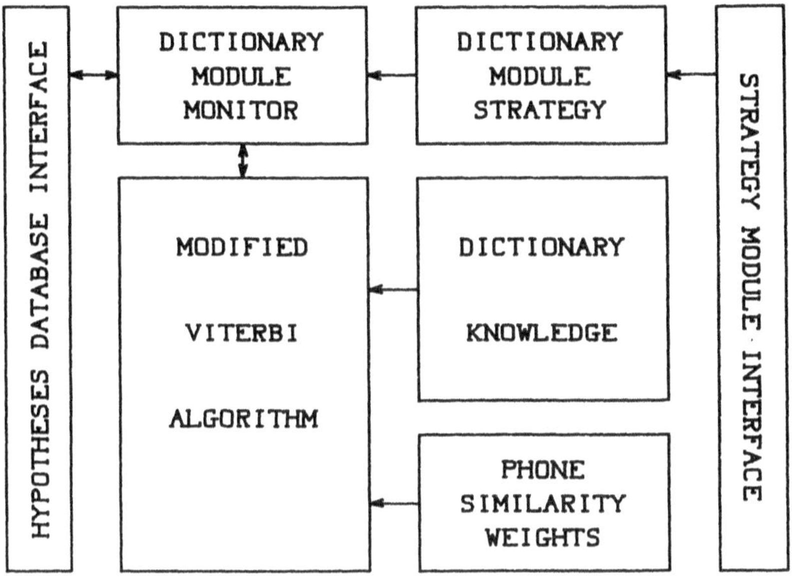

Fig. 5: Dictionary Knowledge Module

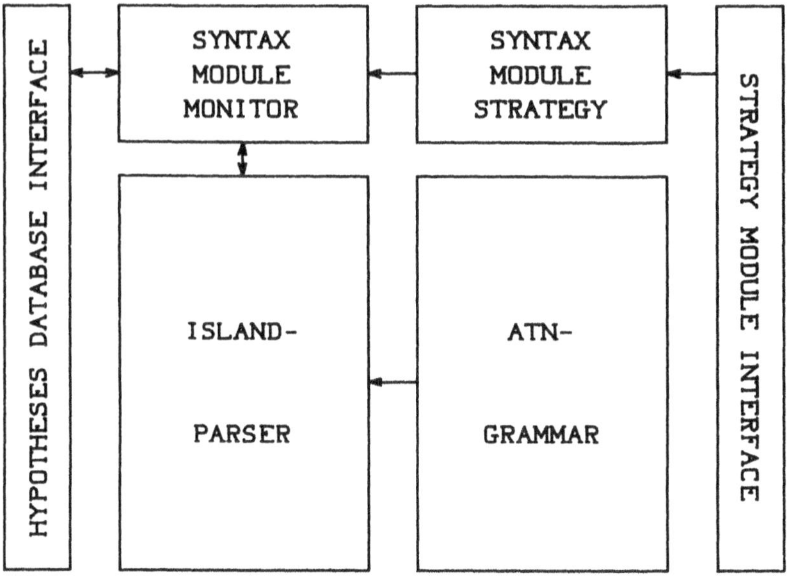

Fig. 6: Syntax Knowledge Module

For the creation and verification of word hypotheses the dictionary knowledge module uses a modified Viterbi algorithm [9] which computes for two given strings of (phone) symbols a distance measure considering the phone similarity weights.

4.3. Syntax Knowledge Module

The knowledge data component of this module (Fig. 6) is an Augmented Transition Network (ATN) grammar. A very "natural" ATN grammar has been developed for the German language [10] which can be allocated fully or partially to a speech understanding system to realize different levels of syntactic refinement for experimental purposes. The syntactic classes of that grammar are linguistic ones, e.g. NOUN or ADJECTIVE.

A direction-independent island-parser [11] enables the handling of partial sentence structure hypotheses (islands). To the left or right of already existing islands the syntax knowledge module using the island-parser also may create top-down new word hypotheses.

4.4. Strategy Module

The strategy module (Fig. 7) is the only system component that

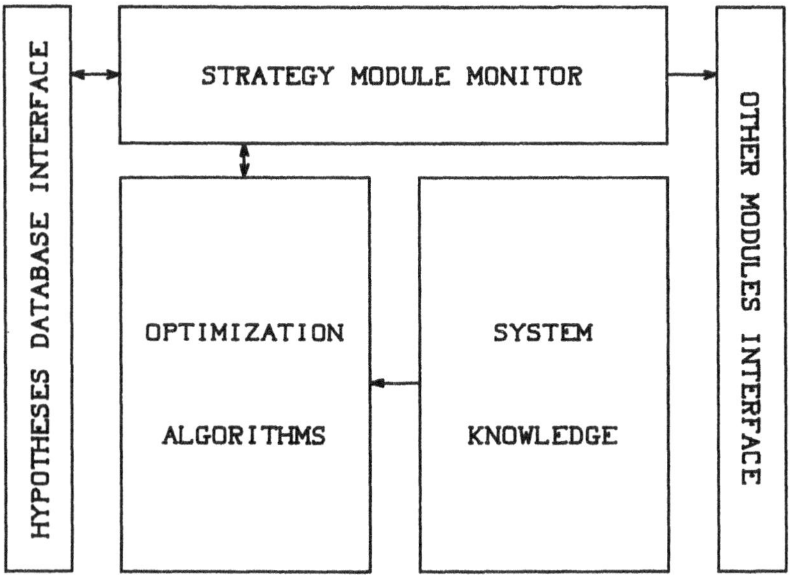

Fig. 7: Strategy Module

knows the actual speech understanding system structure and the contained knowledge modules. It also has the information about the limitations of the system resources and the maximum admissible delay for system reaction.

The strategy watches the database of hypotheses and according to the state of analysis tunes the behaviour parameters of the other knowledge modules. If necessary it also does the system scheduling.

An important task of the strategy module is the computation of general scores for hypotheses of whatever abstraction level. As a basis for the general scoring the strategy uses the single scores, weighted with reliability measures for the respective modules.

With the tuning, the scheduling, and the scoring tasks the strategy module has three different tools to influence a running speech signal analysis. The objective will be to make the understanding process converge rapidly to a hypothesis at sentence meaning level which spans over the whole input speech signal and has a score high enough to be a very likely interpretation.

4.5. Further Knowledge Modules

A semantics and a pragmatics knowledge module are being developed for the speech understanding system, too. With these modules a question-answering application about the German Intercity-Train-System will be realized first. It should enable the processing of sentences being parts of information dialogs concerning connexions, distances, fares etc.

CONCLUDING REMARKS

The main components of the Erlangen Speech Understanding System are implemented with the programming language SIMULA on a CYBER 173, the acoustic-phonetic transcription system in FORTRAN on a PDP11/34.

The construction of speech understanding systems with easy exchangeable dictionary, grammar, and application is intended. The vocabulary should at least include 1000 wordforms and the grammar be as natural as possible.

If good recognition results are reached with the software simulation of a certain speech understanding configuration, an implementation on a parallel processing hardware system will follow.

ACKNOWLEDGEMENTS

The Deutsche Forschungsgemeinschaft (DFG) supported the research concerning the acousic-phonetic transcription.

The development of the semantics and pragmatics modules are granted by the Bundesminister fuer Forschung und Technologie (BMFT).

REFERENCES

[1] Niemann, H., Hein, H.-W.: "A Program System of Parallel Processes for Understanding Continuous Speech". COMPUTING, Suppl. 3, pp. 141-148 (1981)

[2] Hein, H.-W., Niemann, H.: "Expert Knowledge for Automatic Understanding of Continuous Speech". Signal Processing: Theories and Applications (Kunt, M., de Coulon, F., eds.), pp. 647-651. Amsterdam: North-Holland (1980)

[3] Hein, H.-W.: "A System for Understanding Continuous German Speech". GWAI-81: German Workshop on Artificial Intelligence, Bad Honnef, January 1981 (Siekmann, J. H., ed.), pp. 277-286. Berlin: Springer-Verlag (1981)

[4] Niemann, H.: "Pattern Analysis". Springer Series in Information Sciences 4. Berlin: Springer-Verlag (1981)

[5] Erman, L. D., Hayes-Roth, F., Lesser, V. R., Reddy, D. R.: "The HEARSAY-II Speech-Understanding System: Integrating Knowledge to Resolve Uncertainty". COMPUTING SURVEYS, Vol. 12, No. 2, June 1980, pp. 213-253

[6] Lea, W. A.: "Speech Recognition: Past, Present, and Future". Trends in Speech Recognition (Lea, W. A., ed.), pp. 39-98. Englewood Cliffs, N. J.: Prentice-Hall (1980)

[7] Regel, P.: "Automatic Extraction of Phonetic Symbols from the Speech Wave". Pattern Recognition (Research at Lehrstuhl fuer Informatik 5) (Niemann, H., ed.), pp. 65-74. Erlangen: Arbeitsberichte des Instituts fuer Mathematische Maschinen und Datenverarbeitung, Vol. 14, No. 1, February 1981

[8] Regel, P.: "A Module for Acoustic-Phonetic Transcription of Fluently Spoken German Speech". IEEE TRANSACTIONS ON ASSP, to appear.

[9] Haton, J.-P.: "The Representation and Use of a Lexicon in Automatic Speech Recognition and Understanding". Spoken Language Generation and Understanding (J. C. Simon, ed.), pp. 311-335. Dordrecht: D. Reidel (1980)

[10] Brietzmann, A.: "Eine ATN-Grammatik des Deutschen fuer die Automatische Sprachverarbeitung". Lehrstuhl fuer Informatik 5 (Mustererkennung), Universitaet Erlangen-Nuernberg, Diplomarbeit (1980)

[11] Gerlach, M.: "ATN-Inselparser fuer den Syntaxexperten eines Spracherkennungssystems". Lehrstuhl fuer Informatik 5 (Mustererkennung), Universitaet Erlangen-Nuernberg, Studienarbeit (1980)

SEMANTIC AND PRAGMATIC PROCESSING IN CONTINUOUS SPEECH UNDERSTANDING

Jean-Paul HATON, Jean-Marie PIERREL, Simon SABBAGH

CRIN - Université de Nancy I
B.P. 239
54506 Vandoeuvre Cédex, France

The use of speech in man-machine communication is becoming increasingly important and research in this area is now a part of computer science activity, related both to signal processing, pattern recognition and artificial intelligence. The actual applications of speech recognition presently use isolated word recognition systems but the most interesting and promising domains of man-machine dialog will require connected speech recognition and understanding. Although significant results have already been obtained (see for instance (1)) the presently available systems obviously have still to be improved.

In this paper we present an approach to the use of high level constraints (essentially semantics and pragmatics) in continuous speech recognition. This presentation is related to the work carried out by our group during the past seven years in the design of several systems for artificial and pseudo-natural language recognition and understanding. After a brief presentation of these systems and of the distinction between recognition and understanding which appears in them, we successively describe the two processing levels :

- recognition using syntactic and local semantic constraints,
- comprehension and interpretation using global semantic and pragmatic constraints,

that we have implemented. This description will be illustrated by practical examples especially in the framework of the MYRTILLE II system.

J.-P. Haton (ed.), Automatic Speech Analysis and Recognition, 253–268.

1 - RECOGNITION VS UNDERSTANDING

The communication process can be viewed as the interaction between a source and a receiver through a transmission channel (fig. 1). The message emitted by the source is coded, transmitted and eventually altered by the channel, and finally decoded by the receiver.

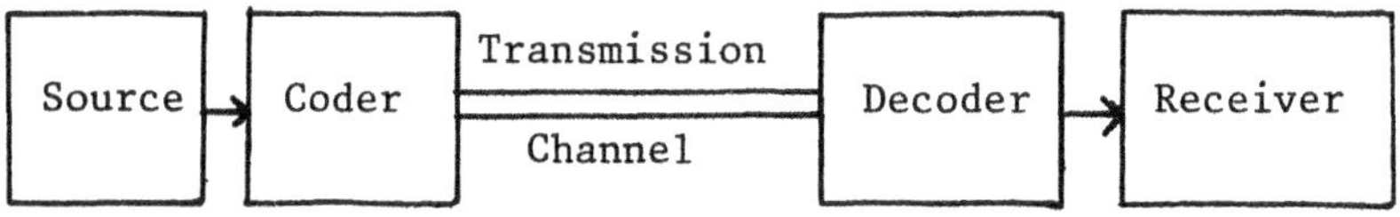

Fig. 1 : A General Communication Model

This communication schema can be either uni- or bi-directional. It implies that the emitter and the receiver share the same code for information transcription and eventually other kinds of knowledge (common references, ...).

Speech communication is essentially a two-way process. The vocal signal is highly encoded since it corresponds to a hierarchy of interrelated coding levels : phonology, syntax, semantics, pragmatics, prosody. The receiver must decode this signal in order to extract the underlying meaning of the message. The distinction between recognition and understanding in speech communication appeared in the 5 year ARPA Project on Speech understanding. This distinction is useful since it does correspond to substantial differences in the architecture and the design of the systems as it will be shown later on.

The goal of speech recognition consists of identifying all the elements (words, phonemes, ...) which constitute the message whereas in speech understanding one aims at finding the meaning of the message even though some elements have been missed or misrecognized. In that sense, a speech understanding system always incorporates a speech recognizer, even if both are not clearly separated. The process of understanding implies an interpretation of the message which extensively uses semantic and pragmatic knowledge.

According to the difficulty of the task (i.e. complexity of the language and of the universe of application, size of the vocabulary, ...) the interpretation will require more or less attention : this will be shown on two systems developed by our group : MYRTILLE I and MYRTILLE II.

MYRTILLE I was designed for artificial language processing with the following characteristics : highly constrained syntax represented by a context-free grammar of reasonable size, vocabulary of less than 100 words. It is a top-down, syntax-driven system in which the recognition process is entirely controlled by the parser which emits at each step word hypotheses according to the context and the grammar (cf fig. 2). These hypotheses are then weighted by a word recognizer and taken into account using a best-first strategy.

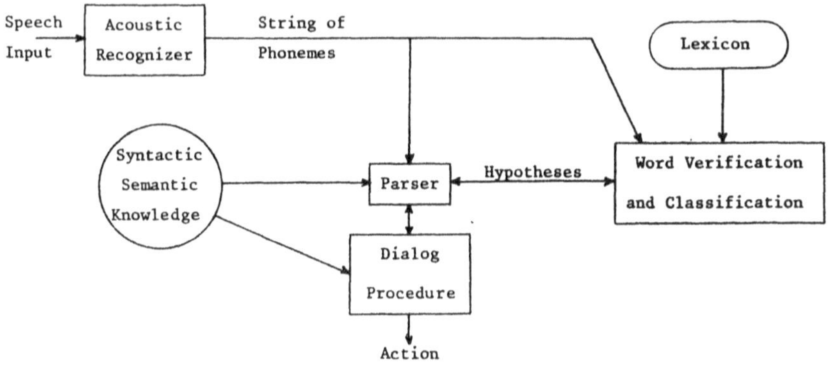

Fig. 2 : Overview of MYRTILLE I

MYRTILLE I was experimented with several languages for instance in the framework of a telephone exchange application (2). With such languages the meaning of a sentence is usually obvious : it is possible to interpret the request of a speaker from the syntactical structure of the sentence given by the recognition process. This system is therefore not a speech understanding system.

The MYRTILLE II system, on the contrary, has to be considered as a speech understanding system (3). It can process pseudo-natural languages corresponding to wide subsets of spoken French with few syntactic restrictions and vocabularies up to 500 words. MYRTILLE II is presently running on a meteorological inquiry application. The difficulty of the task makes it necessary to take into account semantic and pragmatic knowledge in the understanding process at two different levels :

- use of local semantic constraints during the recognition phase : these constraints mainly correspond to word dependancies. They appear both in the syntactic-semantic networks which represent the structures of the language (the so-called procedural networks) and in the hierarchical lexicon. The recognition of a sentence is based upon a hypothesis-and-test paradigm, as summarized on fig. 3. This process will be developed in section 2 .

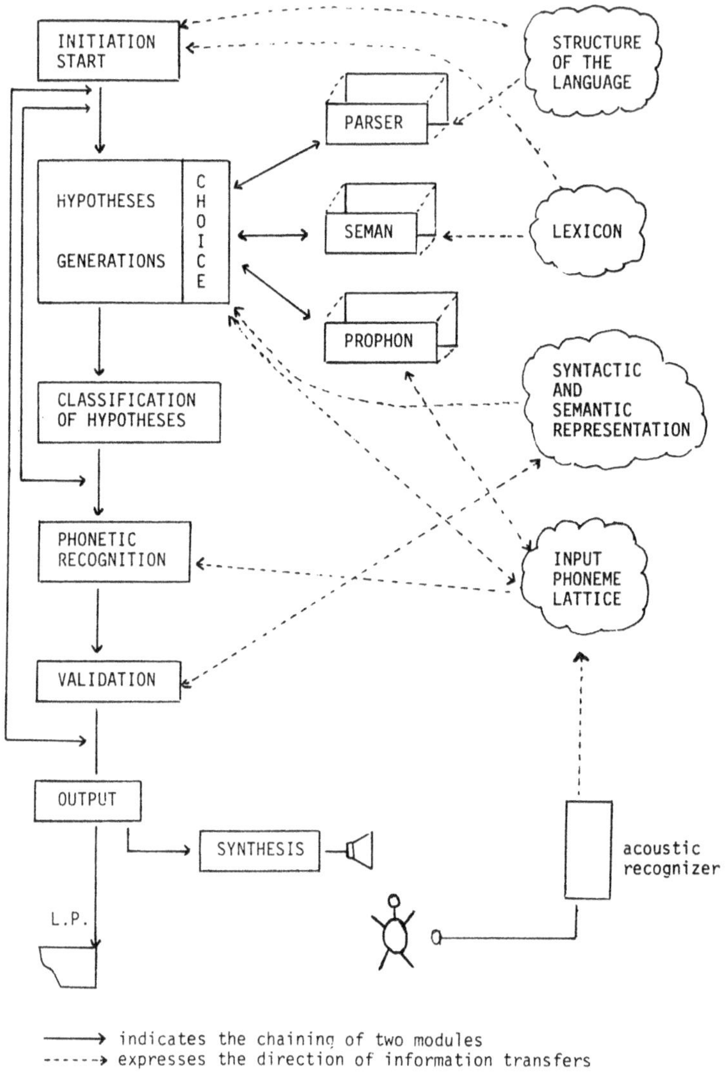

Fig. 3 : Overview of the sentence recognizer in MYRTILLE II

- use of global semantic and pragmatic constraints during the interpretation phase : this process will be detailed in section 3.

The understanding of a sentence in MYRTILLE II can therefore be divided into three successive parts :

- acoustic-phonetic decoding,
- sentence recognition,
- interpretation.

This distinction between recognition and interpretation was choosen for the sake of simplicity and efficiency. The same distinction appears in other systems, for instance KEAL (4). It is also possible to group the two phases as in the semantic networks of the SRI system (5).

2 - USE OF SEMANTIC CONSTRAINTS IN RECOGNITION

2.1. General principle

In the case of natural or pseudo-natural languages, it is necessary to distinguish between two types of syntactic representations :

- semantic grammars,
- general grammars.

In the first case, the semantic information used by the sentence recognizer and the syntactic information are put in a single structure. The principal disadvantage of this model is the difficulty of changing the application : it is necessary to redefine the semantic grammar and the semantic subclasses of terminals at each time. Nevertheless, this model has been used in several systems (1), (6) since the use these semantic grammars are well defined within the formalism of the semantic ATN grammars (7).

In the second case, the language is defined by a general grammar which only determines the main syntactic groups (the terminals of the grammar are equivalent to grammatical units). In this case, if the syntactic procedure is very fast, the hypotheses emitted are too numerous and it is necessary to limit these hypotheses by the use of semantics before the phonetic recognition. Presently the principal solutions for describing semantic informations are :

. procedural representation (8),
. semantic networks (9),
. n-uplets ...

In the MYRTILLE II system, in addition to the structure definition (3), we also need the syntactic-semantic definition of words, as indicated in section 1. Without going directly to the definition of elementary concepts and conceptual dependencies, as Shank has done, we have opted for a tree representation of the vocabulary (4 tree levels):

(i) the first level creates a division of the vocabulary into three large classes :

- the liaison words which appear explicitly in the definition of the syntactic network,
- the words with particular grammatical functions : propositions, conjonctions,... ,
- the general terminals : nouns, verbs, adjectives.

(ii) the second level creates a division according to the grammatical classes : NOUN, VERB, ADJECTIVE...

(iii) the third level distinguishes between sub-classes in terms of the grammatical construction and of the types of conceptual dependencies that they demand. For example, for the verbs, we indicate :

- the list of the types of nouns which can be subject,
- an indicator for the impersonal verbs,
- the list of the types of noun complements,
- the list of the conceptual dependencies,
- the list of the types of grammatical structures of complements,
- etc...

(iv) the fourth level corresponds to the different words or entities. This level details the elementary concepts and the properties attached to each word : specific constructions, acoustic-phonetic representation, type of phonological alterations

With this lexicon, we have defined several procedures of semantic selection , for example :

- selection of adjectives with a given noun and the inverse function,
- selection of noun complements of a given noun and the inverse function,
- selection of subject nouns of a given verb and the inverse function .

For the emission of hypotheses, the procedure SEMAN takes into consideration the word syntactic-semantic definition in the dictionnary. Having taken into account the context, SEMAN limits the number of hypotheses in terms of conceptual dependencies determined by the part of the sentence already processed, as well as of the recognized structure.

For instance in order to limit the hypotheses of adjective type, SEMAN uses the following algorithm :

If the adjective is an adjectival complement (with a copulative verb)
then begin
. search of the subject nouns,
. selection of adjectives which can qualify these subjects with the indicate verb.
end

If the adjective is in a nominal group then begin
if the noun is already recognized
then . selection of possible adjectives which can be attributive adjectives with this noun

else begin
. selection of nouns which can appear in this nominal group,
. selection of possible adjectives which can be attributive adjectives with these nouns
end
end

2.2. Example of recognition (10)

MYRTILLE II is presently implemented with a left-to-right, best-first strategy. The language structure which is used is close to the natural spoken French structure and the vocabulary of the application (meteorological inquiry) contains about 400 words (the term "word" is used here like "root") : 160 nouns, 50 verbs, 70 adjectives and grammatical words.

Let us explain the sentence recognition level in MYRTILLE II with an example which concerns the french sentence : "Est ce que les risques de petites plaques matinales de verglas diminueront demain dans la région de Nancy ?"
(Will the risks of small early morning silver thaw persist tomorrow around Nancy ?).

The figure 4 gives the number of hypotheses emitted at each step of the recognition and the recognized structure (the identifiers used in the tree representation of the sentence are explained in Appendix 1). This recognized structure is given by the system under a linear form :

((est ce que) * INTE (((risques (petites plaques matinales (verglas) * GN) * GN) * GN) * GS (persister) * GV () SGV ((demain) * GC (dans (région (Nancy) * GN) * GN) * GL) * GCØM) * PR) * ENON

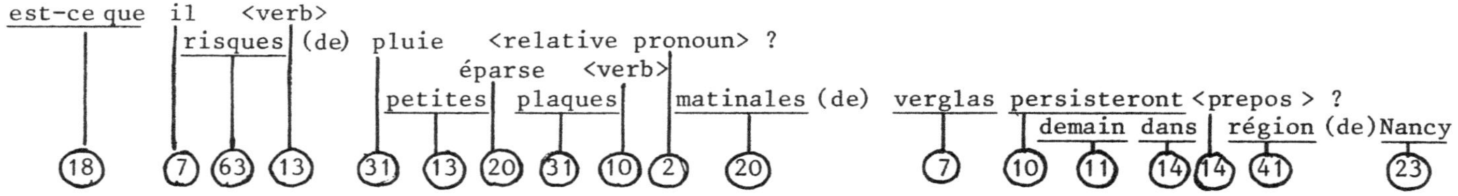

Number of hypotheses emitted at each step

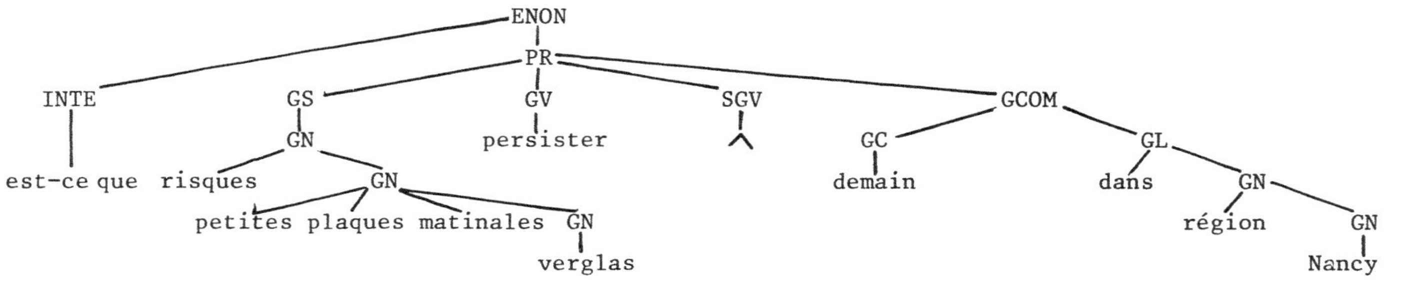

recognized structure

Example of sentence recognition : French sentence : "Est-ce que les risques de petites plaques matinales de verglas diminueront demain dans la région de Nancy ?"
(Will the risks of small early morning silver thaw persist tomorrow around Nancy ?)

Figure 4

We can make several remarks about this example :

* The complexity of the subject noun group ("risques de petites plaques matinales de verglas"). The recognition of a noun group complement is carried out if a voiced plosive was detected (preposition "de" which introduces the noun complement) and a relative if a unvoiced plosive (relative pronoun "qui" or "que").

* During the recognition of a complex noun group SEMAN will emit increasingly restricted hypotheses about nouns and adjectives in the group by taking into account the conceptual dependencies introduced by the already recognized part of the noun group. The number of emitted hypotheses will therefore decrease when the semantic context increases : 63 for "risques", 31 for "plaques" and 7 for "verglas".

* We can see on the recognized structure that the system detected an empty verbal complement (SGV). In the definition of the language a verb is always considered as being followed by a verbal group. Each verb is given an attribute which indicates the kind of verbal group it needs. If a verb does not need any verbal group the system generates an empty one.

* The nominal groups are often composed of a name without determiner : this is quite common in speech recognition since determiners are usually badly recognized especially in French because the determiners are often mono syllabic words : "le", "la", "les", ...

3 - SEMANTIC INTERPRETATION AND DIALOG PROCEDURE

3.1. Introduction :

The purpose of the interpretation level is to check the semantic validity of the sentence which has been recognized ; the dialog procedure is intended to gain complementary information concerning the initial request of the user in order to complete the recognition or to solve ambiguities.

After the description of our linguistic model we will illustrate the way in which the system operates.

3.2. The linguistic model :

Among the different existing models we have choosen the Fillmore case grammar (11). In this model a sentence S is considered as a proposition PROP associated with a modality MOD :

$$S \longrightarrow MOD + PROP$$

The modality carries detailed information : tense, form, etc...

Whereas the proposition represents the deep structure of the sentence the verb is the main element in the sentence and the different noun groups are the arguments of the verb since the verb is considered as a logical predicate. These arguments are called cases ; they are considered to be linguistic universals. However, for a limited domain of knowledge, it is possible to use better adapted cases : for example for an airplane ticket reservation system we can use such cases as departure time, date, city and so on.

Let us give an example of sentence decomposition in the framework of this model :

Sentence : John breaks the window with a ball.
Structure :

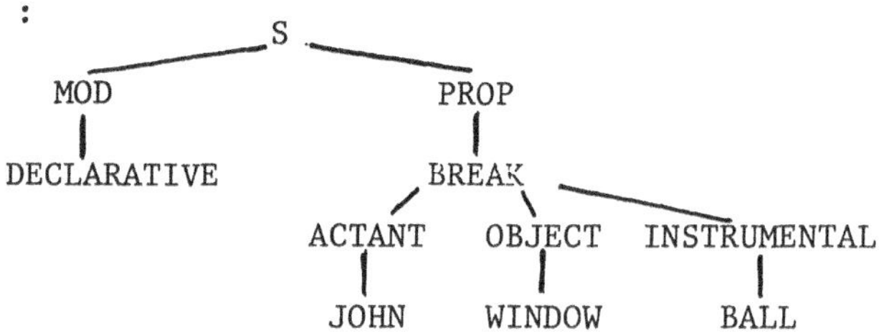

Most of natural language processing systems use case grammars as linguistic model. It is therefore important to precise in what sense we use this model. In fact we only use the case grammar model for the casual verb representation in the lexicon and for the representation of meaning of sentences.

But the semantic analysis procedure doesn't use the selection rules defined by Fillmore : the main reason is that in our case the input is a deep structure (Chomsky's sense) and not the surface structure.

Let us examine the internal representation of the lexicon in the framework of this model. For example to the verb "to increase" (augmenter) are associated several pieces of information in our meteorological application :

To increase :

property	NDGS	NDSGV	SOURCE	OBJECT
value	Source	Object	(QMES, MES)	(UMES)
	OBL	OBL		PREP
	True	NIL		DE

That means that this verb must have for subject group a noun which is the source of the process and the presence of the subject noun is obligatory. This noun must also belong to the sets QMES or MES that are respectively the set of measurable quantities (snow, rain,...) and the set of measurements (thickness, highness,...).

The verbal noun group is the object of the process and its presence is not mandatory ; it must belong to the set UMES of measurement units (fourth column). The proposition DE (in english OF) must preceed the verbal group. Since the recognition of such monosyllabic words is not always pertinent, this preposition was added in the description of the verb in order to obtain a syntacticaly correct string at the end of the generating process.

The internal representation of a sentence also uses the formalism of case grammar. For instance for the french sentence : "De combien de centimètres l'épaisseur de neige a-t-elle augmenté dans les Vosges cette semaine ?"

(How many centimeters did the snow's thickness increase of in the Vosges mountains this week ?)

The representation is as follows :

```
[(ENON (MOD INTER)          (This means that the question
       (QUEST C2))           is about the object C2)

 (VERB (INCREASE)
       (SOURCE C1)
       (OBJECT C2)

 ( C1  (THICKNESS)
       (MOD C3))

 ( C3  (SNOW)
       (PREP OF))

 ( C2  (CENTIMETERS)
       (MOD HOW MUCH)
       (PREP OF))

 (LOC  (VOSGES)              The cases of location (LOC) and
       (PREP IN)             of Time (TPS) are privilegied
                             cases in our application.
 (TPS  (WEEK)                They are deduced from the cir-
       (MOD THIS)]           cumstancial noun groups.
```

3.3. System description and operation

Figure 5 gives an overview of the interpretation process in the MYRTILLE II system. This process is made up of three major parts :

- semantic analyzer,
- dialog procedure,
- generation procedures (synthesis)

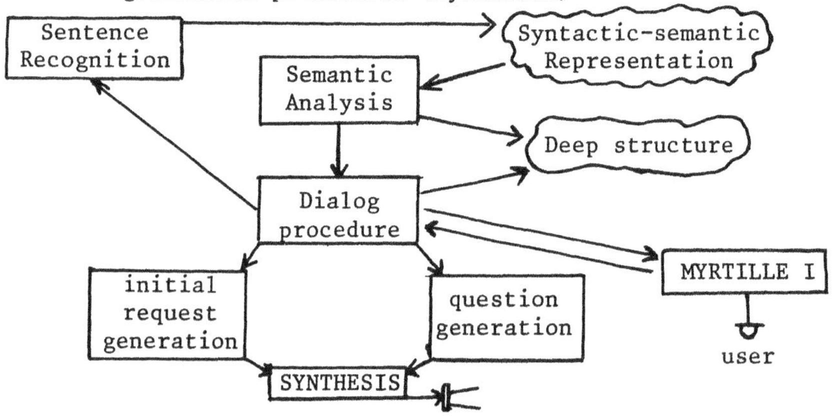

Fig. 5 : Overview of interpretation in MYRTILLE II

Figure 6 shows an actual example of dialog .

(0) System : Centre de renseignements météo, je vous écoute.

(Hello, Weather information center)

Speaker : Je voudrais connaître le temps qu'on envisage demain.

(I want to know what weather you predict for tomorrow.)

(1) System : Vous avez demandé : quel temps envisage-t-on demain ?

(Did you ask : what weather we predict for tomorrow ?)

Speaker : Oui, c'est cela.

(Yes)

(2) System : A Nancy ?

(In Nancy ?)

Speaker : Non, dans la région de Paris.

(No, in the area of Paris).

(3) System : Vous avez demandé : quel temps envisage-t-on demain dans la région de Paris ?

(Did you ask : what weather we predict for tomorrow in the area of Paris ?)

Fig. 6 : A practical example of dialog.

The semantic analyzer has several roles :

- it gives an interpretation of the request,
- it builds a representation which is the deep structure mentioned before,
- it asks questions to the speaker in order to obtain complementary information or to solve ambiguities.

For example after the first request (0) in figure 6, the system builds up the following deep structure :

```
!(ENON (MOD INTER)
       (QUEST C2))

 (VERB (PREDICT)
       (ACTANT (ON))
       (OBJECT C2))

 (C2   (WEATHER)
       (MOD WHICH))

 (TPS  (TOMORROW))

 (LOC  (        ))!
```

The lack of the location case is at the origin of question (2). The analyzer builds a list of questions associated with recognition directives for the answers of the speaker. Most of these responses are predictible as far as their semantic content and syntax form are concerned. The recognition of these answers is carried out by the MYRTILLE I system.

The dialog procedure takes as input the list of questions built up by the semantic analyzer, controls the generation of questions, verifies the speaker's answers and asks for confirmation. This dialog is controlled by the system for sake of effenciency.

4 - CONCLUSION

We have presented in this paper the way in which semantic and pragmatic constraints are taken into account in the MYRTILLE II speech understanding system. The basic idea consists of distinguishing between recognition and interpretation of a sentence.

The recognition phase determines the syntactic surface structure of a sentence by using syntactic and local semantic knowledge. This surface representation is then interpreted by a semantic analyzer and finally a pragmatic dialog procedure asks questions to the speaker in order to solve the remaining ambiguities.

The recognition part of MYRTILLE II is presently running using a centisecond model as phonetic front-end. The connection between the recognition and the interpretation levels is now in progress. Improvements to this system are now being expected both in the acoustic-phonetic trancription of the speech signal and in the strategies used for sentence recognition and interpretation.

APPENDIX 1

List of nonterminals identifiers

ENON	sentence
GC	circumstancial group (time)
GCOM	complement group
GL	circumstancial group (place)
GN	nominal group
GS	subject group
GV	verbal group
INTE	interrogative group
PR	clause
SGV	verbal object group

REFERENCES

(1) W.A. LEA ed. "Trends in Speech Recognition", Prentice Hall, 1980.

(2) J.P. HATON, J.F. MARI, J.M. PIERREL : "Research Towards Speech Understanding Models for Artificial and Natural Languages", IEEE ICASSP, Hartford, 1977.

(3) J.M. PIERREL, J.P. HATON : "Data Structure and Organization of MYRTILLE II System", 4th IJCPR, KYOTO, JAPAN, Nov. 1978.

(4) G. MERCIER et al. : "The Keal Speech Understanding System" in "Spoken Language Generation and Understanding", J.C. SIMON ed., Reidel, 1980.

(5) D. WALKER : "SRI Research on Speech Understanding " in "Trends in Speech Recognition", W. LEA ed., Prentice Hall, 1980.

(6) R. DE MORI : "Recent Advances in Automatic Speech Recognition", 4th IJCPR, KYOTO, JAPAN, Nov. 1978.

(7) W.A. WOODS : "Transition Network Grammars for Natural language Analysis", C.A.C.M., Vol. 13, N° 10, Oct. 1970.

(8) T. WINOGRAD : "Understanding Natural language", Academic Press, New York, 1972.

(9) R.C. SHANK : "Conceptual Information Processing", North Holland, 1975.

(10) J.P. HATON, J.M. PIERREL : "A Model for Natural Spoken Language Understanding", IEEE ICASSP, Denver, 1980.

(11) C. FILLMORE : "Case for Case" in "Universals in Linguitic Theory", BACH, HARMS eds, Holts, Rinehard, Winston, 1968.

THE ARIAL II SPEECH RECOGNITION SYSTEM

PERENNOU Guy

C.E.R.F.I.A. - UNIVERSITE PAUL SABATIER - TOULOUSE

The purpose of this paper is to point out the main ideas and the difficulties when defining and setting up ARIAL II subsystem for phonetical and lexical decoding of connected speech.

It is easy to verify that listeners may transcribe nonsense words, obeying the phonological constraints of French (or another language), with better than 90 % phonemic accuracy.

Our research has always had the aim of obtaining better acoustical input which would permit the use of bottom-to-top methods. However, we shall see presently that a mixed method at phonological level is used in ARIAL II.

INTRODUCTION

ARIAL II is a recognition system for dictated connected speech. It is to be included in a larger system with high linguistic levels. In a first stage we developed ARIAL I (Analyseur lexical pour la reconnaissance de la parole - aide ADI n° 44/79 - rapport final 1979-80).

In the actual project (aide ADI n° 80/444) we introduce a phonological level and an integrated linguistic control with a mixed strategie (bottom-to-top phonological decoding and phonological hypothesis verification). These reinforcements seemed necessary in the applications needing large vocabularies.

J.-P. Haton (ed.), Automatic Speech Analysis and Recognition, 269–275.

1.1. BOTTOM-TO-TOP PHONETIC DECODING

Recent studies have shown that the acoustical level is rich enough in informations to allow an acceptable rate of broad phonetic transcription. In a recent paper, D.H. KLATT (1979) mentioned experiments of COLE, RUDNICKY, REDDY and ZUE (1978) showing that 80 to 90 % of broad phonetic transcription may be obtained by spectrogram readers which are unaware of pronounced words.Furthermore, KLATT mentions experiments of LIBERMAN and NAKATANI indicating that listeners may transcribe nonsense names, obeying the phonological constraints of English, with better than 90 % phonemic accuracy. It is easy to verify that in French we have the same result.

Our research in automatic speech recognition has always had the aim of obtaining better acoustic input which would permit the use of bottom-to-top methods from the acoustical to the lexical level.

We think that speech recognition systems (SRS) will have the capability to transcribe good pronounced utterances with high phonemic accuracy for a large class of speakers (without important adaptation). Therefore, the ideal outputs will contain informations on the input sentence (if there exists one) and moreover on possible pronunciation errors, on the speaker etc.

However, taking into account the results obtained actually at phonetical level, we must limit the vocabulary and try to have well articulated utterances if we want to obtain a good performance.

1.2. MORPHONEME VERIFICATION

Listeners may clearly use linguistic informations (constraints) in order to supply lacking phonetic informations. We introduce also in our system searching guided by linguistic knowledge. If we don't obtain a clear phonological analysis using bottom-to-top decoding, it is possible to supply it by verification of the morphologic unit. Then, we have a mixed method at the phonological level.

1.3. WORDS RECOGNITION

At this level, the decoding process is using a lexical tree of expected morphonologic sequences for all words of the vocabulary.

The choice of an effective set of morphonologic elements is crucial for the size of lexical tree and for the top-to-bottom phonologic recognition.

In the actual ARIAL II implementation, the syllable is generally represented by two morphonologic elements (Fig. 1).

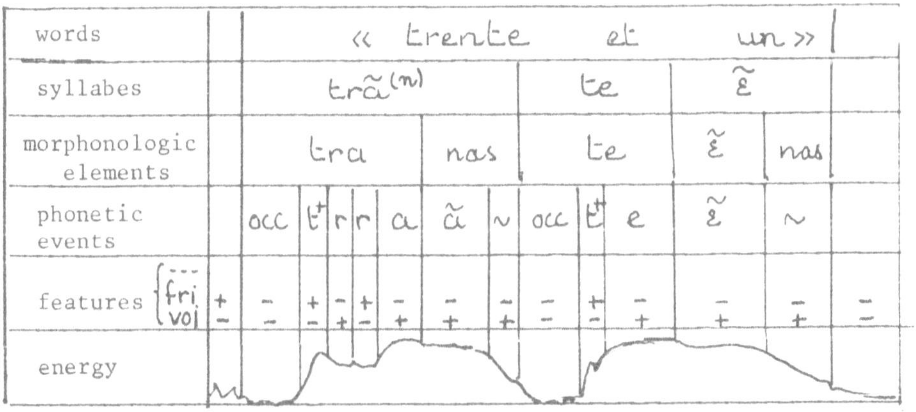

Fig. 1 : Example of multilevel phonological analysis.
The syllable is divided in two parts :
- increasing part and decreasing part
Note that the two last syllables are tending to merge.

As shown in Fig. 3, each arc of the lexical tree represents two consecutive morphonologic elements. Then, for a vocabulary of N words, a rough estimate of the number of arcs is 2N.

2. KNOWLEDGE REPRESENTATION IN ARIAL II

2.1. Phonologic level

An utterance is described as a string of labelled infra-phonemic segments (as shown in fig. 1).

For each segment cues, features and candidates are given. (More details can be found in J. CAELEN (1981)).

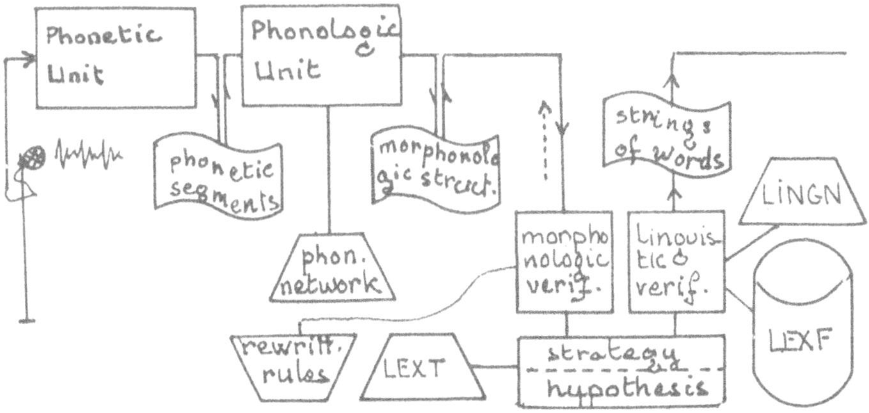

Fig.2 : The ARIAL II organisation

2.2. Morphonological levels

The description of words is based on four sorts of morphonologic elements.

a) Increasing open syllable (IOS)
Example : /tu/, /tre/ etc.. (but not /stri/, which is not increasing, or /sis/, which is not open).

b) Syllabic closing (SC)
Example : /g/ in "segment" or /r/ in "article". If it is necessary features of the previous vowel can be specified : open, nasality, long,.. For instance, in the word "peine" we have IOS → [pe], SC → [op long nə], [pe] + [op long nə] = [pɛ:nə]

c) Final syllabic closing (FSC)
It is a special case of b). Two indices i and j are completing the description. They respectively specify the juncture and the flexional ending rule.

d) Initial presyllabic element (IPE)
Four cases are distinguished : 1) there is a presyllabic consonant (ex : /s/ in "stable"), 2) the initial is the disjunctive h of French (example : "huit" : see following ex.4), 3) the initial is a vowel (see following ex.2), 4) the initial is a consonant (see following ex.3). The morphonologic descriptions of words have the following form :

WORD → (IPE|IOS)(SC|IOS)^n(FSC|i j) n=0,1,2,..

Examples

"samedi"	→	(c-|sa)(mə|di)(# -|01)	("saturday")
"octobre"	→	(-|ɔ)(k|tɔ)(# brə|21)	("october")
"petit"	→	(pə|ti)(# t|12)	("little")
"huit"	→	('|yi)(# t|90)	("eight")
"poignée"	→	(c-|pwa)(neu|ɲe)(long# -|01)	("handful")

2.3. Lexical levels

Two structures embodies the lexical knowledge :
a) The lexical tree LEXT which gives the morphonologic representation of words and for each arc i the set syse (i) of the admissible linguistic categories (as shown in fig. 3).

b) The lexical file LEXF which gives the other informations needed at the linguistic levels.

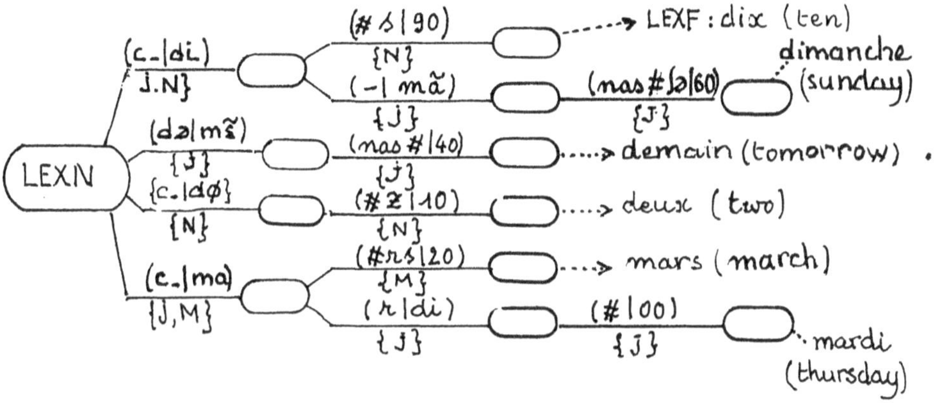

Fig. 3 : A sublexical tree. The morphonologic elements are on the arcs. A set syse(i) of admissible linguistic categories is the arc i (ex : syse(i) = {J,N}).
Each terminal node points to an entry of LEXF.
Notation : J=jour (day), N=numeric, M=month,..
10 :liaison + invariable....

Morphonologic rewritting rules

Rewritting rules are necessary to give all concrete phonetic descriptions of an abstract morphonologic element. We give to them the following form :

$$COND_1 \qquad X \rightarrow Y \qquad COND_2$$

That means : if $COND_1$ is true then X can be rewritten Y and $COND_2$ must be verified in a further stage. If $COND_i$ is not specified, it will be taken as true. The following table is giving a few examples of these rewritting rules.

$COND_1$	X	Y	$COND_2$
i=1 j=0	# Cons	cons	IPE= '-'
i=1 j=1	# Cons	cons	IPE='-';n=sing
i=1 j=0	# Cons	^	IPE='c-"
voy—	cons+	cons	
cons—	cons+	cons+	
	/a/	[ɑ] [a] or [ɔ]	SC=neu ___

2.4. Linguistic level SYSE-1

The SYSE-1 unit is built around a linguistic (syntactic and semantic) augmented transition network (ATN) LINGN. Given a string $S=w_1\ w_2 \ldots w_k$ of k words, SYSE-1 computs for each word w (after a question from the lexical level) the new situation and the predictive constraints SYSE (Sw). Of course, the calculation (L(S),w)→ (L(Sw),SYSE(Sw)) is shorter than (L(Λ),Sw)→(L(Sw),SYSE(Sw)) (Λ is the symbol of the empty string). Therefore, it is advantageous to save L(Sw) if w is accepted. This will happen if SYSE(Sw) is not the empty set.

The predictive constraints like keys, can open arcs of LEXT. Suppose that for a given hypothesis we have to validate an arc i with a set syse (i) of linguistic categories (see fig. 3). This will be possible only if SYSE (Sw)∩ syse(i) is not the empty set Thus, SYSE (Sw) opens a sublexicon for search of w.

3. THE LEXICAL DECODING PROCESS

In our project three main recognition processes can be distinguished
- the phonetic bottom-up decoding process with acoustic, phonetic, phonologic steps
- the lexical decoding process (LDP)
- the linguistic interpretation

In this section, we shall describe the LDP and linguistic levels. At this level, the strategy used by ARIAL is a best-few search moving left-to-right in time. At each time an interval covering one syllable-like unit is examined (from phonologic strings level) with the object to derive from a list H_A of "old hypothesis" (taking in account the previous syllables) a list H_N of "new hypothesis".

3.1. Hypothesis

An hypothesis H has the following form : H=(N,SC,HSYSE,DT,PTST) where : N is a node of the lexical tree LEXN, SC the score of H, SYSE the linguistic constraints, DT the instant of the last phonetic segment validated by H, PTST a pointer to a string of words.

3.2. Creation of an hypothesis

An hypothesis H'= (N', SC', HSYSE', DT', PTST') can be derived, from H in two ways.

a) Let (N,N') be an arc issued from N. If N' is non terminal node in LEXN, HSYSE'=HSYSE∩ syse(i) is a non empty set of linguistic categories, the score SC' after the verification of the morphonological unit PH(i) is greater enough (SC'⩾ Θ) then an hypothesis is created in OLDHYP ,if DT' earlier than the syllabic kernel. Else H' is created in NEWHYP.

b) If N' is a terminal node, corresponding to a word w, a question (PTST,w) is transmitted to the linguistic level. If the answer is

(NON-ACCEPTED) H' is delated. Furthermore, for each answer SYSE, PTST') an hypothesis (0, SCD, SYSE', DT', PTST') is created, 0 being the root of LEXN, SCD an initial score.

CONCLUDING REMARKS

This paper had the aim of giving only the general framework of the ARIAL II project. Instead of designing a specific speech recognition system, we consider ARIAL II as a model for a class of systems. In [4] the main ideas of the project have been given. More details can be found in [1] for the phonetic level, in [5] and [7] for the phonologic level in [6] for the linguistic level.

The lexical unit, with its compiler knowledge, and the SYSE linguistic are actually implemented. They are written in PASCAL on Cii-HB IRIS 80 computer.

The phonetic input is actually simulated on IRIS 80 and we are implementing the phonologic units (bottom-up and verification).

REFERENCES

[1] J. CAELEN and G. CAELEN : 1981,"Indices et propriétés dans le projet ARIAL II". Séminaire Processus d'encodage et de décodage phonétiques - Toulouse - septembre 1981.

[2] R.A. COLE, A. RUDNICKY, R. REDDY and V. ZUE : 1978, "Speech as Patterns on paper", in "Perception and Production of fluent speech", R. COLE (Ed.) Erlbaum.

[3] D.H. KLATT : 1979, "Speech perception : a model of access to phonetic analysis and lexical access". Journal of Phonetics, 7, pp 279-312.

[4] G. PERENNOU : 1980, "ARIAL II : a system for speech recognition". IAPR, Miami.

[5] G. PERENNOU, M. de CALMES, MINH BUI VAN : 1981," Décodage lexical et composante phonologique dans ARIAL II". 12ème Journées d'Etudes sur la Parole - MONTREAL.

[6] G. PERENNOU, M. de CALMES, MINH BUI VAN : 1981, "Décodage lexical et composante syntactico-sémantique dans ARIAL II". 3ème congrès AFCET, Int. Art. et Rec. des Formes. NANCY.

[7] G. PERENNOU et M. de CALMES : 1981, "Le décodage au niveau phonologique dans ARIAL II". Séminaire Galf/Afcet CNRS. Processus d'encodage et de décodage phonétiques. Toulouse.

SPEAKER RECOGNITION: A SURVEY

Patrick CORSI (*)

ABSTRACT

This paper presents a unified discussion of the scientific and practical issues in the field of speaker recognition.
Besides some background on speaker recognition by listening and visual analysis of spectrograms, we survey the computer recognition methods, and briefly discuss some technical aspects of various speaker recognizers, Methods for selecting an efficient set of features, and examples of results of experimental studies are also presented. We then differentiate between the Verification and Identification tasks.
Potential applications of speaker recognition by computer are reviewed.

(*) Obtained a Doct. Eng. degree on speaker recognition at ENSIMAG, Institut National Polytechnique Grenoble, France, in 1977. Since then, a visitor scientist at IBM ISD, Research Laboratory, San Jose, California.
This paper is independent of the current assignment.

J.-P. Haton (ed.), Automatic Speech Analysis and Recognition, 277–308.

I - INTRODUCTION

A person's voice is a complex acoustic signal which encodes various kinds of information, among them a reflect of some of the anatomy and physiology of the speaker. It seems reasonable that, due to the large amount of speaker identity information in the speech signal, a speaker recognition methodology can be carried out by many means. The general question arises then as to find out the most pertinent sounds to select.
The three general methods of talker recognition (namely: by listening, by visually spectrogram inspection, and by computer) will be discussed. Fingerprinting, photographic and antropometric techniques are usual methods of identification. Handwritings and photographs already involve intraperson and interperson variability. But in speaker recognition, the patterns are probably believed to be more varying.

II - THE EARLY WORK IN SPEAKER RECOGNITION.

2.1 INTRODUCTION.

Early researches were in relation with the study of personality and individuality. Among the precursors were: SAPIR, 1927; TAYLOR; 1933; HERZOG 1933; ALLPORT & CANTRIL, 1934; BONAVENTURA, 1935; and principally KAISER, 1939-1944.
Recognition by listening is evidently used in everyday life. It is a mostly subjective process: listener's accuracy and reliability is based upon decisions made on highly personal and unconscious criteria. Recognition by visual comparison of spectrograms has usually been believed to be a more objective method. This is because the graphic features can be interpreted in a more conscious and objective way. Still, the overall decision will rely on personal estimates.
The interest in recognition by means of a machine is an attempt in overcoming such limitations. The current efforts are at enabling such machines, namely computers, to provide better decisions than human ones.

2.2 RECOGNITION BY LISTENING.

This method has been given the earliest research, with McGEHEE (1944). Its importance has been emphasized during the late sixties, when was stressed the need of comparison with other methods: VOIERS, 1964; BRICKER & PRUZANSKY, 1966; HOLMGREN, 1967; STEVENS & al., 1968; CLARKE & BECKER, 1969; ROSENBERG, 1973. Authors have basically employed a common procedure: speakers read a selected material, and the recordings are presented, after audition, to listeners, who will then carry out a recognition task. Each step thus introduces variables that can influence the

resulting performance. The objective of studies in listening is to appraise the likelihood that a listener's judgement might be in error.
Mc GEHEE (1937) studied listening reliability on unfamiliar voices, and found that the reliability of recognition decreases rapidly as the time interval between sessions exceeds two weeks. Vocal disguise has been found to lower recognition scores (REICH & al. (1976) examined the effect of 5 selected disguise kinds on 4 examiners with a 4-weeks training; a 43% overall error rate was observed), as is the effect of increasing the number of speakers heard during the training session. ENDRESS & al. (1971), in the earliest study on vocal disguise, showed results in which speakers shifted the frequencies of the formants by as much as 10 % both above and below their own normal values. PTACEK & SANDERS (1966) attempted an aural age recognition. Mimicry possesses certain features in common with disguise. For instance, it involves a voluntary change of vocal sounding effects. However, in mimicry, a speaker makes the attempt to imitate a specific sounding, usually one of another person. Successfull disguise hides the identity, meanwhile successfull mimicry identifies someone else.
ENDRES & al. (1971) studied spectrographic effects in mimicry by two professional mimics and confirmed the preceding observations. Usually, the formant structure in the imitating speech is different than the one in the speech of the imitated person (cf. also the thesis by HALL (1975)).
The ability of listeners to identify by hearing the voice alone falls far short of 100 percent reliability. But these studies have been worthwile in illustrating the fact that the speech waveform carries speaker dependent information. Still the question about perceptual bases of speaker recognition and their acoustical correlates remains largely unresolved.
The operating mode of aural recognition can be text-independent. Speakers need not to be cooperative. Future development is seen to be quite limited since not much can be done to improve the average individual's ability to recognize speakers by hearing their voice: it is yet a subjective method which does not provide the ability to justify one's conclusions in a court of law, and difficult to perform when several speakers are involved. However, human performance has served as a benchmark of progress for the more automatic methods.

2.3 RECOGNITION BY VISUAL ANALYSIS OF SPECTROGRAMS.

A first study on the "voiceprints" was performed by GRAY & KOFF (1944). Mc GEHEE (1937,44) had to solve legal cases (referred to as the Lindbergh Case, 1938). The operating mode of visual analysis of spectrograms can be text-independent. Speakers need not to be cooperative.
Voice spectrography was developped at Bell Laboratories by POTTER, KOPP & GREEN in 1947. KERSTA made sound-spectrographs adaptable to personal identification. The purpose of the sound spectrograph is

to convert the speech signal into a visual display called the spectrogram. In 1962, KERSTA announced that voice spectrograms could provide a reliable mean of identification (KERSTA, 1962 a,b,c,d). Subjective matchings of eye-brain similarities found in pairs of spectrograms are the foundation of his method, which obtained the unfortunate term of "Voiceprinting": the spectrograph was commercialized by Kersta under the trade name of "Voiceprint". It should not be assumed to be like fingerprinting: the patterns of fingerprint ridges remain unaltered for a same person after the age of 5; moreover, duplication of fingerprints cases have not appeared. The "Voiceprint" had a play-back facility allowing continuous listening of samples prior to be processed. Speakers read selected words or phrases, then spectrograms of different are prepared from them and are presented to a trained observer who attempts to determine whether some utterances were produced by a given speaker. Again, the observers try to perform a recognition task; they match spectrograms that represent the same speaker and are instructed to examine features such that:

- similar mean frequencies of vowel formants,
- formants bandwidths,
- gaps and type of vertical striations,
- slopes of formants,
- durations,
- characteristic patterns of fricatives and interformant energies.

KASHYAP (1976) indeed affirmed that speaker dependent featured are not to be found in a limited range of sounds or categories of sounds. As when listening to voice, each step introduces variables that can affect performance. The fallibility of the observer is a crucial issue in the legal use of this method (BORDERS, 1966; LADEFOGED & VANDERSLICE, 1967; McDADE, 1968; BOLT & al., 1970). Many claims in favor of this method have been made (essentially by KERSTA, 1962 a,b,c; however often based on unpublished results). The methodology of "voiceprint" comparison is poorly documented, and the claims concerning the unalteration of recognition performance by the aging process and some minor disabling of the vocal organs and also the disguising of the voice, could not be considered as established. Kersta claimed that spectrograms of several utterances of the same words by a given speaker always contain more similar spectral features than when produced by different speakers. The remaining dissimilarities are assumed to be a result of intraspeaker variability. Underlying these claims is the hypothesis that the probability that two persons have similar enough vocal-tract dimensions and articulatory patterns to produce undistinguishable spectrograms is extremely small. This hypothesis is a belief that has not been fully tested yet. We will briefly discuss the sources of voice variation in the next paragraph.

Other investigators have tried to duplicate the favourable results, but with higher error rates than KERSTA's (STEVENS & al.,

1968). TOSI (1967, 1968) could only conclude that the "Kersta method shows promise", suggesting the need for specific independent studies. Eventually, the scientific community could not fully appraise the design of these experiments and their results.

BOLT & al. (1970) evidenced the need of further studies to assess the feasability of this method of identification since it has obvious potential in forensic applications. The "spectrographic" method does not seem to have achieved its full capability: there is a lack of study to determine optimal procedures for discriminating among speakers. But it is unlikely that it will ever lead to a general way of identification. There is, in fact, wider agreement on the ability of humans to recognize speakers by listening alone than by inspection of spectrograms alone.

The Michigan State Police study on recognition by spectrogram examination, reports studies based on 9 clue words. The grand mean percentage of errors from the 17500 trials was 8.9%. One third of the errors are for false identification. Contemporary tests result in better recognition scores (95% vs. 88% for non contemporary spectrogram matching). Context of the clue words is important (96% for isolated words, 93% for fixed contexts, and 87% for a random context). Awereness of examiners when the trials are open lead to better scores too (94.5% vs. 90% for open trials). In other words, the upper limit for the range of errors is found in open trials, non contemporary tests, and random context for clue words.
A second set of experiments was made to obtain information concerning the effect of a reduction in the number of clue words on the accuracy of the examiners. Little difference was found. Main results suggest that experienced examiners can identify or eliminate one unknown speaker from among as many as 40 known speakers. Professional examiners can use "voicegrams" with a frequency range of 80-4000 Hz, with a useful frequency of field less than 3KHz. But the effects of noise and spectral characteristics on visual identification performance remain largely onknown. Loss of bandwith in higher frequencies may affect visual identification.
The opinion of some phoneticians and linguists, as of STEVENS & al. (1968), who feel that recognition by listening is more accurate than by analysis of spectrograms, seems to be a valid hypothesis; however some results suggest that the performance of machines sometimes exceed that of human listeners. STEVENS & al. (1968) give the following comparison of error rates (approximated):

sample length	error rate % by listening	error rate % with spectrograms
1 syllabe	12	32

2 syllabes	9	24
1 sentence	9	18

A recent evaluation committee of sound spectrograms by BOLT & al. besides recommending particular techniques, has concluded that "the technical uncertainties are so great that forensic applications should be approached with great caution".

2.4 CONCLUSIONS.

It is not clearly established as to know how much would the conditions in practical cases (as legal cases for instance) degrade the recognition rates. The following factors should be considered:
- the size of speaker population (the known speakers); extrapolation cannot be made for the whole population of which the statistical sample is the known population,
- openness of the known population; however the inter and intra-variability could be extrapolated in some applications,
- availability of clues,
- type of decisions the examiners are asked to make, the decision time limit, and the possibility of alternate decisions (positive identification or elimination, versus doubt).

2.5 RECOGNITION BY MACHINE.

The preceeding 20 years have brought about an expanding use of voice processing technology while numerous related scientific experiments have been undertaken. One approach is to have the machine generate and examine amplitude - frequency - time matrices of specific speech samples. The other approach has been to have the machine extract speaker dependent parameters from the signal, and analyse them. The last method proves to be a promising one and follows the development of pattern recognition and other contemporary techniques.
To achieve high performance, there is a need to properly identify and measure the speech parameters. The analysis and decision processes must also be incorporated into the machine. Major research on automatic speaker recognition has been made by ATAL, BUNGE, DAS, DODDINGTON, FURUI, PRUZANSKY, ROSENBERG, SAMBUR, WOLF, and others.
Identification by machine is reaching the level of auditive accuracy. Questions regarding the most appropriate speech parameters, the optimal decision rules, still can be considered as open. A totally reliable machine has not been built and is difficult to be predicted. But even if it comes of age, there will still be a need of human experts trained in phonetics and spectrography to select the proper samples from the known and unknown voices for the machine or to discuss the appropriateness of an alternative method.

The area of speaker recognition can be divided into a large number of subareas depending on such factors as the:
- nature of the task (namely Verification and Identification),
- nature of the parameters used,
- population size,
- phonetic context (sounds, words, sentences, etc.),
- nature of the classifiers and the decision rules,
- recognition rate obtained, etc.

The paper by ATAL (1976) presents a general framework for the study of the subject along with a survey of the progress achieved. The detailed review by ROSENBERG (1976) specializes in speaker verification and presents results of practical systems. Among other surveys is the one by BEEK & al. (1977).
PRUZANSKY (1963) was among the first to initiate modern research, by using 17 filter banks, and correlating two digital spectrograms for a similarity measure. PRUZANSKY & MATHEWS (1964) improved this technique; so did LI & al. (1966) by using linear discriminators.
DODDINGTON (1971) replaced filter banks by formant analysis; since then it is accepted that the higher formants represent tinier phenomenas which rather characterize the speaker. For instance, the first two formants are enough to phonetically define a vowel, meanwhile the following two are mainly speaker specific. However, the higher the formants, the more unprecise their measure.
The polish shool (CZAJKA, GROSZKA, GUBRINOWICZ, JASSEM, KOSCIELAK, KRZYSKO, FRACKOWIAK-RICHTER, STEFFEN-BATOG, etc.) has published quite extensive researches on speaker recognition from 1970 to 1973, including the study of the short-term distribution of the fundamental and the use of discriminant functions. Its more recent papers (BASZTURA & MAJEWSKI, 1978; BASZTURA & JURKIEWICZ, 1978) especially deal with short and long-term zero-crossing methods.

Frequency domain methods tend to be independent of the phonetic context by averaging over a long enough duration or by extracting statistical or predictive parameters. They are important for the study of transition phases. Among the tools used in the recent years, we find the following: averaged auto-correlation (BRICKER & al., 1971; YEGNANARAYANA & al., 1977); instantaneous spectra covariance matrix (LI & HUGUES, 1974); spectrum and fundamental frequency histograms (BEEK & al. 1971; HUNT et al. 1977); linear prediction coefficients (SAMBUR, 1976 a,b; GRENIER, 1977); long term averaged spectra (FURUI et al. 1972). Predictive coding has been successfully used (SAMBUR, 1972; ATAL, 1974; ROSENBERG & SAMBUR, 1975; GRENIER, 1977,78). WOLF (1972) and SAMBUR (1975) studied the spectrum for some nasals and vowels. Long term averaged spectra (BUNGE at al.; 1977) gives a high degree of security in Verification (false acceptance error rate was 0.8%, false rejection rate was 1.6%, both with distance and tolerance region classifiers).
However, it is known that the vowels in all languages are sensitive to coarticulation effects, i.e. to the inter-phonemes dynamic

pollution (the acoustical properties of nasal consonants are also believed to be very speaker dependent).

Time-domain methods are more recent, and have been considerably less studied (ATAL, 1972,b, 1974; ROSENBERG & SAMBUR, 1975; SARMA & YEGNANARAYANA, 1976; VIDALON & al., 1977). Temporal methods are heavily dependent on the phonetic and linguistic context. They also introduce the need for a reliable segmentation, that does not eliminate the speaker's differences. They are nonetheless important for the study of stable segments. Some attempts measure the voice onset time for intervocalic stops like /k/, /t/, and /p/. An early study by BRICKER & PRUZANSKY (1966), which has been corroborated by others, shows that the number of sounds is more important than the sample duration.
LIN & PILLAY (1976) used zero-crossing methods in specific frequency- bands, GOLDSTEIN (1975) a formant computation by linear prediction. SAMBUR (1972) selected acoustic parameters out of a set of 92, including the structure, transitions, and formant frequency bandwidth for vowels and nasal consonants, the fundamental and various durations.
Motivation for using the pitch is also due to the possibility of measuring it under poor signal-to-noise ratio and high spectral distorsion conditions. However, when over telephone lines, two obvious major problems are raised: voicing detection and substitution of the fundamental frequency. It should be noted that spectral parameters suffer a priori more than temporal ones of the telephone lines various distorsions. ROSENBERG (1976) has extensively evaluated the degradations due to a telephone system, KRAUSE (1976) the modifications occuring in the spectrum.

Modifications of the voice have been of early interest: Whispered voice by WILLIAMS, 1964, RAMISHVILI, 1966, and SCHWARTZ & RINE, 1968; vocoder distorsions by SCHEARME & HOLMES (1969). The major contribution of voicing to speaker recognition, when sample duration is over 1 second, has been stressed by POLLACK & al. (1954), who compared normal speech against whispered speech. A few experiments were directed towards the study of speakers members of a same family (KERSTA, 1965 b), especially twins (ALLPERT & al. (1963). Women's spectrograms have been believed to be easier to recognize (KERSTA, 1962), however the general difficulty of recognition depends upon the speaker. Differences between men/women/children have been evidenced in the spectrum (SOTTER & STEINBERG, 1950), but more in the fundamental (HOLMGREN, 1963,67; MATSUMOTO & al., 1973; RAMISHVILI & TUSHVILI, 1976; CLARKE & BECKER, 1969); SCHARTZ (1968) and INGEMAN (1968) used /s/ and /f/ isolated. In an early experiment involving 5 speakers, YOUNG & CAMPBELL (1967) obtained 78.4% recognition with isolated words (words: "you", "it", etc.), versus 37.3% in the continuous case (isolated words tend to be longer coarticulation effects less

important). Cf. also: SCWARTZ & RINE (1968), and WEINBERG & BENNET (1971) on the same subject.
Emotional state of the speaker affects, besides the value of the fundamental, the position of the formants, (and especially in the first syllabe of the words). It can be observed by tracking the speaking rate and the frequent occurence of meaningless words (KRAUSE, 1976) in extemporaneous speech.
Resistance to imposture has always been of concern since any system is faced with the number of types of impostors. LUMMIS & ROSENBERG (1972) have improved a verification system using 4 well-trained impostors imitating 8 speakers. DODDINGTON (1974 b) reported a degradation of the Texas Instrument system he designed, when 2 professional impostors are imitating 6 talkers. It has also been shown that the vocal dynamic can be easy to imitate. Success is increased by immediate replication of an heared utterance. HAIR & REKIETA (1972) used a professional imitator and showed that imposture could be eliminated by using more parameters for a Verification experiment.
Other studies were by ENDRES, BAMBACH & FLOSSER (1971), HAIR & REKIETA (1972), FURUI, LUCK, etc.
Changes in the fundamental have been largely addressed (MYSAK, 1959; PTACEK & al., 1966; PTACEK & SANDER, 1966; SHIPP & HOLLIEN, 1969; ENDRESS & al. 1971; RYAN & BURL, 1972; HARTMAN & DAN HAUER, 1976; FURUI, 1973-78). According to ATKINSON (1976), its intra-speaker variability is almost comparable to inter-speaker variability. This variation of the pitch has been considered to be the only major obstacle to its generalized use, despite its high value when discriminating between sexes.

The fundamental frequency of the voice has nonetheless been the most studied parameter. Its resistance to imitation is however not high (LUMMIS, 1971), but its efficiency in discriminating speakers is always slightly greater when compared to many temporal parameters (CORSI, 1979).
Other comparative studies have assessed the superiority of the long-term spectrum versus the averaged fundamental frequency in a same experiment (CLARKE & BECKER, 1969; also PRUZANSKY, 1963)).
WILLIAMS (1970) and LEVIN & LORD (1975) considered it as an indicator of the emotional and physiological state of the speaker.
Long-term variations of voice due to the aging process (vocal register gets narrower and maximal value of that first formant of vowels is found lower (PTACEK & al. 1966)).
Long-term spectra distributions stabilize after less than one minute of continuous speech. Correlation of spectral characteristics produces about 1 percent overlap between intraspeaker and interspeaker variation for a particular 30 speakers population.

But, due to the fairly large number of factors affecting each experiment, an inter-systems comparative analysis of the efficiency of the parameters is likely to be hazardous.

III - PRINCIPLES OF SPEAKER RECOGNITION.

3.1 - SOURCES OF VARIABILITY IN THE VOICE

There are two acoustical differences among speakers: the anatomical and physiological aspect which are relatively known, and the individual use of one's vocal system which generates individual realisations of the phonological system. According to FRANCESCATO (in STEVENS, 1971), one should replace the distinction:

"inter-speaker variations" versus "inter-speaker variations"

by the dichotomy:

"inherent(permanent) variations" versus "variations conditioned by situation"

The lattest expression refers to the situation recognition domain and the former one to the speaker recognition domain itself. Four fundamental physico-acoustical mechanisms account for the variations in voice: the lungs, the pharynx, the mouth cavity, and the nasal cavities. LOWERRE (1977) distinguished three kinds of variations in continuous speech: the dialectical and contextual ones, which introduce changes in the pronounciations of the words, and the acoustical one which introduces modifications in the vocal tract. GARVIN & LADEFOGED (1973) introduced the organic and learned differences. As organic differences, we find the properties of the lower breathing system (cavities, vocal chords, etc.). They introduce, for instance, variations in pitch, in intensity and duration of phonatory pauses. Due to intonation, pauses distribution, accentuation and context, these variations are greater for continuous speech than for isolated words. They are conditioned by three external factors: the heredity, the sex, and the age. The learned differences result from learned neural commands (muscles controls, and articulatory gestures (these could be defined as a temporal organisation of the vocal behavior). As external factors, we see the speaker's geographic origin which accounts for dialectical differences, and some social and cultural factors.

For STEVENS (1971), the larynx is the most important factor, besides the part of the vocal tract between the glottis and the lips. It particularises the vocal cords elasticity, shape, asymetries and humidity. As it is well known, the vocal tract filters the vocal source and individualizes the formant frequencies. Other sources are to include the endocrine glands functionning, the health status of the laryngyal source and all the illnesses influencing the vocal channels: allergies, colds, etc. Intra-speaker variations increase with fatigue, stress, fear, and the aging process involves many changes (for instance the decrease of the speaking rate).

3.2 GENERAL PRINCIPLES

The definition of HECKER (1971) for speaker recognition is still valid: it consists of any decision process based upon the inter-persons variations. All methods of speaker recognition are based on the fact that a given word or phrase tends to be uttered differently by different speakers. The parameters, when combined together or individually, should therefore provide distributions for different speakers that are concentrated at widely different locations, in the parameter space. For a single measurement (or parameter) we commonly define the following ratio as the F-ratio:

$$\frac{\text{variance of speaker means}}{\text{average of intra-speaker variances}}$$

Suitability of sound is measured by its efficiency in discriminating between speakers, measured in turn by the error rate.

When considering a set of parameters, one will generalize the above definition with a multi-dimensional measure of effectiveness, called the divergence D (cf. ATAL, 1968; 1972 b):

$$D = \mathrm{Tr}\ W' * B,$$

where W' is the inverse of the intra-covariance matrix, and B is the inter-speaker covariance matrix (i.e. a measure of the inter-speaker dispersion). Hence, the inter-parameter correlation is taken into account. D is maximized to find a set of parameters that linearly transform the original measurements vector into a space of maximum speaker separation. As a particular case, the unidimensional F-ratio would lead to an ordering of the coordinates by their inter-speaker separation amounts.

The decision is dependent on the set of parameters used and a more objective measure can be found in averaging different recognition rates with different parameters.

A trial is by definition an attempt to perform a recognition and describes the matching of a "known sample", against an "unknown sample". Three kinds of error are then possible in the identification task:

- a match exists and a wrong one has been selected (false identification),
- a match exists but none has been found (false rejection),
- no match exists, althought one has been selected (false identification).

Let us consider the case of the "elementary discrimination task". A match will be by definition the simple binary task of verifying the matching of a reference and a sample test. Probabilities associated with a match (y) and a non-match (y') can be represented by the associated probabilistic density functions (pdfs), if defined. The decision matrix of the task is as follow:

HOW WELL THE SYSTEM PERFORMS:	TRUE STATE: y (sample matches)	y' (sample does not match)
Y sample matches	$P(Y\|y)$ correct acceptance	$P(Y\|y')$ incorrect acceptance
Y' sample does not match	$P(Y'\|y)$ incorrect rejection	$P(Y'\|y')$ correct rejection

The column sums are 1, so there are only two independent variables. The user is concerned about the related probability:

$$P(y|Y) = \frac{P(y) * P(Y|y)}{P(Y)}$$

which is the conditional probability of match given the decision of a match (P(y) is the probability that a match a priori exists, and P(Y) the probability that the system favors the decision of a match; the latter can be estimated by empirical relative frequency of match decisions. If decision costs and a priori costs are defined, an optimum decision criterion can be derived, which maximizes the overall expected value of a decision.

In the early progresses, measures were made manually to ensure a better precision. But pattern recognition techniques have gradually been adopted. Two aspects are involved in automatic speaker recognition: the creation of reference patterns and the determination of similarity between such a reference pattern and a test pattern. Thus, the main steps to be performed in a speaker recognition system are:

- data acquisition,
- feature selection /extraction,
- classification for training / decision for testing.

Inputs to a basic system are voices, either on or off-line from a data base, and a model of each known speaker. Outputs are graphs of the features vectors, statistical data relative to the decision criteria, recognition rates and analysis of errors false acceptance and reject; speaker proximities and inter-speakers confusion matrices are classic research tools.
The parameters are available in digital form and provide the way to reduce the huge amount of data which comes from the digitization of speech in a given limited band limit. The data transmission speed is as least twice the band limit, thus allowing a sufficient dynamic range.
Two major tasks are distinguished, depending on the application field: Although these two problems have a lot in common, the recognition procedures employed in each case can be very different.

Verification task. Two inputs are available: the claimed identity of the speaker by himorherself (by means of a badge, a keyed code, etc.) on one hand, a sample of the speaker's speech on the other. A prior template of the claimed speaker must be made available. Then the Verification task is simply the binary decision: is the claimed speaker the actual one who has spoken ? Here the number of speakers can be assumed to be fixed, and the error probability is independent of the population size. Authentication is often used as a synonym to Verification.
Identification task. Here, only one input is available: a sample of an unknown and unclaimed speaker. Given a reference population of N speakers, the Identification task refers to the determination of the "closest speaker" from the population, to the input one. Common Synonyms for Identification are: Detection, Determination. Two errors are to considered in the Identification task:

1- false acceptance (false recognition),
2- false alarm (false reject).

Here, decisions are made on reference pattern similarity. Assuming there are N speakers known to the system, this task is a N-class decision problem. When it is assumed the actual speaker belongs to the referenced population, the task is said to be closed, otherwise open. In the former case, the machine decides which known speaker produced the sample, and a match may always exist, possibly with a low probability, therefore the only possible error kind is the selection of a wrong matching (false identification).
In the case of Identification by exhaustive Verification of all speakers we see that N comparisons should be performed. Then, the error probability will increase with the size N, since there is a finite non null probability of incorrect decision for each comparison. However, the reader will observe this is true in general.
Likewise, applications of practical interest involve a large number of speakers. Methods that recognize a limit on the number of speakers must be devised (by using proper tree-structured algorithms for instance), (BRICKER & al., 1971); more research is here needed whatsoever.

It is important (and complex) to determine accurate decision thresholds. Usually, in the Verification task, false acceptance is less acceptable than false reject for protection purpose. However, that may depend upon the application. Very often, one seeks to minimize the probability of false acceptance under the constraint the false reject is upper bounded. The symbolic figure below shows the errors dependence upon the rejection threshold.

```
Error
rates
 %  ____________________________
   |                E(fa) *     |
   | . E(fr)             *      |
   |  .                 *       |
   |    .             *         |
   |      .         *           |
   |        .*                  |
   | *  *          .    .    .  |
   |____________________________|
      Threshold values
  low                      high
```

E(fr): false reject error).
At a low threshold, many speakers are falsely rejected, due to the variability of their speech. This case corresponds to high security requirements.

E(fa): false acceptance error.
At a high value of the theshold, speakers with similar voices are falsely accepted. Security requirements must not be stringent.

The overall goal is to minimise the error rate, which is the total of each kind. The frequency of reject of the input sample is also of importance in practical situations. The correspondence between error rate and reject rate is known in the bayesian case, for a minimum risk decision rule. We will later introduce a well-established procedure to evaluate the performance on that correspondence.

Verification errors may be attributed to three causes:

1- variations in speech by the same speaker, including laryngitis and cold, and also variations in speaking rate, voice quality and articulation),
2- similarities in speech between speakers, (this includes mimicry and tape recording of another speaker),
3- problems in automatic measurement, comparison and time registration (including variations in the speech signal environment, differences between microphones, and distances mouth-microphones, errors in time alignment).

The speaker is said to be an impostor when the attempt to modify his voice is purposely made. The label of "casual impostor" (ROSENBERG, 1972) refers to non attempted and unexpected modifications of the voice. Common assumption are that the impostor comes from the same population, and that all speakers have equal imposture likelihood. In fact, as ATAL (1977) points out, it may be that the higher the acceptance probability for a speaker, the higher the probability to have him as an impostor.

We finally observe the specificity of the Verification task, in which:

- all speakers are supposed to be cooperative whenever they are trained.
- the phonetic content of both the training and the test can be the same and not vary among speakers.
- the speakers can be asked to replicate his attempt,
- the acoustic environment can often be controlled,
- the parameters can be adapted to the (claimed) speaker.

These conditions have made several laboratory Verification systems possible. However, the degree of cooperativity in a given speech sample cannot realistically be directly addressed.

3.3 METHODS FOR RECOGNITION.

It is desirable to select for speaker recognition those acoustic parameters of speech that show low intraspeaker but high interspeaker variability. It is further possible to reduce the intraspeaker variability by requiring that the text of the unknown and reference utterances of a speaker be the same. In the early progresses, measures were manually made; pattern recognition techniques have gradually been adopted.
Choice and selection of pertinent parameters has been a definite concern (PRUZANSKY & MATHEWS, 1964; DAS & MOHN, 1971; MOHN, 1971; DAS & al., 1972 a,b; WOLF, 1971; HAIR & REKIETA, 1972; ROSENBERG & SAMBUR, 1975; SAMBUR, 1975). WOLF (1972) has listed the ideal properties for speaker recognition parameters. Essentially, they should be efficient, easy to measure, stable over time and speaking environment. They also should not be susceptible to mimicry, and be frequently occuring in speech.
A large number of experimental studies deal with text-dependent format , yet studies by ATAL (1974), FURUI & al. (1972), etc., have described methods suited for both text-dependent and text-independent situations. Text-independent is more difficult since it includes additional variability due to the differences in the text of the reference and the unknown utterances.
Methods of feature extraction include statistical analysis, analysis of selected speech segments, contour analysis. There are two general ways for talker recognition. First, to make use of statistical parameters which are independent of the phonetic sequence (for instance, the long-term averaged speech spectrum). Second, to measure parameters on preselected speech events, and then time-normalize them. (an example is contituted by measurements made on specified phonemes). This method has a limited value when considering the long-term variations of the speech.
Statistical analysis. Short-time representation is especially useful in the case of non-stationary signals. Short spectral values are here computed over a range of 0 - 10 KHz, for instance, and with a suitable window spanning 20 to 30 ms. Features can be obtained by averaging the time-dependent behaviour of the parameters. Averaging short-time spectra produces the so-called "long-time spectrum", which describe a mean position of the vocal tract, and allows a text-independence; the empirical variance of the short-term spectra estimates the deviation of the vocal tract from the neutral position. Common statistical measures include average values, normal deviation, and correlation coefficients between parameters and evaluation of the dynamic changes in speech parameters. Statistical analysis is often performed in real time

(such as the averaging of short-time spectra and the detection of local features in the frequency domain).
To list some other method, we must include the spectrum homomorphically filtered, inverse logarithmic power spectrum with repeated FFT (homorphic filtering), logarithmic power spectrum with linear prediction (inverse filtering) cepstrum coefficients with double FFT, cepstrum calculated by FFT and from the predictor coefficients, the parcor coefficients (partial autocorrelation), the intensity (computed as an integral of the squared signal over an averaging interval), etc.
The approximation of the probability density is another method: it estimates, for instance, the probability for a formant to fall into a specific frequency band.
Analysis of selected segments. Ways for locating a defined speech event in an utterance are to align by time normalisation the test and reference utterances. It is in some way the same to align the speaking rates. Other methods use non-linear warping of the time-axis; they require much computing time. Segmentation of the utterance provides another way of locating speech events.
Contour analysis. Contour analysis is based on the fact that the temporal structure of the acoustic parameters is fairly speaker-dependent. Moreover, in case of telephone transmission, the temporal structure of the parameters remains unchanged. Contour analysis requires time normalization, because the dimensionality of the vector depends on the duration of the utterance. Thus, this method is not suited for text-independent speaker recognition.

Various classifiers have been evaluated by comparison of their performances. The Bayes classifier, for instance, compares the local value of the probability density function:

$$p(x,w(i)),\ i=1,\ \ldots,\ K,$$

and selects the class for which it is maximum, if a priori probabilities are equal and the cost of making an error is the same for all speakers. For high dimensional feature vectors, the estimation of the pdf is expensive, unless an independence of the classes is assumed. Then, the pdf is described by the product of the marginal densities, which can be estimated either by the histogram approach, either by a Parzen window approach (which led BUNGE & al. (1977) to a 20 times faster method), or rectangular or Gaussian windows, etc. The minimum distance classifier computes a reference for each class, for instance its mean, and assigns the closest reference to the unknown sample.
In the Verification task, a similarity threshold has to be given, which is a rejection threshold for the distance measure or for the pdf. The sample claimed to (assumed to) be from speaker w(i) is only verified if the distance is less (the pdf is greater) than the reject threshold. The previous two classifiers are optimal when the features are independent. In practice, a decorrelation procedure is operated to reduce the frequent dependencies. An Euclidean minimum distance classifier with a Karhunen-Loeve

transformation is equivalent to the so-called Mahalanobis minimum distance classifier. Such a transformation is required if the speech parameters show significant inter-correlations (or have widely different intra-speaker variances).
Procedures of various complexity have been tried. The nearest-neighbour classifier and the minimum distance classifier were used by BUNGE et al. (1977) with various distance measures and various data normalisations (including Euclidian distance and unweighted modified cross-correlation distance. Weighting by inverse variances is quite common. BUNGE et al. (1977) used a pattern normalisation according to equal maximum value or equal pattern energy. The Mahalanobis classifier has often been reported to help improving the recognition rate versus other classifiers. Its disadvantages are the larger storage and computing time requirements. These constraints are reduced with the Bayes classifier.
The relation between the value of the rejection threshold and the error rate for false acceptance and false rejection has been seen: the trade off is in terms of the security of the system (small false acceptance rate) and customer convenience (small false rejection rate). Decision theory provides a classical procedure for decomposing the performance in two independent components: the ROC (Receiver Operating Characteristic) curve is a graphical representation of the available trade-offs that are available given a level of performance of the system. To improve the latter, more data (training), and more powerful measurements will be needed. The ROC curve is basically the following plot:

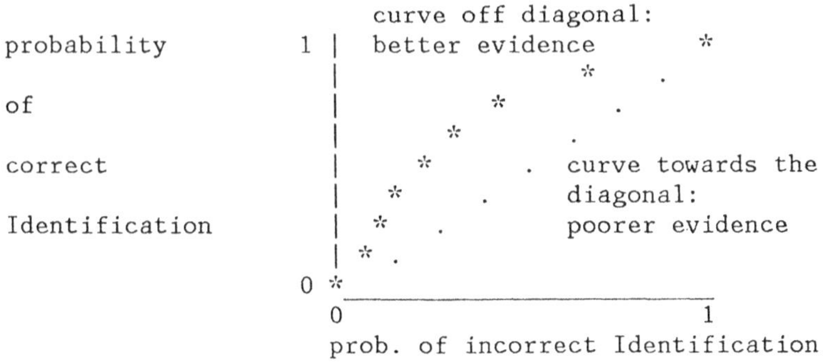

The diagonal line is the limit case where correct and incorrect identifications are equally probable. If we can increase the capacity of discrimination, then we can increase the probability of correct identification for any value of incorrect identification probability. If not, we must tolerate a higher probability of incorrect identification, to still increase the correct identification one.
BUNGE & al. (1977) have compared various methods for automatic evaluation of the rejection threshold. The distribution of the

error rate among individual speakers can be different for distinct utterances (as observed by FURUI & ITAKURA (1973), who improved the recognition accuracy using a combination of words). The old idea that speaker's sex indication leverages the recognition shows promises: DILLER & SIEBENAND 1981's set of experiments tend to establish that the speaker sex can be predicted in 80-95% of the cases, by simple procedures (sex of the best matching spectral pattern); furthermore, the word recognition rate is slightly better when based on both sex-specific lexicon and sex discrimination function, although, according to the same authors, there are not many other speaker sex prediction methods, besides pitch detection, and frequency-amplitude characteristics of voiceless sounds.

IV - SPEAKER RECOGNITION SYSTEMS.

Implementation of a speaker recognition system involves a combination of speech processing techniques and pattern recognition methodology. Let us note that the present focus on the front-end algorithms is partly explainable by the still rapid changes of the hardware.

Usually, the front-end steps of speech processing are mainly carried out in hardware. The digitized data is stored on digital tapes, which are used as input to a software system. Segmentation of the utterances is done at an early stage, and proper alignment of the utterances allows the extraction of the desired features.

The following questions should be answered when designing a speaker recognizer (adapted from SARMA & YEGNANARAYANA, 1975):

THE ENVIRONMENT:

1 - WHICH TASK ?	- Verification or/and Identification
	- errors tolerance
	- which environment (quiet, public place)
2 - WHAT PHONETIC CONTEXT ?	- short utterances (sounds, single words
	- longer (> 2s.) utterances
	- fixed material (speaker independent)
	- arbitrary material, speaker dependent (then language is of concern)
	- one language only, alternatives
3 - WHICH MEDIUM ?	- quality microphone
	- telephone (analog, digital, PCM,vocoders)
4 - WHICH SPEAKERS AND HOW MANY ?	- native speakers only
	- few speakers (< 10), many (< 30 ?)
	- speaker independent system for the Identification
	- cooperative, casual
	- are impostors to be handled ? How many ?
	- both sexes ?

THE SYSTEM:

1 - TRAINING DESIGN.	- number of samples and replications per speaker, how are they stretched time-wise - training/test data splitting method - allowable training session duration - allowable frequency of new trainings
2 - WHAT ORDER OF PROCESSING CAPACITY ?	- large/small computer - on line availability of templates and data access methods - what computation time (vs. real time)
3 - WHAT KIND OF ERRORS ?	- miss, false alarm - error probability estimation

Small systems seem to be more adapted to the Verification task, but their implementation faces some obstacles:
- the uncertainty about parameters reliability,
- the difficulty of a good segmentation of speech,
- the amount of storage needed,
- the required response time.

Real time systems seem to be affordable only when considering a small number of parameters together. Many systems are hybrid concerning the use of both temporal and frequential parameters. For instance, ROSENBERG (1976) described a Verification system based on some description of the fundamental frequency (average, standard deviation, range), the intensity of the first three formants, and some linear prediction coefficients.
A general observation is that the systems that are not based on filter bank techniques, generally make use of pitch, intensity, formant, or LPC analysis.
There are not many systems operating on a significantly large population, or under "real world" conditions. Since large population sizes are expectable in most applications, a multiple classifier scheme implemented on a decision tree has been devised by DANTE & SARMA (1979). The features used at each stage were found by minimizing a cost function. Their number depends on the outcome of each feature measurement, which includes the formants extracted by a 10th order linear prediction analysis on the nasal sound /n/, and the vowel /a/ for 60 male speakers. The number of classes with which the matching is performed is reduced to a minimum. Accuracy was reported to be 97.2% with 900 test samples.
We will say that performance evaluation is a recent field in pattern recognition. No general study seems to have been made. The obvious main components of the system are:
- performance in discriminating among speakers
- computational effort involved
- computation time involved
- storage requirements.

The cooperative west-german AUROS project (Philips and Heinrich Hertz Institute) is a field experiment which goal is an automatic recognition based on code sentences for the verification task and highly speaker-characterictic sounds. The text-independent signal analysis is performed in two stages: the statistical measurements in time and frequency domains, and various classifiers (linear, Mahalanobis, distribution-free tolerance regions) based on several distance measures, and computed on several pre-processings (correlation matrices, similarity matrices, variability coefficients). Features include the relative maxima/minima in short term spectra, the use of various linear transformations (FFT, Walsh-Hadamard, inverse filtering, etc.) for the evaluation of off-line spectrum, formant frequencies and bandwidth, pitch, and long-term spectra. Typical results show a 99% correct acceptance on code words and text- text-independent features, on 2 databases of 50 speakers each (2500 utterances). For Verification, both error rates are under 1%. Reports by BUNGE, HOFKER & JESORSKY have been written.

The SASIS program conducted by Rockwell International aims at a computer-aided speaker identification. It carries out studies involving a recruitment of over 250 speakers, a large set of 162 parameters reduced to 30 features (cf. PAUL & al. (1974)).

The Texas Instruments entry control system is the first fully automatized large scale talker Verification system which provides high operational security. Verification is based upon a four-word randomized utterance built from a set of 16 monosyllabic words. The analysis provides spectral amplitude information on precise sections: 14 channels with digital filter banks (4-pole recursive bandpass filter) uniformly spaced within the range 300-2700 Hz. The decision strategy is sequential and may require up to 4 utterances for the trial (1.35 utterances on the average, requiring 5.6 s of transaction time). The 1976 results reached a true speaker acceptance of 99.94%, with 0.82% impostor acceptance. Storage requirements were about 9400 bits per speaker, and included the on-line update of reference data whenever the verification is labelled as successful. DODDINGTON (1976), reports that a frequent usage is desirable to ensure a constant high performance level. At this date, probably several millions of tests have been made over a period of 6 years, for several hundred of speakers (87% male). (See HAIR & REKIETA (1972), for one report).

The Bell Labs telephone system is primarily intended to work over dialed-up telephone lines. Hereby, measurements were chosen to be insensitive to the phase or spectral amplitude distorsions of the telephone system (pitch, overal intensity, formant frequencies, linear predictor coefficients). It was found that both reject customer and accept impostor error rates stabilized at around 4% from high initial values around 15%, after a period of some 20 trials. This result shows the need of speaker's reference refreshing (cf. ROSENBERG, 1976, b).

Small size systems have been reported by BRODERICK & al. (1975), EL CHAFEI's thesis (1978), etc.

V - PRACTICAL APPLICATIONS OF SPEAKER RECOGNITION BY COMPUTER.

Many modern man-machine and man-man communication techniques emphasize the acoustic medium. Any security device must then work with the vocal signal. However, speech is already obviously going through the telephone and satellite channels, and teleconferences have come of age. In the recognition and systhesis areas, speech can bring specific advantages which have been widely discussed by scientists as well by private companies task forces. Speech and speaker recognition applicability are strenghened when used together. On the other hand, the availability of home terminal's opens the way to daily remote financial, and other operations over one's private telephone line.

Practical application in the field of crime detection is viewed as a more complicated task, which still requires much fundamental research. This being due to the wherever possible disguise of the voice, to the inadequacy, low quality or inavailability of speech samples of specific content.
Automatic processing is an essential feature of a practical Verification system. Since the Verification task offers specific advantages discussed above (see also BEEK, NEUBERG & HODGE, 1977), it has been the first task to be implemented for practical applications. The verification of any legitimate person should be successful within one or two attempts: the false-reject rate measures this performance. Secondly, non-legitimate persons (impostors) should be refused: this is measured by the false-acceptance rate.
Let us illustrate with an example. When automatised, a banking work station requires less trained people to verify documents (signatures, etc.). Long distance financial applications become possible, throught telephone, radio, or computer networks links. So far, verifying a person exactly means to experience a visual matching of two documents provided by the person (one of which may be the actual body of the person). So far a person has to carry many different proofs of his/her identity since security requirements are common. The error probability is however never null, and follows the same general rules as in the automatic case. False documenting, loss of documents, and document duplication, are common problems.
Before a technique is to provide law enforcement agencies with a reasonably valid investigational tool, it must meet certain

criteria. As an example, experiments tend to demonstrate that non-contemporary samples make the match difficult.
Before designing a system, it is a critical issue to determine the statistical affordable error rate for the given application. It should usually be very low, however there exist applications which may afford a high error rate. For instance, phone credit card verification number, phone orderings in which a rough first verification is what is required to begin processing the client's order, before receiving payment.
System architecture may be affected by sex-dependent components (like sex-dependent phone-lines or lexicons) since it may be evaluated easier and better to discrimitate among homogeneous populations, at the expense of predicting or asking the speaker's sex.
It is questionable that voice security systems would replace most systems, however it is advisable today to scan the fields which might be concerned and in what extend, both for Identification and Verification tasks. The diagram below summarizes the potential applications sectors. Economic feasibility is not discussed in this paper, nor is an evaluation of the user's cost per speaker.

VI - CONCLUSIONS.

The effectiveness of automatic talker verification has been demonstrated over the recent years. In more instances, the performances of machines exceed those of human listeners and spectrograms analysts.
The problem of Verification is easier to solve and has led to high recognition rates, using a large number of speakers, under laboratory conditions. Work is currently done in field experiments which provide more realistic conditions, and with telephone channels.
The high recognition rates achieved in speaker verification systems were possible under restrictive conditions, such as a cooperative speaker, a fixed text, a non noisy speech signal. While some authors emphasize the good results obtained by specific methods in laboratory environments, others point out that these are still achieved under these limited conditions. It is still true that questions remain concerning the effect of noise, bandwith truncations, spectral distorsion on systems performance. Research is still needed, for instance in the effects of the professional mimicry, the long-term modifications in the human voice, as well as the finding of the perceptual bases of speaker recognition. However, one can realistically say that speaker verification has already emerged as a practical technology. The systems have approached the point where they can be used in practical situations. Improvements will depend not only on the reduction of error rates, but also in the increase of accuracy with which the rates can be quantified; here, human factors appear to be of importance.
That the market is real for many transactional services which require authorized users seems plain, however profitability has not, in many instances, been assessed yet.

FIELD CLASSIFICATION OF POTENTIAL APPLICATIONS FOR SPEAKER RECOGNITION.

FIELD	EXAMPLES	TYPE	PERSONEL	SPECIFIC ADVANTAGES OBSERVATIONS
Access control to physical processes	-controlled zones (military sites, lockers, development labs. computer rooms storage rooms)	Verif.		-military security -industrial protection -nuclear power plants
Selective access to information (various authorization levels). Entry control	information centers -(documentation, archives, data bases (medical, public, banking, scientific, ...), -account numbers access, credit cards validation, acoustical passport systems.	Verif. and Ident.		-is one solution to the data bases confidentiality problems -is made possible by phone -can integrate the three tasks: -Verif./Ident. -word recognition -vocal response
Validation of financial, and banking operations, ...	-external and internal "work-stations" -remote bill phoned payments -telephone credit cards verification -seats reservation -remote voting -phone ordering	Verif.	operator may be needed depending on case	-can be done over telephone lines (no mail nor physical transportation remote homes, travel -50% accuracy can be enough to reduce current abuse costs -over phone lines
Assignment of confidential and nominal tasks	-handling of dangerous or remote material	Verif. only	automatic	-assignment of confidential and nominal tasks -industrial
Home devices and applications	-home protection (electronic lockers garages, doors, cars	Verif. only		-one solution to personal goods protection
Teleconferences	-reserved access to the meeting (signed or not messages -make listeners aware of the current speaker -channels security inspection	Ident.	automatic	-will not reduce line listening
Criminology	-location of criminals	Ident.	trained personel still needed	-still requires fundamental research (ex: disguises, noise)
Side applications	-emotional state & illness detection			

REFERENCES

(1) ALPERT M., KURTZBERG R.L., PILOT M. & FRIEDHOFF A.J. (1963) Comparison of the Spectra of the Voices of Twins ; J.A.S.A., 35, 1877 (A).

(2) ALLPORT G.W. & CANTRIL H. (1934) Judging Personality from Voice ; J. Soc. Psychol., 5, 37-55.

(3) ATAL B.S. (1968) Automatic Speaker Recognition Based on Speech Contours ; Ph. D. Diss. Polytech. Inst. Brooklyn, Brooklyn, N.Y., June.

(4) ATAL B.S. (1972 a) Text-Independent Speaker Recognition ; J.A.S.A., 52, 181 (A), 83th ASA, New York.

(5) ATAL B.S. (1972 b) Automatic Speaker Recognition Based on Speech Contours, J.A.S.A., 52, 1687-1697.

(6) ATAL B.S. (1974) Effectiveness of Linear Prediction Characteristics of the Speech Wave for Automatic Speaker Identification and Verification; J.A.S.A., 55, 6, 1304-1312.

(7) ATAL B.S. (1976) Automatic Recognition of Speakers from Their Voices ; Proc. IEEE 64, 460-475.

(8) ATKINSON J.E. (1976) Inter and Intra-Speaker Variability in Fundamental Voice Frequency, J.A.S.A., 60, 2, 440-455.

(9) BASZTURA C. & JURKIEWICZ J. (1978) The Zero-Crossing Analysis of a Speech Signal in the Short-Term Method of Automatic Speaker Recognition, Archives of Acoustics, 3, 3, 185-196.

(10) BASZTURA C. & MAJEWSKI W. (1978) The Application of Long-Term Analysis of the Zero-Crossing of a Speech Signal in Automatic Speaker Identification, Archives of Acoustics, 3, 1, 3-15.

(11) BEEK B., GRECH J. & MEEKER F. (1971) Automatic Speaker Recognition System ; Rome Air Development Center, report.

(12) BEEK B., NEUBERG L.P. & HODGE D.C. (1977) An Assessment of the Technology of Automatic Speech Recognition for Military Applications ; IEEE Trans. on ASSP, 25, 310-322.

(13) BOLT R.H., COOPER F.S., DAVID E.E. Jr, DENES P.B., PICKETT J.M., & STEVENS K.N. (1970) Speaker Identification by Speech Spectrograms : A Scientist's View of its Reliability for Legal Purposes ; J.A.S.A., 47, 597-612.

(14) BONAVENTURA M. (1935) Ausdruck der Personlichkeit in der Sprechstime und im Photogramm ; Arch. Geo. Psychol. 94, 501-570.

(15) BRICKER P.D. & PRUZANSKY S. (1966) Effects of Stimulus Content and Duration on Talker Identification ; J.A.S.A., 40, 1441-1454.

(16) BRICKER P.D., GNANADESIKAN R., MATHEWS M.V., PRUZANSKY S., TUKEY P.A., WACHTER K.W. & WARNER J.L. (1971) Statistical Techniques for Talker Identification ; B.S.T.J., 50, 1427-1454.

(17) BRODERICK P.K., PAUL J.E. & RENNICK R.J. (1975) Semi Automatic Speaker Identification System.

(18) BUNGE E. (1975) Automatic Speaker Recognition By Computers ; Proc. Carnahan Conf. on Crime Countermeasues, Lexington, Kentucky.

(19) BUNGE E. (1976) Automatic Speaker Recognition By Computers ; 1976 IEEE International Conference On Acoustics, Speech And Signal Processing, Philadelphia, Apr. 12-14, p. 738 (abstract).

(20) BUNGE E. (1977) Speaker Recognition by Computer ; Philips Technical Review, 37, 207-219, 8.

(21) BUNGE E., HOFKER U., JESORSKY P., DRIENER B. & WESSELING D. (1977) Statistical Techniques For Automatic Speaker Recognition ; IEEE Int. Conf. on ASSP, Hartford, May 9-11, pp 772-783.

(22) CLARKE F.R. & BECKER R.W. (1969) Comparison of Techniques for Discriminating among Talkers ; J.S.H.R. 12, 747-761.

(23) CORSI P. (1979) Reconnaissance automatique du Locuteur : Présentation générale, Méthodologie et Expérimentation, Perspectives d'Application ; Doct-Ing. Thesis, Grenoble Univ., INPG, Oct.

(24) DANTE H.M. & SARMA V.V.S. (1979) Multistage décision schemes for speaker Recognition ; IEEE Conf. ASSP, 797-800.

(25) DAS S.K. & MOHN W.S., WILLETT S.S. & CHAPMAN W.D. (1972) Two Speaker Verification Experiments ; IEEE AFCRL, Conf. Speech Comm. & Process, 275-278.

(26) DAS S.K. & MOHN W.S. (1971) A Scheme for Speech Processing in Automatic Speaker Verification ; IEEE Trans. AU-19, 32-43.

(27) DILLER T.C. & SIEBENAND J.F. (1981) Speaker-Independent Word Recognition Using Sex-Dependent Clustering ; IEEE Conf. ASSP, 180-183.

(28) DODDINGTON G.R. (1970) A Method of Speaker Verification ; Ph. D. Dissert., Univ. Wisconsin, Dept EE.

(29) DODDINGTON G.R. (1971) A Method of Speaker Verification ; J.A.S.A., 49, 139 A.

(30) DODDINGTON G. (1974 a) Speaker Verification ; Tech. Rept. RADC-TR-U1-963 700 F. Rome Air Development Center, Griffis afb, N.Y.

(31) DODDINGTON G. (1974 b) Speaker Verification ; Final Report, Tech Rept. RADC 74-179, Rome Air Development Center, Griffis AFB, N.Y.

(32) DODDINGTON G. (1976) Speaker Verification ; Tech. Rept., TR-76-262, Rome Air Development Center, Griffis AFB, N.Y.

(33) EL CHAFEI C. (1978) Un système de reconnaissance automatique de locuteurs sur mini-ordinateur ; Doct-Ing. Thesis, Paris-Sud Univ., Orsay.

(34) ENDRESS W., BAMBACH W. & FLOSSER G. (1971) Voice Spectrograms as a Function of Age, Voice Disguise and Voice Imitation ; J.A.S.A., 49, 6(2), 1842-1848.

(35) FURUI S., ITAKURA F. & SAITO S. (1972) Talker Recognition by Long Time Averaged Speech Spectrum ; Electronics and Communications in Japan, 55-A, 54-61.

(36) FURUI S. & ITAKURA F. (1973) Talker Recognition by Statistical Features of Speech Sounds ; Electronics and Communications in Japan, 56-A, 62-71.

(37) FURUI S. (1974) An Analysis of Long-Term Variation of Feature Parameters of Speech and its Application to Talker Recognition ; Electronics and Communications in Japan, 57A, 34-42.

(38) FURUI S. (1978) Research on Individual Information in Speech Wave ; Ph. D. Dissert., Tokyo Univ., Faculty of Eng., Math. Eng. and Inst. Phys. Dept.

(39) GOLDSTEIN U.G. (1975) Some Speaker-Identifying Features Based on Formant Tracks ; US Navy Office of Naval Research, Contract N00014-67-A-0204-0069.

(40) GRENIER Y. (1977) Identification du Locuteur et Adaptation au Locuteur d'un Système de Reconnaissance Phonétique ; Doct. Ing. Thesis, ENST, Paris.

(41) GRENIER Y. (1978) Speaker Identification from Linear Prediction ; 4th IJCPR, Kyoto, Nov.

(42) HAIR G.D. & REKIETA T.W. (1972) Automatic Speaker Verification Using Phoneme Spectra ; J.A.S.A., 51, 131 (A).

(43) HALL M.E. (1975) Spectrographic Analysis of Interspeaker and Intraspeaker Variabilities of Professional Mimicry ; Unpublished Master Thesis, Michigan State University.

(44) HARTMAN D.E. & DANHAUER J.R. (1976) Perceptual Features of Speech for Males in Four Perceived Age Decades ; J.A.S.A. 59, 712-715.

(45) HECKER M.H.L. (1971) Speaker Recognition, an Interpretive Survey of the Literature ; American Speech & Hearing Assoc., Monograph 16.

(46) HERZOG H. (1933) Stimme und Personlich Lichkeit ; J. Psychol. 130, 300-379.

(47) HOLMGREN G.L. (1963) Speaker Recognition in Speech Communication Systems ; Rept. 13-73801-13, Rpt. AFCRL-63-119, Air Force Cambridge Research Laboratories, Office of Aerospace Research, Bedford, Mass.

(48) HOLMGREN G.L. (1967) Physical and Psychological Correlates of Speaker Recognition ; J.S.H.L., 10, 57-66.

(49) HUNT M.J., YATES J.W. & BRIDLE J.S. (1977) Automatic Speaker Recognition for Use Over Communication Channels ; IEEE Conf., ASSP, 764-767.

(50) INGEMAN F. (1968) Identification of the Speaker's Sex from Voiceless Fricatives ; J.A.S.A., 44, 1142-1144.

(51) KAISER L. (1939/40/41/42/44) Biological and Statistical Research Concerning the Speech of 216 Dutch Students, I/II/III/IV/V . Archives Neerlandaises de Phonétique Expérimentale, 15, 1-76/16, 77-136/17, 143-211/18, 1-58/19, 37-78.

(52) KASHYAP R.L. (1976) Speaker Recognition from an Unknown Utterance and Speaker-Speech Interaction ; IEEE Trans. ASSP, 24, 481-488.

(53) KERSTA L.G. (1962 a) Voiceprint Identification ; Nature, 196, 4861, 1253-1257.

(54) KERSTA L.G. (1962 b) Voiceprint Identification ; J.A.S.A., 34, 725 (A).

(55) KERSTA L.G. (1962 c) Voiceprint Identification ; J.A.S.A. 34, 1070 (A).

(56) KERSTA L.G. (1962 d) Voice spectrograms for Unique Personal Identifications ; Bell Labs Record, 40, 214-215.

(57) KRAUSE H.J. (1976) Possibilities of Identification by Voice and Limits ; Archiv fur Kriminologie, 157, 5 & 6, 1976, pp 154-164 (in German).

(58) LEVIN H. & LORD W. (1975) Speech Pitch Frequency as Emotional State Indicator ; IEEE Trans. SMC, 5, 2, 259.

(59) LI K.P., DAMMANN J.E. & CHAPMAN W.D. (1966) Experimental Studies in Speaker Verification using a Adaptative System ; J.A.S.A., 40, 966-978.

(60) LI K.P. & HUGHES G.W. (1974) Talker Differences as they Appear in Correlation Matrices of Continuous Speech Spectra ; J.A.S.A., 55, 833-837.

(61) LIN W.C. & PILLAY S.K. (1976) Feature Evaluation and Selection for an On-Line Adaptive Speaker Verification System IEEE Conf. ASSP, 734-737.

(62) LUCK J.E. (1969) Automatic Speaker Verification using Cepstral Measurements ; J.A.S.A., 46, pp 1026-1032.

(63) LUMMIS R.C. (1971) Real-Time Technique for Speaker Verification by Computer ; J.A.S.A., 50, 106 (A).

(64) LUMMIS R.C. & ROSENBERG A.E. (1972) Test on Automatic Speaker Verification Method with Intensively Trained Professional Mimics ; J.A.S.A., 51, 131-132, A.

(65) Mc GEHEE F. (1937) The Reliability of the Identification of the Human Voice ; J. Gen. Psychol., 17, 249-271.

(66) Mc GEHEE F. (1944) An Experimental Study in Voice Recognition J. Gen. Psychol., 31, 53-65.

(67) MATSUMOTO H., HIKI S., SONE T. & NIMURA T. (1973) Multidimensional Representation of Personal Quality and its Acoustical Correlates ; IEEE Trans., AU-21, 428-436.

(68) MOHN W.S. (1971) Two Statistical Feature Evaluation Techniques Applied to Speaker Identification ; IEEE Trans. C-20 979-987.

(69) MYSAK E.D. (1959) Pitch and Duration Characteristics of Older Males ; J.S.H.R., 2, 46-54.

(70) PAUL J.E., RABINOWITZ A.S., RIGANATI J.P. & RICHARDSON J.M. (1975) Semi-Automatic Speaker Identification System (SASIS) - Analytical Studies ; Final Report, Rockwell International Report N° C74-11841501, Prepared for the Aerospace Corporation.

(71) POLLACK I., PICKETT J.M. & SUMBY W.H. (1954) On the Identification of Speakers by Voice ; J.A.S.A., 26, 403-406.

(72) POTTER R.K., KOPP G.A. & GREEN H.C. (1947) Visible Speech ; Van Nostrand, New York (reprinted by Dover).

(73) PRUZANSKY S. (1963) Pattern-Matching Procedure for Automatic Talker Recognition ; J.A.S.A., 35, 354-358.

(74) PRUZANSKY S. & MATHEWS M.V. (1963) Talker Recognition Procedure Based on Analysis of Variance ; J.A.S.A., 35, 1877A

(75) PRIZANSKY S. & MATHEWS M.V. (1964) Talker Recognition Procedure Based on Analysis of Variance ; J.A.S.A., 36, 2041-2047.

(76) PTACEK P.H. & SANDER E.K. (1966) Age Recognition from Voice ; J.S.H.R., 9, 273-277.

(77) PTACEK P.H., SANDER E.K., MALONEY W.H. & JACKSON C.C.R. (1966) Phonatory and Related Changes with Advanced Ages ; J.S.H.R., 9, 353-360.

(78) RAMISHVILI G.S. (1966) Automatic Voice Recognition ; Eng. Cyber., 5, 84-90.

(79) RAMISHVILI G.S. & TUSHIVILI H.A. (1976) On the Connection of some Time Characteristics of the Speech Signal with the Individuality of the Voice ; IEEE Conf., ASSP, 730-733.

(80) REICH A.R., MOLL K.L. & CURTIS J.F. (1976) Effects of selected vocal disguises upon spectrographic speaker identification ; J.A.S.A., 60 : 919-925.

(81) RABINER L.R., LEVINSON S.E., ROSENBERG A.E. & WILPON J.G. (1979) Speaker-Independent Recognition of isolated words using clustering techniques, IEEE Trans. ASSP-27, 336-349

(82) ROSENBERG A.E. (1972) Listener Performance in Speaker Verification Tasks ; Proc. IEEE Conf on Speech Comm. & Process., Air Force Cambridge, Res. Lab., Bedgord, MA, 283.

(83) ROSENBERG A.E. (1973) Listener Performance in Speaker Verification Tasks ; IEEE Trans. AU-21, 221-225.

(84) ROSENBERG A.E. & SAMBUR M.R. (1975) New Techniques for Automatic Speaker Verification ; IEEE Trans. ASSP-23, 2, 169-176.

(85) ROSENBERG A.E. (1976a) Automatic Speaker Verification : A Review ; Proc. IEEE, 64, 475-487.

(86) ROSENBERG A.E. (1976b) Evaluation of an Automatic Speaker Verification System Over Telephone Lines ; B.S.T.J., 55, 6, 723-743.

(87) ROSENBERG A.E. & SHIPLAY K.L. (1981) Speaker Identification and Verification Combined with Speaker Independent Word Recognition ; IEEE Conf. ASSP, 184-187.

(88) RYAN W.J. & BURL L.W. (1972) Predictors of Age in the Male Voice ; J.A.S.A. 53, 345 (A).

(89) SAMBUR M.R. (1972) Speaker Recognition and Verification Using Linear Prediction Analysis ; Ph. D. Dissert., M.I.T., Cambridge.

(90) SAMBUR M.R. (1975) Selection of Acoustic Features for Speaker Identification ; IEEE Trans. ASSP-23, 176-182.

(91) SAMBUR M.R. (1976 a) Speaker Recognition using Orthogonal Linear Prdiction ; IEEE Trans., ASSP-24, 283-289.

(92) SAMBUR M.R. (1976b) Text-Independent Speaker Recognition using Orthogonal Linear Prediction ; IEEE Conf. on Acoustics Speech and Signal Process., 727-729.

(93) SAPIR E. (1927) Speech as a Personality Trait ; Amer. J. Soc 32, 892-905.

(94) SARMA V.V.S. & YEGNANARAYANA B. (1975) A Critical Survey of Automatic Speaker Recognition Systems ; J. Comput. Soc. India, 6, 1, 9-19.

(95) SARMA V.V.S. & YEGNANARAYANA B. (1976) Cascade realization of Digital Inverse Filter for Extracting Speaker Dependent Features ; IEEE Int. Conf. Acoustics Speech and Signal Process., 723-726.

(96) SHEARME J.N. & HOLMES J.N. (1956) An Experiment Concerning the Recognition of Voice ; Language and Speech 2, 123 131

(97) SCHWARTZ M.F. (1968) Identification of Speaker Sex from Isolated Voiceless Fricatives ; J.A.S.A., 43, 1178-1179.

(98) SCHWARTZ M.F. & RINE H.E. (1968) Identification of Speaker Sex from Isolated ; Whispered Vowels, J.A.S.A., 44, 1136-1737.

(99) SHIPP T. & HOLLIEN H. (1969) Perception of the Aging Male Voice ; J.S.H.R., 12, 703-710.

(100) STEVENS K.N., WILLIAMS C.E., CARBONNEL J.R. & WOODS B. (1968) Speaker Authentification and Identification : a comparison of Spectrographic and Auditory Presentations of Speech Materials ; J.A.S.A., 44, 1596-1607.

(101) TAYLOR H.C. (1933) Social Agreements on Personality Traits as Judged from Speech ; J. Soc. Psychol., 5, 244-248.

(102) TOSI O. (1967) Evaluation of the Voiceprint Method ; Report to the Michigan Dept. of the State Police.

(103) TOSI O. (1968) Speaker Identification Trough Acoustic Spectrography ; Proc. 14th Int. Cong. on Logopedics And Phoniatrics, Paris.

(104) TOSI O., OYER H., LASHBROOK W., PEDREY C., NICHOL J. & NASH E. (1972) Experiment on Voice Identification ; J.A.S.A., 51, 2030-2043.

(105) VIDALON M., SHRIDHAR M., CANAS M. (1977) Speaker Verification using Composite Reference ; IEEE Conf., ASSP, 758-760.

(106) VOIERS W.D. (1964) Perceptual Bases of Speaker Identity ; J.A.S.A., 36, 1065-1075.

(107) WEINBERG B. & BENETT S. (1971) A Study of Talker Sex Recogntion of Esophageal Voices ; J.S.H.R. 14, 391-395.

(108) WILLIAMS C.E. (1964) The Effects of Selected Factors on the Aural Identification of Speakers ; sec. III in "Methods for Psychoacoustic Evaluation of Speech Communication Systems". Dept. ESD-TDR-65-153, Electronic Systems Division, Air Force Systems Command, Hanscom Field, Mass.

(109) WILLIAMS C.E., STEVENS K.N. & HECKER M.H.L. (1970) Acoustical Manifestations of Emotional Speech ; J.A.S.A., 47, p 66 (abstract).

(110) WOLF J.J. (1972) Efficient Acoustic Parameters for Speaker Recognition ; Ph. D. Dissert., Dept. EE, M.I.T.

(111) WOLF J.J. (1972) Efficient Acoustic Parameters for Speaker Recognition ; J.A.S.A., 51, pp 2044-2056.

(112) WOLF J.J. (1971) Choice of Speaker Recognition Parameters ; Q.P.R., M.I.T., 97, 125-133.

(113) YEGNANARAYANA B., SARMA V.V.S. & VENUGOPAL D. (1977) Studies on Speaker Recognition using a Fourier Analysis System ; IEEE Conf. ASSP, 776-779.

Voice Identification Research ; Submitted by the Dept. of Michigan State Police, East Lansing, Michigan (1972).

COMPLEXITY OF ALGORITHMS AND PATTERN RECOGNITION

J.C. SIMON

Institut de Programmation, Université P. & M. Curie,
4 place Jussieu, 75230 Paris Cedex 05, France.

The Pattern Recognition (P.R.) problem of the identification of an object from a representation data may be termed as a reduction in the complexity of the representation. Pattern recognizers have to find algorithms which escape the untractability of combinatorial explosion, without being defeated by errors. Thus the recent results obtained in the field of computational complexity are of interest to P.R.. An overview is given, with an attention to the class of NP-complete problems, and polynomial problems of interest to P.R.; some hints are given how to escape untractability.

AN OUTLINE OF THE PATTERN RECOGNITION PROBLEMS.

Nowadays, Automatic Pattern Recognition (P.R.) is a very wide field of similar techniques, applied to comparable problems. Data being given by sensors (images, speech, more generally sampled signals) some algorithms, implemented by programs on computers, are supposed to perform an "identification".

In other words let

$$X = (x_1, x_2, \ldots, x_n)$$

be an *initial data*, i.e.
a representation given by some sensors;

$$\mathrm{X} = (\mathrm{x}_1, \mathrm{x}_2, \ldots, \mathrm{x}_n)$$

be a *variable data*,
belonging to $\mathcal{X}$, *a representation space;*

$$\Omega = \{\omega_1, \omega_2, \ldots, \omega_p\}$$

be a set of *names*.

J.-P. Haton (ed.), Automatic Speech Analysis and Recognition, 309–329.

An *identification* is a mapping $\mathfrak{L}$ from the representation space unto the set of names:

$$\mathfrak{L} : \mathcal{X} \to \Omega \qquad (1)$$

Of course (1) has to be implemented "constructively", and this is all the problems of P.R.:

- find an algorithm Å, which would perform (1) properly,
- implement Å by a program, written in a computer language

But an identification is not the only interpretation looked for the representation of an object, of a scene. For instance:

- *a feature* is the result of a *partial identification* (a phoneme in a spoken word , a segment in a letter, a contour or a texture in an image).
- *a fuzzy identification.* The result of the process is to associate *a membership value* f_i to the name ω_i.
- *a symbolic description*, such that:
  - the most representative object of a class,
  - a representation of a class,
  - a concept, described by a "type",
  - a statement, an expression, a logical proposition, etc..

What are the informations, specific to a particular P.R. problem, which enable to build up an algorithm performing (1) properly? J.C. SIMON (1). They consist of:

- A *training set*, T = $\{(X,\omega)\}$, a set of couples of representations and names, samples of (1).
- The properties of the representation space $\mathcal{X}$ such as *the similarities or the distances* between X and X'.
- The use of specific operators (arithmetic, logical, syntaxic) and the evaluation of their performances.

The most simple algorithm to perform (1) would be to extend the "training set" T to the entire domain of $\mathcal{X}$, and to record the results in the memory of a computer. Then whenever an X appears, a simple search in the memory would give the proper name ω.

Feasible in principle, this process is completely impossible in practice. On the first hand the domain of $\mathcal{X}$ may be infinite; but even if supposed finite it is always of an exponential nature, and the number of recorded couples would have to be of "astronomical" nature.

Thus the amount of representation requested for a second of speech is 10 to 50Kbits. A "phoneme" duration being of the order of 10ms or above, as the duration of the pitch period, one is tempted to record all possible signals of 10ms duration. But this request 100 to 500 bits; and the number of possible X would be 10^{30} to 10^{150}, which are truly astronomical numbers.

Images make things even worst. A TV image requires well above 1 million bits, a printed letter of the order of 200 bits.

To keep in a memory all the binary templates requires for 3x3, 512, 5x5, 10^7; 8x8, 10^{19}!

On the other hand the number of names, the cardinality of Ω is never very large. In speech, the phonemes are approximately 40; the number of words to be recognized goes from a few tenths to a thousand.

An identification can be considered as a reduction of the complexity of the representations. Thus to operate this reduction or *projection* of many orders of magnitude, we need to find algorithms, which would perform in a reasonable amount of time, in other words which would not be "too complex" (the aim of this paper is precisely to make clear what is meant by the complexity of algorithms).

A first approach to the reduction of complexity is to divide the problem of identification in *sub-problems*. Clearly this is what is done, when a word is decomposed into syllables or phonemes, when *elementary features*, such as a contour, a texture, is searched in an image. The signal is decomposed in sub-signals, the image in sub-images. The general P.R. problem is decomposed in multi-level problems. Identifications are done at each levels; but each of which treats a problem of less complexity.

But still we are left with the necessity to find proper algorithms for each of these levels of decision; and the "memory template" type does not seem appropriate, even though the complexity of representation has been drastically reduced. Thus the P.R. litterature is overwhelmed by description of P.R. algorithms and of their performance on such and such problem. As the efficiency, not only errorwise, but also in time (in number of elementary operations) is essential for any real P.R. problem, it seems useful to point out to the P.R. community the recent progresses done in the domain of the complexity of algorithms.

This is what is done in the following paragraphs, keeping in mind that we adress to an audience of P.R. specialists in the domain of speech.

DEFINITION OF ALGORITHMS AND EFFECTIVELY COMPUTABLE FUNCTIONS.

An algorithm may be loosely defined as "a specific set of rules for performing a task". Equivalent words frequently used are *finite procedures* or *effectively computable functions*.

Several definitions of finite procedures were based on methods very different from each others (CHURCH, GODEL, HERBRAND, KLEENE, POST, TURING, etc...) using the *closure operations, formal symbol manipulation systems, idealized mechanical computers*. All the definitions turned out to be equivalent, in the sense that they all "compute" the same class of functions: *partial recursive functions*, which is exactly the class of effectively computable functions (the CHURCH-TURING thesis). M. MATCHEY & P. YOUNG (2).

Primitive recursive and partial recursive functions.

An algebraic definition of effectively computable functions on a set of words, by closure operations.

Let $A = (a_1,...a_k)$ be a k-alphabet, and A^* the infinite set of words x, which may be formed from A (including the empty word ε).

Let us consider the elementary functions of domain $(A^*)^n$ and of range A^*: *erase (or null)* $E(x) = \varepsilon$; the jth *successor* $S_j(x) = x\,a_j$; the *projection* P_j^n .

A *primitive recursive function* is formed from these elementary functions, using the *composition and the (primitive) recursive operations.*

However, it is possible to show (2) p. 25, by the Cantor diagonal argument, that some effectively computable functions are not primitive recursive. This does translate in reality by the fact that not every algorithm produces an output (i.e. halts) on every possible input; sometimes they "loop". When a *partial function* is not defined on an argument, it is assumed to be non terminating (looping). These partial recursive function are defined through a minimization recursive operation instead of a primitive recursive operation, the rest being unchanged.

Every primitive recursive function is shown to be a total recursive function, i.e. a partial recursive defined everywhere.

Random access machines and programs.

A Random Access Machine (RAM) is an idealized computer, with a potentially infinite memory (or set of registers), which may store any of the words of A^*. Seven "instructions" are available in the RAM; Y and Z being register names:

add$_j$ Y : it adds a_j to the right end of the word in register Y.
del Y : it deletes the first letter of the word in Y.
clr Y : it replaces the word in Y by the empty word ε.
Y $\leftarrow$ Z : it copies the word in Z into Y, leaving Z unchanged.
jmp X : jump instruction
Y *jmp*$_j$ X : conditional jump executed if the first letter of the word Y is a_j.
continue : does nothing; may be considered as a "halt" instruction

A RAM program is a finite sequence of instructions. It halts if it reaches a continue instruction.

Turing Machines

A Turing Machine TM is an idealized tape automata. It contains a two-way (potentially) infinite ribbon, divided into squares and a finite control device with a read-or-write head, which moves along the tape. The tape symbols are the letters of the finite k-alphabet A, one of them being considered as "blank". There are p "internal states". A typical instruction of the control device would be interpreted as : "if in state i looking at the symbol a_j on the tape, replace it by a_n, move one square in a direction along the tape, and go to state m".

Markov algorithms

Markov algorithms are formal symbol manipulation schemes which might be considered as deterministic versions of formal grammars. A Markov Algorithm (MA) is a finite ordered sequence of *productions*, or rules rewriting a word p into another word q, both belonging to A^*. Some productions are "terminating" ones. A partial function is MA-computable if some Markov algorithm computes it.

Theorem

The following classes of partial functions are all the same:

- *partial recursive functions,*
- *Random Access Machine computable functions,*
- *Turing Machine computable functions,*
- *Markov Algorithm computable functions.*

Moreover, these equivalences are effective in that, given a program for a partial function in any one of these four formalisms, we can effectively translate it into a program for the same function in any of the other formalisms.

Remarks

- All of the functions used in practice in P.R. algorithms are of the class of primitive recursive functions.
- Restricted to primitive recursive functions, the demonstrations of the above theorem are rather simple.
- It seems important to underline that no modern computer has "more power" that a simple RAM or TM. Of course it is a lot more easy to use, specially with the high level languages.
- The computer has been developped for fast arithmetic computation, but its use is far more general. Today, most of the work done on a computer is word manipulation rather than proper computation.

The fact that a computer has such an universal value is well reflected by the concept of algorithm or effective procedure, which was sketched above.

We now turn our attention on the *computational complexity* of an algorithm (program), in other words, the use of computational resources such as time or memory space, which are required by the execution of the algorithm. As this last item depends upon the input data, the computational complexity (in brief the complexity) depends also of the inputs.

A HIERARCHY OF COMPUTATIONAL COMPLEXITY.

First let us be more precise about he input data. P.R. problems have usually simple encoding schemes; for instance the input data is a list of comparable numerical samples. It is not true generally for a problem solvable by an effective procedure. For instance let us consider a problem on a graph. There are several techniques to encode the graph itself: a matrix, a list of vertex and a list of "successors", etc... But one may obtain one from the other through a "polynomial" manipulation.

We will admit that the input data is a "reasonable" encoding of the data of the problem: a list of elements taken in an alphabet (the symbols are not necessarily comparable, i.e. they are not part of a space). This finite alphabet is encoded by binary numbers. Thus an input data will be a string of binary digits, or bits. But of course its decoding may not be straightforward. However this string may always be defined by a length n.

The algorithm, or effective procedure, solving the stated problem will require a number of operations, or elementary steps, which will be a function of n. If n is "large" this measure of the computational complexity is estimated by the largest term. For instance, if a polynomial by the term of largest degree; if the sum of a polynomial and and exponential term, by the exponential term. Usually a multiplicative constant is not mentioned in the currently used notation, such as $O(n^2)$ or $O(3^n)$.

Remark

It seems interesting to point out that an identification is nothing else but a "clever" new encoding of the input data. But of course this clever encoding is made by the human recognizer. And what is asked to the machine is to do the same for any input data, which is far from easy...

An algorithm is in fact labeled by its complexity. The algorithm, best for a problem, is the one which has the lowest complexity function. Here comes a difficult question:

Is it possible to prove that there does exist (or not) a better algorithm that the one we know at a given time?

For example is it possible to prove that no polynomial algorithm exists, or in other words (usually) that the best algorithm is of exponential nature?

Some variations exist around this question, according to the data: some algorithms change of class (for example from exponential to polynomial) for a small variation of the input data. Thus one should distinguish:

- the most favourable,
- the worst,
- the average complexity.

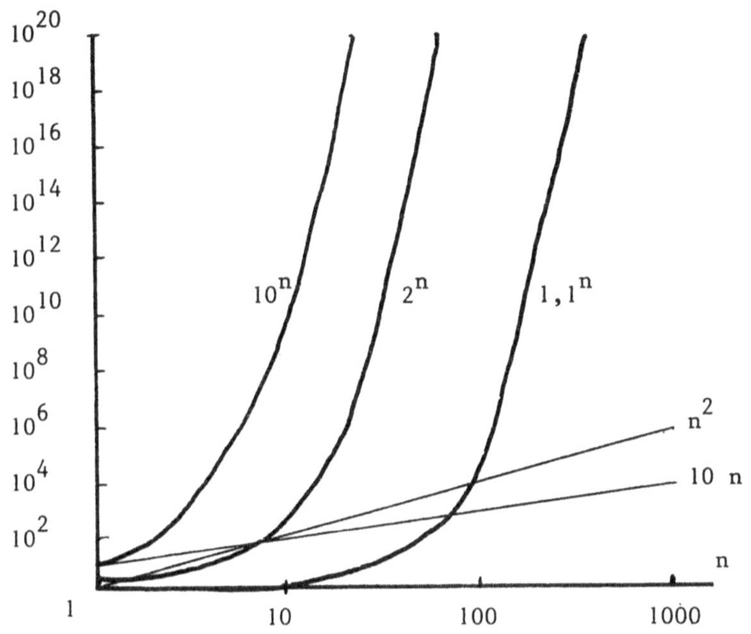

Fig. 1, variations of polynomials and exponentials.

Fig. 1 shows clearly why everything which is not polynomial, in particular the exponentials, is qualified of *intractable*. To illustrate this "intractability", let us see what is the increase in performances that a computer 10 time or 1.000 time faster would bring to algorithms with different complexities. Let N be the number of operations that are performed by the original computer, on different complexities. The following Table shows the increase in performances that a new computer 10 or 1.000 time faster would bring.

computing power	n	nLn	n^2	n^5	$1,1^n$	2^n	10^n
x10	10N	10N	3N	1,6N	N+24	N+3	N+1
x1000	1000N	390N	32N	4N	N+72	N+10	N+3

Table 1, increase in performances versus complexity

A quick glance on this Table 1 shows that most of the time an increase, even large, in computing power is useless for exponential algorithms, with not small N.

Thus it is a lot more interesting to find a polynomial algorithm, if this is possible, rather than using an exponential one and buy a very powerful computer!

Let us now describe a hierarchy of complexity.

Undecidable problems

It has been shown since the years 30-40 by theoreticians like GODEL, POST, TURING, that some problems, very well defined, in the algorithmic sense are *undecidable*; in other words that *no algorithm could be found at all to solve them.*

For instance:

- whether or not an arbitrary program would halt on an arbitrary input(TURING).
- there is no algorithm for deciding whether or not arbitrary sentences of the language of arithmetic are true in the usual interpretation of the natural numbers, under addition and multiplication (GODEL).
- HILBERT's tenth problem: the solvability of polynomial equations in integer (MATIJASEVIC).
- the POST correspondance problem, which is quite a simple problem: Let A be a finite alphabet, x_i, y_i belong to A^*. Let $\{(x_i, y_i)\}$ be a set of couples, called "correspondances". $1 \leq i \leq p$. Wether or not is there a solution consisting of a finite sequence of integers $i_1, i_2, \ldots i_q$, such that:

$$x_{i_1} x_{i_2} \ldots x_{i_q} = y_{i_1} y_{i_2} \ldots y_{i_q} \qquad (2)$$

A consequence of the undecibility of the POST correspondance problem is that wether or not two context free languages have an empty intersection is undecidable.

However a small restriction, which would be encountered in practice, does there exist a finite sequence of length bounded by N? is a decidable problem, exponential in nature of course, but definitely decidable.

If we assume that the brain of man is of algorithmic nature, then of course all the P.R. problems are decidable, as they seem to be solved by man. Even more so they should not be of exponential nature as the time of solving seems so short...

Inherently exponential or higher.

Some problems have been shown to be inherently exponential. This is true of course if it is asked to obtain an exponential number of answers. For instance:

- testing of a program for all possible data;
- extending the training set on the set $\mathcal{X}$;
- find all the cliques or subtrees of a graph,etc...

Some problems have been proved to be higher than EXP. For instance PRESSBURGER arithmetic is arithmetic reduced to addition. M.O. RABIN (3) has proved that, though is is a decidable theory, the complexity of a proof is of the order of: $2^{2^{cn}}$, c being a constant.

A sub-class of primitive recursive functions (2) p. 53,

are called by the theoreticians m-*elementary*, and their complexity is of the type 2 exponentiated m time, to the exponent n.

$$O(n) = \underbrace{2^{2^{2^{\cdot^{\cdot^{2^{n}}}}}}}_{m}$$

NP problems.

This is a class of problems, which are not shown either polynomial or exponential by nature, but for which no polynomial algorithm has been shown. Moreover, they are *polynomialy equivalent* in the sense that if only is solved by a polynomial algorithm, then all are polynomial problems. But this happy event is most unlikely...

This rather strange class is of a great practical importance, as we will soon see. More than 300 quite interesting problems have been prove to belong to that NP class, M. GAREY and D. JOHNSON (4). The abrevation NP, does not mean at all non-polynomial, but "*non-deterministic polynomial*".

Polynomial problems.

The "good" problems indeed, for which a polynomial algorithm has been found. Fortunately many usual problems in computer science are polynomial. For instance:

- the sorting and searching problems;
- the digital filters in signal and image processing;
- dynamic programming;
- the transitive closure of a graph;
- the shortest path in a graph between two nodes;
- deterministic syntactic analysis, etc...

Let us examine now in more details the class NP.

THE CLASS NP-COMPLETE

Let us give an example from LAURIERE (5) of an NP-problem. Let R be a logical expression $R(p_1,p_2,...p_n)$, well-formed with the variables $p_1,p_2,...,p_n$ and the operators $\vee$ (or), $\wedge$ (and), $\ulcorner$ (not). Let us write an algorithm, which finds the correct values of the variables so that R is T. (R and the variables may either be T or F)

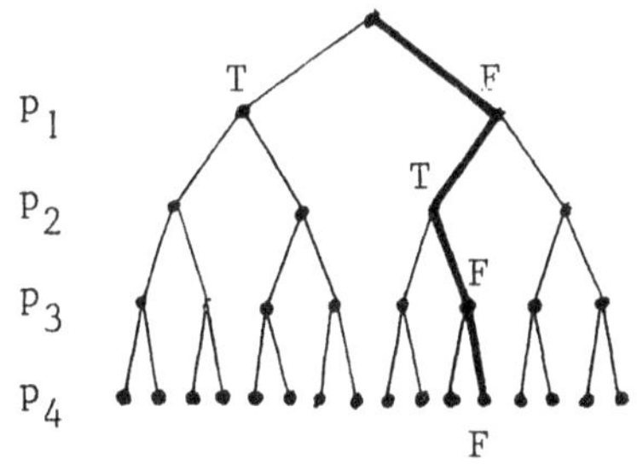

Fig. 2, decision tree

```
REPEAT FOR i = (1:n)
 | pi := CHOICE(T,F)
 |    IF R(p1,p2...pn) = T
 |       THEN "success" END;
 |       ELSE "failure"
ENDOFREPEAT
```

Let the function CHOICE(X) be an intruction which creates as many copies of the current state, as there are elements in X.

For instance let R be:

$$R = (p_1 \vee \lceil p_2 \vee \lceil p_3) \wedge (p_1 \vee p_2 \vee p_4) \wedge (p_3 \vee \lceil p_4) \wedge (\lceil p_1) \quad (3)$$

The solution for R = T, is $(p_1,p_2,p_3,p_4) = (F,T,F,F)$ which corresponds to the underlined path in the decision tree of the Fig. 2. The algorithm REPEAT does 16 copies of itself, the number 12 is the good one. At each step it copies again the formula (3), with the appropriate values of p_i.

The abstract model for such a computation is a Turing Machine, which is able to duplicate itself at each step. It is a Non Deterministic Turing Machine NDTM, which is able to make 2^i computation, at each step; each of which requesting in fact a Deterministic TM. An NP-problem is a problem whose solution ask for a NDTM, solving it in *polynomial time.*

An interesting remark is that, if we could have had a "guessing machine" (a little bird?), which would tell us which branch of the decision tree we should take, a single DTM would suffice, and the computation would be done in polynomial time, (in linear time even, in this example). But without this divination machine, the number of outputs, which have to be checked, is obviously exponential.

In other word this is typically a problem for which we use a DTM and, without a divination machine, we have to make an exponential number of "backtracks", or we have a NDTM and we conduct an exponential number of computations in parallel.

The irritating problem is that the problem is not inherently exponential because of the possibility of this guessing machine, but in reality exponential because it is not in fact realistic to think that one would be able to discover a good guessing machine.

Definitions

A problem Q is *reducible* to a problem R, iff, for all solutions s of R, there exists a polynomial function g(s), such that g(s) is a solution of Q. We write then:

$$Q \propto R \quad (4)$$

It is clear that the relation $\propto$ is "transitive"; but not evidently symmetrical. If:

$$R \propto Q \quad (4') ,$$

then with (4) and (4'), Q and R are *polynomially equivalent.*

NP-hard. A problem R is said NP-hard iff every NP-problem is reducible to R.

NP-complete. A problem R is said to be NP-complete iff every NP problem is reducible to R, *and* if R is NP (NP-hard + NP).

Historically the first NP-complete problem is precisely the problem of the above example; the SATISFIABILITY (SAT) problem of a collection of logical clauses over a set of boolean variables. It is COOK (6) who, in 1971, did present the interesting concept of NP problems and proved that SAT is NP-complete.

Curiously, SAT becomes polynomial if the number of variables in each clause is two or less. Thus SAT is usually quoted as 3-SAT.

A host of other interesting problems have been proved to be polynomially equivalent to 3-SAT. They form the NP-complete class of problems; for which, if a polynomial solution were found for one, all of them would become polynomial, instead of exponential. But it is quite unlikely. However we will come back on this question in the last paragraph, and explain how *heuristic algorithms* may, in a sense, play the role of the guessing machine; and thus reduce NP to P, or at least reduce the exponentiality.

We refer the reader to the basic review of GAREY and JOHNSON (4) for the demonstrations of the NP-completeness of the following algorithms, which are going to be quoted in relation with their interest in P.R. (the most interesting being pointed by a ★).

Graph theory. Graph $G = (V,E)$, positive integer $K \leq |V|$ or $|E|$

Vertex cover. Is there a vertex cover of size K or less for G, i.e. a subset V' of V, with Card(V') smaller or equal to K, such that for each edge of E at least one of the nodes belongs to V' ?

Dominating set. Is there a dominating set of size K or less for G, i.e. a subset V' of V, with Card(V') smaller or equal to K, such that for all node u belonging to V - V', there is an arc (u,v) belonging to E, such that v belongs to V'?

★ Graph K-colorability. G is K-colorable if there exist a function f of K values, such that $f(u) \neq f(v)$, (u,v) is an arc of E. This is polynomial for K = 2. A graph is 2-colorable iff all the cycles are even, which may be polynomially tested.

Minimum maximal matching. Is there a subset E' of E, with Card(E') smaller than K, such that E' is a maximal matching, i.e. no two edges in E' share a common endpoint and every edge in E - E' shares a common endpoint with some edge in E' ?

★ Partition into isomorphic subgraphs.

★ Partition into forests. Partition of the set V of nodes, in V_i with i smaller than K, each graph induced by V_i is a forest, i.e. does not contain any circuit.

★ Partition or covering by cliques (complete graphs).

★ Clique of size K or more.

★ Induced subgraph with a given property Π. Examples: G is a clique, G is planar, G is bipartite, G is a forest, etc...

Degree bounded connected subgraph. Is there a subset E' of E, with Card(E') greater than K, such that the subgraph is connected and has no vertex with degree greater than d ?

Planar subgraph, with Card(E') greater than K ?

★ Minimum K-connected subgraph. Does there exist a subgraph G' = (V,E') which cannot be disconnected by removing less than K vertices?

Hamiltonian path or circuit. (a path through all vertices)

★ Subgraph isomophorphism. Does G contains a subgraph isomorphic to H?

★ Largest common subgraph. Does there exist a subgraph G' of G, and a subgraph H' of H, with a number of vertex greater than K, which are isomorphic?

★ Graph contractibility or homomorphism. Is it possible to obtain a graph isomorphic to H from a graph G by suppressing arcs and/or vertex?

★ Special properties spanning trees, such as degree constraint, maximum number of leaves, shortest total path length, bounded diameter, isomorphic to a given tree, etc...?

Graph partitioning. A graph G with weights w for each u belonging to V. Is there a partition of V into subsets V_i, such that Σ w(u) $\leq$ K, and such that if E' is the subset of edges belonging to E, and having weights 1 (for each e), then Σ 1(e) $\leq$ J?

★ Kth shortest path. Are they K or more simple path between two vertices s and t of V, each having a total length L or less?

Sets and partitions

★ Set packing. A collection C of finite sets. Does C contains at least K mutually disjoint sets?

Set splitting. Let C be a collection of subsets of a finite set S. Is there a partition of S in two subset S' and S", such that no subset in C is entirely contained in either S' or S"?

★ Minimum cover. Let C be a collection of subsets of a finite set S. Does C contains a subset C', with Card(C') $\leq$ K, such that every element of S belongs at least to one member of C'?

★ Partition. Finite set A of elements a, weighted by s(a). Is there a subset A' of A such that $\Sigma_{a \in A'} s(a) = \Sigma_{a \in A-A'} s(a)$?

★ Subset sum. Is there a subset A' of A such that the sum of the weighted elements of A' is exactly M?

Numerical matching with target sums.

Minimum sum of squares. Can A be partitioned into K disjoint subsets, such that $\Sigma(\Sigma s(a))^2 \leq M$.

★ Kth largest subset. Are they K or more distinct subsets A' of A, such that the sum of the weights of the elements of A' does not exceed M?

Many P.R. problems imply at a stage or another a problem on a weighted set or graph. We have just seen that many NP-complete problems may be relevant to P.R. We will see some more.

Compression and representation.

Let A be a finite alphabet, R be a finite set of strings from A^* and K be a positive integer.

Shortest common supersequence. Is there a string w of length less than K, such that each string x of R is a subsequence of w? : $w = w_0\ x_1\ w_1\ x_2\ w_2\ \ldots . x_k\ w_k$.

★ Shortest common superstring. Is there a w of length less than K, such that each string x of R is a substring of w?: w=w'xw".

Longest common subsequence.

Bounded Post correspondance problem.

Hitting string. Let A be a set of binary strings a of length n. Is there a binary string x of length n such that for each string a of A, there is an integer i, positive and smaller or equal to n, for which the symbol of rank i of a and of rank i of x are identical?

★ Consecutive one matrix. A matrix m x n of bits A, integer K . Is there an m x K submatrix A' of A, such that the columns of A' can be permuted so that in each row all the 1's occur consecutively?

★ Consecutive ones matrix partition. A is an m x n matrix of 0's and 1's. Can the rows of A be partitioned such that the resulting m_1 x n, m_2 x n matrices have each the consecutive 1's property?

Consecutive ones matrix augmentation. A is an m x n matrix, integer K. Is there a matrix A', obtained from A by changing K or less 0's into 1's, such that A' has the consecutive 1's property ?

★ Consecutive block minimization. A is an m x n matrix of 0's and 1's, K a positive integer. Is there a permutation of columns of A , such that the result is a matrix A' having at most K blocks of consecutive 1's?

★ String to string correction . Let A be finite alphabet, x,y, two strings formed on A. Is there a way to derive y from x by a sequence of K or fewer operations of single symbol deletion, or adjacent symbol interchange?

Grouping by swapping. A string x formed on a finite A. Is there a sequence of K or fewer adjacent symbol interchanges, which convert x into a string y, in which all the occurences of each symbol a of A are in a single block? (no configuration aba).

★ Rectilinear picture compression. An m x n matrix A of 0's and 1's. Is there a collection of K or fewer rectangles that cover precisely the 1's in the picture of the array of 0's and 1's?

★ Clustering. Let χ be a finite set, a distance d(X,X'). Is there a partition of χ in K classes or less, such that d(X,X') is smaller than D in each class?

★ Decision tree. Let $\mathcal{X}$ be a finite set of objects X, and $T = (T_1, T_2, \ldots T_m)$ be a set of binary tests. Is there a decision tree for $\mathcal{X}$, using the tests T_i, such that the length of the path form the root to the leaf is smaller or equal to K? (length of the questionnaire)

Syntactic problems

Finite state automata inequivalence. Two finite state automata on the same input alphabet. Do they recognize (or generate) different languages?

Linear bounded automaton acceptance. Does A accept x?

★ Non erasing stack automaton acceptance. " " "

Finite state automata intersection. A set of deterministic finite automata on the same input alphabet. Is there a string accepted by each automaton?

★ Regular grammar inequivalence.

Non LR(K) context-free grammar. Is the context-free grammar G *not* an LR(K) grammar?

★ Context-free programmed language membership. Is the string x generated by the context-free programmed grammar G?

★ Context-sensitive language membership.

This terminates our list of NP-complete problems, which in reality are in much larger number, even today. We made an attempt to get interested in the subject the P.R. specialists, who should refer to GAREY & JOHNSON (4), for more details.

Remark on complementary problems.

Let us call complementary problems, the two problems corresponding to the two questions: Given I, is X true for I ? is X false for I ?

If one can be solve in polynomial time on a deterministic machine, it is obvious that it is the same for the other.

But it is not at all the same for all problems solved in polynomial time on a non-deterministic machine. Let us consider the "Traveling Salesman" problem: Given a set of cities and their interdistances, does there exist a tour of all cities which is bounded by L? This problem has been proved NP-complete. But the complementary problem: does there exist no tour of all cities bounded by L? is not solvable in polynomial time on a NDTM. In fact to obtain the answer, we have to check all the possible tours.

If a problem is NP-complete or NP-hard, it is easy to show,(4) p. 114, that the complementary problem must be NP-hard.

Remark on optimization problems

In the phrasing of the NP-complete problems, the reader has probably noticed that the search for the "best" solution is carefully avoided. For example the traveling salesman problem answers to the existence of a tour bounded by L; and this is an NP-complete problem. It is shown that the optimization problem, "find the best tour" is in fact NP-hard.

POLYNOMIAL PROBLEMS

The main objective of a human problem solver is to find first the "best" algorithm for his problem, and obviously to escape the intractability of exponential algorithms (though sometimes, exponential algorithms are usable for small n...). They look for a polynomial algorithm, if they can find it!

An interesting situation is that many different algorithms may solve a given problem. For instance: "Is the first player the winner in a game of NIM?" is apparently an NP-problem. But a O(n) algorithm has been found which insures the win of the first player!

However polynomial algorithms are in small number, and may exist under different formulations:

An explicit formula. For instance the digital filters, currently used in the first step of P.R. algorithms.

A recursive formulation.

An iterative formulation.

If the recursive formulation produces simple programs, it may require vast amounts of memory at the execution. It is thus preferable to find an iterative form. The transformation from recursion to iteration has been a topic very much studied in the recent years.

We will now review some polynomial algorithms, useful in P.R. problems.

Graph problems.

Let G be a graph, with m arcs and n vertices.

Euler problem: find a path passing once and only once through all the arcs of G. Algorithm in O(m).

Minimum spanning tree. Alg. in O(m.Log m).

Shortest path between two vertices. Alg. in O(m.n).

Dynamic programming. (best path).

Connected subgraphs. Alg. in O(m).

Transitive closure. Alg. in $O(n^3)$.

Maximum of flux. FULKERSON algorithm in $O(n^3)$.

Maximum coupling. Find the greatest subset of arcs, such that they do not meet in any vertex. Alg. in $O(n^{5/3})$.

Planarity of a graph. In 1930, alg. in $O(n^6)$, before 70, $O(n^2)$, in 70 O(n Log n), in 74 HOPCROFT & TARJAN found one in O(n)!

Sorting and searching.

Sorting a list of values. The elementary algorithms are in $O(n^2)$. A lot of more refined algorithms work in $O(n \text{ Log } n)$; see for instance KNUTH (7).

Searching in a memory. If the memory is properly organized, in $O(\text{Log } n)$, even in $O(1)$: the Hash-coding.

Find a word of length a in a text of length n (all the occurences). Alg. in $a.O(n)$.

Signal and image transforms

S. WINOGRAD (8) declares that algebraic complexity of computation is concerned with determining the smallest number of arithmetic operations which are needed to perform a computation. As the addition (ADD) of two numbers of n digits requires $O(n)$ operations and the multiplication (MULT) $O(n^2)$, many efforts are made to minimize the number of MULT.

Most of the signal and image processing transforms can be defined as t quantities ψ_k :

$$\psi_k = \sum_{i=1}^{i=n} f_{k,i}\, y_i, \qquad \text{with} \quad 1 \leq k \leq t, \tag{5}$$

Also written under matrix form as:

$$\Psi = A(F)\, y^T, \tag{5'}$$

A is a matrix, y a line n-vector. Let $\rho_r(\, A(F)\,)$ be the row-rank of A(F); $\rho_c(\, A(F)\,)$ being the column rank. WINOGRAD shows the following results on $\mu(\psi_1, \ldots.\ \psi_t)$ the minimum number of MULT.

i) If $\psi_1, \ldots\ \psi_t$ are given by (5'), then:

$$\mu(\psi_i, \ldots \psi_t) \geq \min\,(\rho_r, \rho_c) \tag{6}$$

ii) If $\rho_r = t$, then $\mu(\psi_1, \ldots \psi_t) = t$, iff there exists a non singular matrix C, such that $\rho_r(c\ A(F)) = 1$ for every row c of C.

iii) Moreover if $\psi_0, \psi_1, \ldots. \psi_{m+n}$, are given by :

$$\sum_{k=0}^{m+n} \psi_k u^k = \Big(\sum_{i=0}^{m} x_i u^i\Big)\Big(\sum_{j=0}^{n} y_j u^j\Big) \tag{7}$$

Then $$\mu(\psi_0, \ldots \psi_{m+n}) = m + n + 1, \tag{8}$$

This result is applicable to the Fourier Transform, in particular.

Results are also obtained on *bilinear forms* $\sum_k \sum_j g_{ijk} x_j y_k = \psi_i$.

<u>Digital filters</u> or finite impulse response filters. They give m answers Q_i:

$$Q_i = \sum_{j=0}^{n-1} x_{i+j}\, y_j \qquad (9)$$

Let F(m,n) be the sequence of $Q_0, Q_1, \ldots Q_{m-1}$ of m consecutive answers. WINOGRAD states the results:

i) every algorithm for computing F(m,n) uses at least m + n - 1 MULT.

ii) m + n - 1 is the minimum number of MULT. needed to compute F(m,n).

<u>Discrete Fourier Transform</u>. Let the field of constant in (7) be the field of rational numbers, then the minimum number of multiplications needed to compute the cyclic convolution of two n-dimensional vectors is 2n - d(n), d(n) being the number of divisors of n. If p is an odd prime number, then:

$$\mu(\mathrm{DTF}(p)) = 2p - d(p-1) - 3. \qquad (10)$$

$$\mu(\mathrm{DTF}(p^n)) = O(2p^n). \qquad (11)$$

and $\mu(\mathrm{DTF}(2^n)) = O(2^{n+1})$.

<u>Fast Transforms</u>. Leaving the problem of MULT versus ADD, most of the usual transforms FOURIER, KARHUNEN-LOEVE, HADAMARD, Cosine, require n^2 operations. But special organisations of the computation (iterative), allow to obtain the results in O(n Logn). FFT has been the first proposed by COOLEY and TUKEY in 1965. By now all the preceding transforms have "fast" transforms, i.e. in nLn. The fast HADAMARD transform is specially interesting, as it requires only ADD. (as the ordinary one...).

A case study: the successive algorithms to find the convex hull.

The problem of finding the convex hull in d-dimensional spaces finds its application in many domains, classification, clustering, and of course image processing. TOUSSAINT (9) has specially studied this problem, and found some interesting algorithms.

First let us give some formulas useful later. It has been shown by RENYI that the number of point in a plane convex hull (CH) is in O(Log n) if they are uniformly distributed, in $O((\mathrm{Log}\ n)^{1/2})$ if the distribution is gaussian. In d dimension, BENTLEY has generalized the formula for the uniform distribution and has shown that it is in $O((\mathrm{Log}\ n)^{d-1})$.

The points are defined by their coordinates stored in the memory. Finding the point of smallest i-th coordinate is in O(n).

CH0, (a bad algorithm...) It is possible to propose an intractable algorithm. Let A be the set of n points:
- choose a subset A' of A,
- build up all the polygons on A',
- if one is convex and contains all the points of A', it is the CH looked for! Evidently exponential.

CH1, (better), in $O(n^3)$
- consider all the couples of points in the plane,
- if all the n - 2 other points are on the same side of the line x,x', then these points are in the CH.

CH2.
- find the point in the plane of smallest ordinate, let it be x_0
- from x_0 rotate a half line, initially horizontal, the first point met is called x_1; the points x_0 and x_1 are in the CH.
- from x_1 rotate in the same sense than before a line starting from x_0, x_1. The next point met by the line x_2 is part of the CH. Etc.. If h is the number of points of the CH, the algorithm is in O(nh). Thus:
- the average is in O(nLn), but
- the worst case is in $O(n^2)$.

CH3. GRAHAM.
- choose a point a of A inside A, O(1);
- sort the points according to their distances to a, O(nLn).
- use the greatest distances to find the points in the CH.

CH3 is in O(nLn) in average, and worst case is in O(nLn).

CH4, TOUSSAINT.
- select the upper point x_0, the most right x_1, the lowest x_2, the most left x_3. They are part of the CH.
- eliminate all the points which are not outside the quadrangle x_0, x_1, x_2, x_3.
- start again by finding the points which are outmost of the lines of the quadrangle. Four points are obtained. They are part of the CH. Eliminate the points which are in the triangles so formed.
- start again with the lines of the polygon previously formed, etc.

This algorithm has a complexity:
- in average of O(n),
- in the worst case of O(nLn).

This case study shows us:
-that many different algorithms may exist for a given problem, and that there is no general recipe to find the best. However:
- the "divide and conquer" method is advantageous; getting rid of the points which could not belong to the CH is the cause of the success of CH4. Thus we have reduced the dimensionality of the problem, drastically after the first step (except in the worst case).

This terminates, but does not exhaust, the review of the polynomial algorithms.

HOW TO RELAX INTRACTABILITY

Intractability, i.e. (hyper) exponential time computation or memory allocation, has to be avoided if we want to obtain any practical results. It is the main reason for the existence of P.R. algorithms, exact template matching being intractable for image or signal recognition.

The decrease of the order α of the polynomial in $O(n^{\alpha})$ is also of a great interest in P.R., n being often very large, of the order of 10^4 or much more.

Assuming that the processes of the brain are algorithms, Nature provides us with astonishing performances: the recognition of speech in real time, of very complex scenes. Certainly the brain hardware is in very large quantity. It may be modeled by a graph of the order of 10^{11} vertices (the neurones), and 10^{14} branches (the dendrites and synapses). But it is far from enough to accomodate exponentially intractable combinatorial explorations, even if the time response of a neurone would be smaller that 10^{-3} second. The obvious parallelism of the brain is not enough either to explain its success in the domains of perception and memory retrieval, with intractable algorithms. It seems likely that the functionning of the brain escapes any combinatorial explosions. Probably this is why the human brain is able to perform so well in the fuzzy and the undefined... As RABIN (3) says: " The human problem solver is not infaillible. He argues, makes deductions, often solves problems, but not infrequently produces wrong solutions. *Could it be that our success is, in part, attributable to the laxity of our procedures?*...Trying to double check every step and make sure that we are right before we proceed to the next step, can impede us to the extent that we never finish".

Relax the conditions of the problem.

Separation between P and NP. Sometimes a reduction in the conditions of the problem is enough to pass from NP to P. As an example let us recall the SATisfiability problem, which is NP in general except if two logical variables only exist in each clause, problem which is P.

Reduce the dimensionality n, by looking at subgoals, subcases, subproblems. Looking for elementary features is a typical approach in P.R.; also for a reduced dimension of the initial data .

Find a suboptimal solution. As an example in P.R. , the Dynamic clustering algorithm, DIDAY & SIMON (10) finds suboptimal clusters. Many efforts have been made along this line for the "traveling salesman" or the "bin packing" problems GAREY & JOHSON (4) chapter 6. Even evaluation of the performance of the suboptimal solution is made versus the optimum.

Admit errors. This is done deliberately, when a polynomial filter is used instead of a template to find a feature in P.R.

Here the polynomial nature of the operator is decided at the start. Instead of finding the "best" algorithm, a tractable algorithm is chosen and adjusted so to give the "best" performances, hoping that the errors will also be tractable... Such an approach is quite frequent in P.R.

Define natural measures based on the semantics of the problem.

In the representation space. Similarity or distances are found between two objects, having the representations X and X'. A "neighbourhood search" is performed in $\mathcal{X}$, from a "good" solution. (for instance the Nearest Neighbour techniques).

In the goal space. Artificial Intelligence tries to approach the "oracle" or "divination" machine, which we have alluded to in the description of the NP processes, using the properties of the goal space. For example in the chess programs, the branches of the tree are reduced from 40 to 5 by considerations on the nature of the game of chess. More generally "winning strategies" in games are supposed to be some sort of oracle machine telling you what to do at each step. But it is well known that they may be defeated, sometimes by a beginner who ignores them: they are not 100% safe, i.e. produce errors.

Heuristic programs implement sets of decision rules, which allow to act in the absence of theoretical results. They are based often on the way human beings act in a complicated situation of problem solving (for example theorem proving). It is clear that traveling salesmen are able to organize their tours without the help of mathematicians, that a teacher knows better how to solve a problem that his students: he has a number of "rule of thumb", from which may be deduced operational heuristics.

Unfortunately, P.R. is a set of process which is quite subconscious. It seems impossible to use the way we operate to understand speech or images to find good heuristics. Moreover our ideas on the subject may be quite misleading...

CONCLUSION.

We hope that this fast overview of the late complexity results will be of some help to the P.R. specialists, making them more conscious of the real issues of the P.R. problems, which in fact are nothing but a fight against intractability, without being defeated by errors...

REFERENCES.

(1) SIMON J.C. *Recent progress to a formal approach of pattern recognition and scene analysis*. Pattern Recognition Jr. 7, pp. 117-124, 1975.

(2) MATCHEY M. & YOUNG P. *An introduction to the general theory of algorithms*, The computer science library, a book by Elsevier-North Holland, N.Y. 1978.
(3) RABIN M.O. *Theoretical impediments to artificial intelligence*. Information Processing 74 (IFIP-74) pp. 615-619, North Holland publ. Cy 1974.
(4) GAREY M.R. & JOHNSON D.S. *Computer and intractability*. A book by W.H. Freeman and Co. San Francisco, 1979.
(5) LAURIERE J.-L. *Notes de cours sur les problèmes NP*, Institut de Programmation, Univ. P. et M. Curie 1981.
(6) COOK S.A. *A hierarchy for non deterministic time complexity* J. Comput. System Sc. 7, pp.343-353, 1973.
(7) KNUTH D.E. *Sorting and searching*, volume 3 of "The art of computer programming", Addison-Wesley publ. Co, Philippines, 1973.
(8) WINOGRAD S. *Application of complexity of computation to signal processing*. in SIMON J.C. "Digital image processing" A NATO serie C book, REIDEL edit. Dordrecht, Holl. 1982.
(9) TOUSSAINT G.T. & AKL S.G. *Efficient convex hull algorithms for P.R. applications*. Proceedings of the 4th Int. Joint Conf. on P.R. KYOTO, Japan, Nov 1978.
(10) DIDAY E. SIMON J.C., *Clustering analysis*, in FU K.S. "Digital Pattern Recognition", a book by Springer Verlag, 1976.

SELECTING, DESIGNING, AND USING PRACTICAL SPEECH RECOGNIZERS

Wayne A. Lea

Speech Science Publications

ABSTRACT

An expanding public awareness of the value of voice command of machines has made it critical that the previous trial-and-error methods for buying, using, and designing speech recognizers be replaced by systematic procedures. A methodology has been developed with Honeywell for identifying frequently-occurring discrete tasks in high-workload situations for which manual machine control is difficult and the utility of voice input is high. The feasibilities of voice in such tasks have also been assessed, based on the confusabilities of the vocabularies, the noise conditions, and other factors. The detailed specifications of current devices that permit recognition of isolated words and restricted word sequences have been compiled, and criteria defined for selecting the best device for each purpose. Recently, initial side-by-side comparisons of recognizers have been made, and tests have been designed for systematic testing of devices under a variety of representative conditions. The design of better recognizers should begin with consideration of features that assure competitiveness, special facilities that aid the user, and algorithms that are not critically dependent upon word boundaries, equal attention on all acoustic data, or simplistic template matching. It is now possible to advance the state of the "art" of speech recognition to more of a systematic "science" for effective buying, trying, applying, and modifying of speech recognizers.

1. WHAT'S WRONG WITH SPEECH RECOGNITION TECHNOLOGY?

The trends in speech recognition are in many ways quite promising. A variety of commercial devices are available for recognizing isolated words, phrases, and some restricted forms of connected speech, with high reported accuracies and decreasing costs.

J.-P. Haton (ed.), Automatic Speech Analysis and Recognition, 331–368.

Major research projects have indicated that the available isolated word recognizers may at future times be replaceable by advanced generations of recognizers that handle complex phrases and sentences, so that this need not be a dead-ended technology. Human interaction studies (2) and modality studies with actual computer input devices (i.e., a recognizer of isolated spoken words, a keyboard, and a graphical pen (11)) have shown that voice is the fastest, most natural modality for complex tasks, and frees the hands and eyes for effective accomplishment of other tasks. Practical successes in a variety of real-world applications verify such theoretical and experimental arguments. As recent trends bring recognizer prices down to values competitive with keyboard entry, widespread attention has been focused on the expanding role of voice input. Virtually every manufacturer of computers and peripheral devices in the United States and Japan (and, presumably, in Europe as well) has a project dedicated to voice interactions with machines. Manufacturers of toys, appliances, calculators, and other consumer products see large potential markets for voice-actuated devices. Some of the largest corporations in the world (e.g., Exxon, General Motors, Honeywell, IBM, ITT, Marconi, Matsushita, Nippon, Sharp, Toshiba) have joined the long-established small companies in the market place, seeking leadership opportunities in speech recognizer design, production, and application.

Thus, there is an upward swing towards improved capabilities, an expanding market, and widespread acceptance of the technology. Yet, there are serious gaps in this technology, some of which are not being given adequate attention. If the growing success of this technology is going to be more than a passing fad, critical research needs to be done, major changes in technical philosophy seem to be needed, and systematic studies of user needs and practical applications must be undertaken. Looking ahead, critical issues arise: How long can the excitement and growth continue, and what will sustain that advancing thrust? How will this field affect, and be affected by, other technological trends? What role will further research play in this awakening field? What theoretical or practical factors will limit future progress? Who will (or should) be paying the bills for future advances? In what directions do we turn for further breakthroughs?

I deal with these and other critical issues in workshops I regularly conduct for researchers, designers, planners, and users (8), and here I want to focus on methodologies that should assure future advances in recognizer selection, design, and use in practical situations. Using a framework that summarizes a variety of needed projects in speech recognition (Section 1.1), I will focus on critical needs of actual users of recognizers (Section 1.2), their need for thorough evaluations of recognizers (Section 1.3), and arguments for why there is a need for major changes in recognition techniques (Section 1.4). We will consider how to systematically select tasks that warrant voice input (Section 2). It is surprising how much work has been done regarding design and sales of recognizers without ample consideration of these critical questions of WHY voice input is needed and exactly which tasks really warrant voice input. Once we know what we want recognizers for, then we can reasonably consider how to select a device which would be best suited to the intended tasks. We shall briefly consider the spectrum of available devices and the primary criteria for selecting a device (Section 3). One gap in current information about available devices concerns their relative accuracies in various task conditions, but progress is being made on thorough procedures for side-by-side comparisons of recognizers, using common databases and standard tests (Section 4). From the work summarized in Sections 2 to 4, then, we have procedures for effective selection of tasks, devices, and testing procedures, so that we can clearly know what can currently be done. Another critical need concerns how to design new recognizer algorithms that will permit both dependable (low-risk) improvements in capabilities and longer-range advances based on research (high-risk) ideas. We shall see that advanced designs should focus especially on phonological, prosodic, and linguistic models (Section 5).

As will be summarized in Section 6, this work on selecting and assessing tasks, devices, performances, and new ideas for advanced algorithms should contribute much towards turning the "art" of recognizer design and use into a "science" of stable growth and increasing value. Further justifications of arguments here are presented in the references (Section 7).

1.1 A Framework for Describing Current Issues

There are so many unresolved issues, unsolved problems, and limitations in current technology that an unstructured listing becomes overwhelming and minimally informative. To structure the assessment of current technology and focus on key issues, the framework of Figure 1 has been found very useful. Projects that deserve high priority are shown by capital lettering and solid circles, while less crucial topics are shown with small lettering and open circles in Figure 1. This paper is primarily concerned with the TASKS THAT REQUIRE VOICE I/O, the RECOGNITION TECHNIQUES (and associated software, electronics, and research needs), and PERFORMANCE EVALUATION. However, some of the needed work on the related issues of HUMAN FACTORS, LANGUAGES, and CHANNEL AND ENVIRONMENTAL CONDITIONS deserve a brief mention here.

A critical human factors issues that will relate to the adequacy of future devices concerns whether continuous speech is really needed or useful, and exactly what it contributes above that provided by isolated word recognition. Regarding the value of continuous speech, we are about at the level of knowledge we

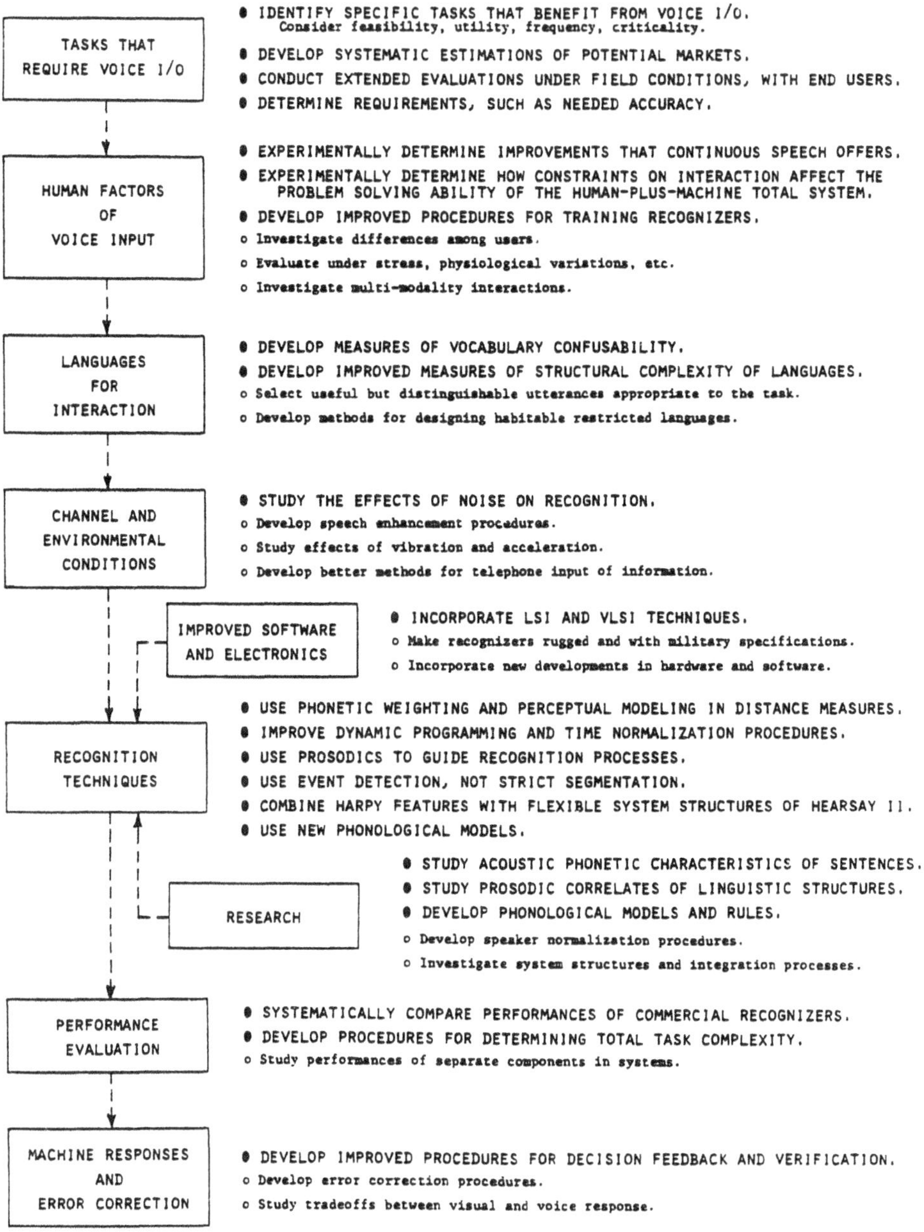

Figure 1. Extensive work is needed in all stages of the communication of spoken messages to machines.

were with the general issue of the value of voice I/O about ten years ago. We need evidence that our intuitive expectations are to be borne out by practical benefits in real applications. A related human factors issue concerns how much the performance and utility of a voice input channel is affected by various ways of constraining interactions with machines. How is total problem solving performance affected by changes in vocabulary size, branch ing factors, stilted versus natural syntax, etc. (5,8).

In the LANGUAGE area for interaction, research is needed on how to measure and control the confusabilities of vocabularies and how to properly select vocabularies that match recognizer capabilities. Vocabulary selections are a critical aspect of task and device selections, and thus will be discussed further in Sections 2 and 3. Further work is also needed on adequate measures of the structural complexity of each language for interaction. Branching factors, entropy, and perplexity are only a beginning in these areas (5). We need to understand how to maximally constrain a recognizer's task while minimally reducing the effectiveness of interactions. Later work must consider what makes a restricted language habitable, or easy to correctly use.

CHANNEL AND ENVIRONMENTAL factors like vibration, stress, microphone changes and movements, telephone effects, and noise will all affect the selection of tasks, and the design of enhanceable algorithms, as will be discussed in Sections 2 and 5. Other needed studies listed in Figure 1 are discussed elsewhere (5,6).

With all this important work that cries out to be done, it is difficult to understand why so many are entering this field with an inherent intention to reinvent the wheel, using previous technology in their own special way, to carve out for themselves a piece of the market, without any substantial new contribution. One goal of this chapter is to suggest some ways to prevent such waste, and to help maximize the directed contributions from future work.

1.2 Critical Needs of Actual Users

Figure 1 illustrated needs from the perspective of the total technology. However, the end user of speech recognizers is not necessarily interested in supporting and advancing the fascinating new field of speech input to machines. He or she wants an effective alternative to manual entry of data or commands, and particularly needs to know the value and feasibility of voice for each specific task in the total discourse of possible interactions in the intended application. An objective methodology for selectively defining when speech is useful for each task is needed. Guidelines are needed for how to interface voice with the rest of a total system, so that the user knows when to use voice input, when voice output will help, how the recognizer decisions can actually control machine actions, and how to redesign discourses of human-machine interactions to exploit characteristic advantages of voice input.

Once a user establishes that speech recognition is needed, he or she then needs procedures for informed selection of the best recognizer for the intended applications. A full compilation of the specifications of each device is needed, along with guidelines regarding the primary selection criteria for choosing the best device. In addition to such assessment of current devices, users may need to know what forms of advanced devices might be expected to appear within a few years. In this competitive field, users can (and should) demand that manufacturers provide thorough performance characteristics for each device, and some proof of claimed capabilities, based on standard tests. We shall next elaborate on this need for testing recognizers.

1.3 The Need for Thorough Evaluations

Little has been learned about why recognizers sometimes succeed and sometimes fail. Currently available information does not provide sufficient, unbiased performance evaluation results, so that a user of recognizers does not yet have all the necessary information to make a fully-informed selection from the many devices. What is more, the user cannot fully determine from available data under what conditions a recognizer will work best, or what causes its errors. Independent testing under a variety of controlled conditions is needed.

Also, very little has been written about how to compare the performances of recognizers, except perhaps that they ought to be tested on a common task with identical conditions. No such direct comparison tests have been published. Also, even under identical task conditions, questions remain about the relative significance to be attached to various limitations of each of the available devices. What if device A is better than device B for noisy speech conditions, but, in contrast, B handles speaker differences better than A? There are many factors that affect the accuracy of recognition for any recognizer. Such factors, which may be called "dimensions of difficulty" of the recognition task (4, 9), include: the form of speech (isolated words, connected sequences of digits, formatted word sequences, sentences, etc.); the population of speakers and their characteristics (sex, dialect and experience); rate of speaking; training procedures; vocabularies and structures of interaction; channel bandwidth, noise level, type of noise, and spectral distortions; environmental conditions like amplitudes, concurrent manual tasks, stress, fatigue, vibration, acceleration, etc.; required recognition accuracy; etc. We currently do not have any quantitative method for determining the overall difficulty of each multi-dimensional task. Nor do we currently know how to compare one recognizer's performance of say 60% recognition on a difficult task with another recognizer's (or the same recognizer's) 90% accuracy on a much easier problem. Until we have some sort of ideal measure of task difficulty, such that the difficulty metric directly predicts recognition error rate under various conditions, controlled tests need to be conducted under identical conditions for all recognizers.

Evaluation of recognizers thus involves much more than simply accepting published or claimed recognition accuracies. Accuracies are generally promised to be "over 99%", but field experiences reported to me by various users vary from over 99% to as low as 80% or even lower, for simple vocabularies like the digits. No public reports had reported independent verifications or testings of such claimed accuracies until recently (cf. limited results reported in references 2,13,15). Performances in practical field conditions have often been substantially poorer than in ideal laboratory conditions, with error rates sometimes increasing by an order of magnitude (e.g., from 0.3% to 3.0%). When someone has wanted to assess a recognizer, the standard procedure has been to buy it and try it out for the specific application. Such a "buy and try" philosophy seems to have been responsible for the purchases of a substantial number (indeed, a large majority) of all speech recognizers that have ever been sold. In the absence of "standard tests" for assessing devices, users have had to resort to incomplete, highly specialized, and uncontrolled testing that is often duplicative of other previous tests and not fully informative about performances in various task situations.

As long as the selection, purchase, and use of speech recognizers is dominated by a "buy and try" philosophy, and testing is confined to specialized pilot tests of small size and little generality, there will be uncertainty, waste, poor choices, unsystematic results, and a "field mysticism" that separates field results from laboratory tests. The reliability and user confidence that are commonplace with consumer products, computer peripherals, and laboratory or factory equipment will not be associated with speech recognizers until standard tests have been defined, conducted, and widely publicized and discussed. The ideas in Section 4 are intended to help bring about such needed progress.

1.4 Major Changes Needed in Recognition Techniques

Another critical need is to develop more advanced recognizers that can deal with the growing demands for versatile voice interactions with machines. Frankly, most commercial speech recognizers are based on ten-or-twenty-year-old technology that was known at the time of its initial development to be inadequate and suitable only as a temporary expediency. What is more, the theory behind such techniques is, in this author's opinion, seriously inadequate and destined to limit progress in accurate recognition of complex spoken commands.

In my WORKSHOPS ON SELECTING, DESIGNING, AND USING SPEECH RECOGNIZERS (8), I have repeatedly stressed a variety of serious inadequacies in current recognition techniques, which are illustrated in Figure 2. Recognizers usually first detect word boundaries from whether energy is above or below a threshold amount. Then they attempt to identify the acoustic pattern between those boundaries, by associating the pattern with the closest matching template for expected word pronunciations. They thus begin analysis using the most poorly articulated and variable portions of the word or utterance. The assumptions are usually made that each abstract unit (phoneme, syllable, or word) is associated with a well-defined stretch of the speech wave (linearity) and a unique defining set of acoustic features is to be evident for each such segment (invariance). Words like rider (/rɑɪdɝ/) versus writer (rɑɪtɝ/), where the phonemic distinction is in the fourth segment, but the phonetic difference is only reliably found in the length of the second segment [ɑ·], yielding [rɑ·ɪɾɚ] versus [rɑɪɾɚ], illustrate the common violation of linearity and invariance assumptions. Such violations occur frequently enough in speech (even in isolated words) to set an upper bound on the accuracy of any recognition based on such linearity and invariance assumptions. Recognizers also frequently depend too much on the location of strict boundaries between segments, in accordance with the linearity and invariance assumptions.

Recognizers are also quite susceptible to amplitude variations and other acoustic variations such as noise, speaker differences, and rate of speaking. Amplitude and time normalization procedures are usually based on strict acoustic criteria that hardly take into account the fact that the signal being processed is speech. Indeed, most speech recognition techniques give no such acknowledgement of the distinctive nature of speech, and could be as readily applied to any signal matching. Phonotactics and phonological constraints are rarely used, and prosodic information is noticeably absent from most recognition processes. Most speech recognizers are based on an ignorance model of the speech that gives equal attention to all the acoustic details, considering essentially no knowledge that would suggest focusing on specific message-distinguishing aspects of the signal. The speech science evidence for "robust" features and "islands of reliability", such as vowel and sibilant regions, are not considered in simple acoustic template matching. Without linguistic or speech science knowledge, the reasons for recognition errors are often not evident to those who view speech recognition as straightforward signal processing, and algorithmic improvements do not then directly fall out from error analysis, as they often

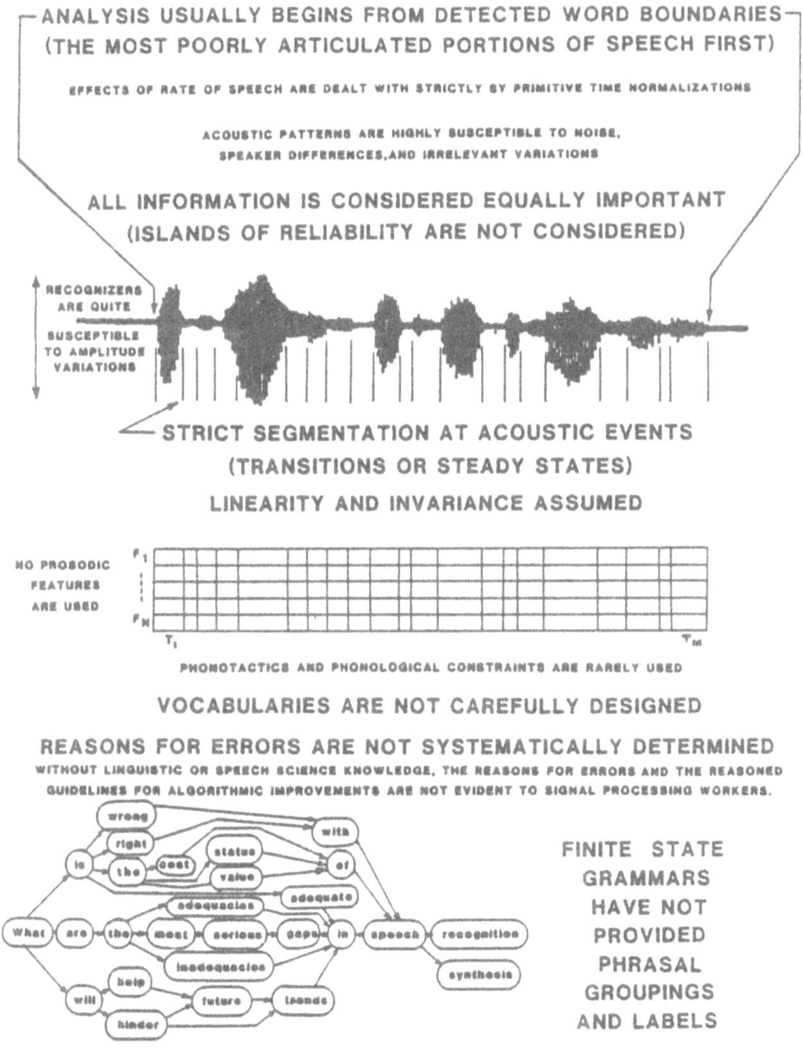

Figure 2. A summary of what is wrong with current speech recognition technology, with the size of lettering indicative of the significance of the associated limitation of current methods.

do for competent speech scientists.

Prudence suggests that recognition vocabularies be designed to avoid readily confusable words, but little is done to design vocabularies for specific tasks. There currently is no systematic procedure for assessing vocabulary confusabilities and replacing confusable words with easily distinguished words. Linguistic models still play only a primitive filtering role in recognizers, whereby constraints on allowable word sequences are used to reduce combinatorial explosions of possible word sequences that could be hypothesized by a recognizer. The ability to use phrasal groupings of words to identify subsequences in sentences or discourses, and to label those phrases with syntactic categories, have not been fully used in recognition schemes. One reason is that finite state grammars, which have dominated the syntactic analysis in recognizers, do not automatically provide a characterization of such phrasal units. It is possible to build "layered finite state grammars" which, like the more advanced augmented transition network grammars, can characterize phrases as well as allowed word sequences. It is also possible to use prosodic cues to the phrasal bracketing of sentences to reinforce some word sequence hypotheses and to guide both word matching and parsing. Little is being done to incorporate prosodic features in commercial recognizers.

In short, speech recognition as it is currently reaching the user is usually antiquated, based on inadequate and erroneous theory and openly acknowledged technical expediencies, restrictive in its focus on general signal processing and neglect of speech science, ignorant in its equal attention to all acoustic details, susceptible to irrelevant acoustic changes like noise and speaker differences, and incomplete in its disregard for prosodics, phonological structures, and linguistic units. These are rather severe indictments, but we will suggest in Section 5 how such limitations can be overcome (cf. 8).

One reason for the persistence of these limitations in recognition techniques has been the severe restrictions placed on tasks addressed by most previous speech recognition efforts with commercial goals. As long as the tasks addressed are close to trivial, adequate progress can appear to be achieved. However, as future work demands the use of voice in more difficult task environments, inadequacies of current techniques will be revealed. Thus, as we begin in the next section to inquire about complex tasks that benefit from voice input, we open the door to evidence of current limitations and crucial needs for future work.

2. SELECTING TASKS THAT WARRANT VOICE INPUT

Not all interactions with machines need or can benefit from voice input. Prospective users of speech recognizers need to

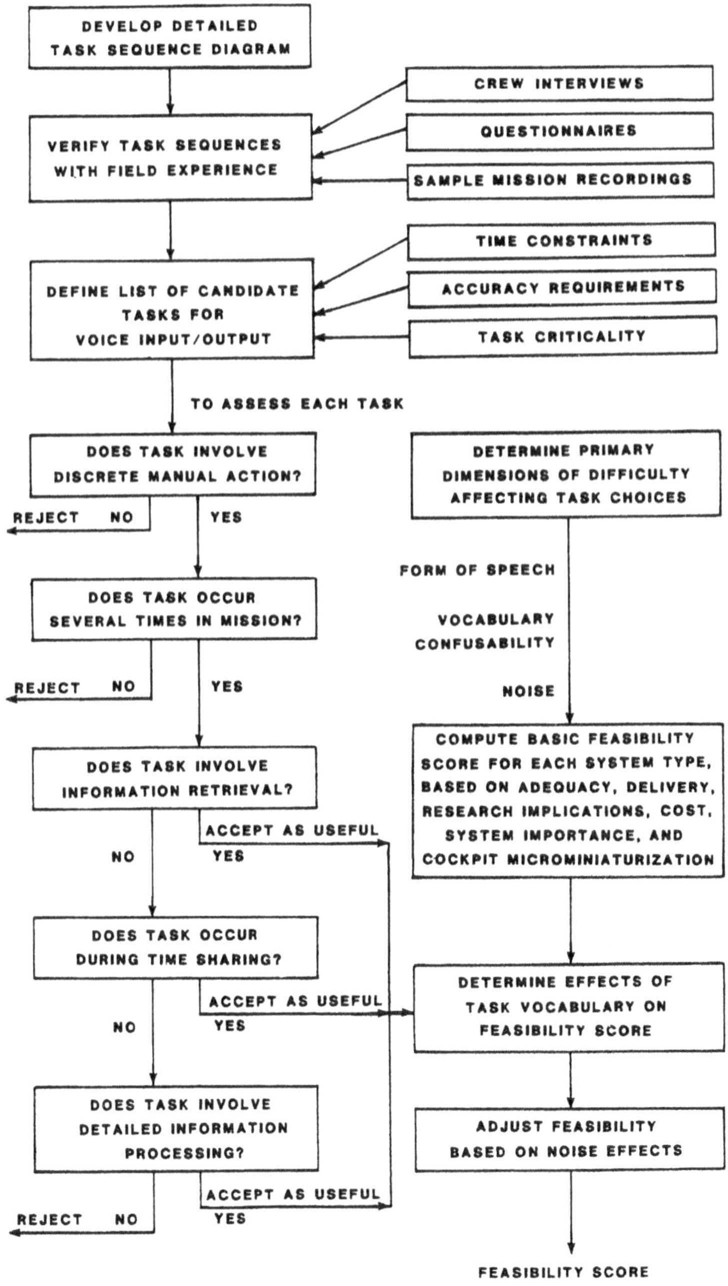

Figure 3. Procedures for Selecting Tasks for Which Voice I/O is Useful and Feasible.

address three questions:

(1) Exactly what tasks are expected that might involve voice or other interactions with a machine; and what must the worker be doing at each moment in such interactions?

(2) Which of the interactive tasks are ones for which voice input would seem to promise real benefits, such as freeing the user's hands and eyes, increasing the communication capabilities of the worker, or speeding data entry?

(3) Are those tasks for which voice is useful also ones for which adequate voice input capabilities are feasible?

These questions suggest a three-stage effort in assessing tasks for voice input: task analysis, to define what must be done by the total human-machine system; rating the utility of voice for each task; and rating the feasibility of voice input for each task. We next consider these three topics of what tasks must be performed, why use voice, and when adequate voice input can be achieved.

2.1 Task Analysis Procedures

Figure 3 illustrates the total process of analyzing an application to determine which tasks are most appropriate for voice input. First, one must know precisely what work is currently being done, or planned. It is surprising how few of the governmental and industrial jobs with computers have been analyzed sufficiently to fully detail the step-by-step processes done by each worker, such that planners can decide which tasks might be adapted to use of speech recognizers. As a result, most previous purchases of recognizers admittedly have not been fully justified by defined needs. Apparently, only the military has defined detailed mission scenarios, with well-defined duties for each worker or crew member, and step-by-step procedures described in training manuals. The most advanced American work in task analysis has thus been done for military airborne applications, under support of the Wright Patterson Air Force Base (WPAFB) and the Naval Air Development Center (NADC), and related agencies. The procedures shown in Figure 3 were developed for WPAFB by Robert North and the staff of the Man-Machine Sciences branch of the Honeywell Systems and Research Center, Minneapolis, Minnesota, with this author as a consultant.

A detailed analysis of the step-by-step actions of the worker can identify periods during which workload is high, involving a number of simultaneous tasks such as manual controls and keyboard entries at the same time as (or close in time to) when visual displays have to be monitored, or the worker's eyes are busy measuring dimensions, watching packages on conveyor systems, or navigating an aircraft. Such high-workload, hands-busy, eyes-busy situations are suitable for voice as a replacement modality or added mode of interactions. It is not merely the number of simultaneous tasks that affect workload; some tasks functionally interfere with each other and are anthropometrically or visually difficult to accomplish simultaneously. On the other hand, if the worker is already engaged in a voice task, voice entry to a computer may be impossible, difficult, or confusing, so conflicts with other voice tasks can discourage voice in some tasks.

Honeywell uses a valuable method called "operational sequence diagrams" to specify all the interactive steps being taken at each short interval of time, including the information flow to and from each worker, the modalities currently used for each interaction, and a measure of the amount of time sharing required by simultaneous tasks of various difficulties. High-workload periods are expected to be the highest-priority times for incorporation of voice input.

2.2 Assessing the Utility of Voice Input

After identifying periods of high event frequency and computing workload indices in each period, this workload description can be provided to experienced workers, via questionnaires and interviews, to subjectively verify the objectively derived task sequences and workload assessments. This total process thus provides the information for filtering of tasks, to isolate high-priority tasks for which voice is useful. As detailed on the left side of Figure 3, suitability criteria for selecting voice-interactive tasks include special attention on replacing discrete manual tasks with discrete voice inputs. Discrete data entry (for example, by isolated words separated by pauses) is not only more in line with technically feasible voice interactions with current or soon-forthcoming devices, but discreteness is intrinsic to the usual processes of goal-directed changes, measurements, and information retrieval involved in human-to-machine input tasks. Other criteria for selecting tasks that are suitable for voice input include the frequency of occurrence of the task within a mission, and how much complex information processing and information retrieval is needed in the task performance. Honeywell (13) found that pilots particularly identified information retrieval, such as looking up detailed instructions in a manual, as a process that could be automated and then benefit from voice interactions.

A variety of utility criteria, shown in Figure 3, can be considered in assessing the value of voice for each task. If time constraints are severe, so that normal times to speak each command and wait for a response make voice inappropriate, or if the accuracy requirement is extreme (at 99.9% correct, for example), then the task may not be suitable for voice. Related to accuracy, if the task is critical to a mission, such as dropping bombs or avoiding terrain in an aircraft, error-prone voice input will be inappropriate. Voice also may not be very useful if the work - load and attention requirements are minimal, no difficult tasks are involved, or if the worker always needs immediate feedback of machine decisions. If a task frequently will be interrupted by another task requiring the worker to respond vocally, it may be unsuitable for voice.

From such assessments, tasks can be identified for which voice input can be useful.

2.3 Assessing the Feasibility of Voice Input for Each Task

It is not enough to know that voice would be advantageous in any specific task; we must also know that the form of speech recognition capability needed is possible, and, indeed, feasible within a reasonably short time. We must address the issue of when we can implement each voice input task, given the present capabilities and the projected future capabilities. The assumption is that feasibility is determined by: (1) Relative difficulties of various tasks, based on all dimensions of recognition task difficulty; and (2) Current state of technology, coupled with projected advances to come from algorithm developments and research and development.

Speech recognition technology has been able to make what progress it has exhibited due to the careful constraining of the difficulty of the task, so that limited algorithms have some possibility of achieving accurate decisions about utterance identities. Thus, most successes have been with isolated word recognition, with a small vocabulary , and near-ideal environmental conditions. History, as well as common sense, thus verify that feasibility is associated with simplification of the task to be performed. Yet, we do not have any comprehensive theory of what makes a task simple or difficult, and we do have extensive evidence that many factors affect task difficulty for speech recognizers. We must identify those factors that affect the difficulty of recognition tasks, and offer some general ideas about how to scale the difficulties of various tasks. To convert difficulty into feasibility assessments, we must also describe the current and projected capabilities of recognizers, in terms of what task difficulties they are able to successfully handle with reasonable accuracy.

The overall difficulty of a speech recognition task, D, is a function of many parameters:

$$D = f(SP, VC, N, BW, AV, SR, FS, EC, AR)$$

where SP is the variation due to speaker population; VC is the confusability of the vocabulary; N is the noise level and effects of various types of noises; BW is the restriction on the frequency bandwidth of the signal; AV is the amplitude variability; SR is the speaking rate; FS is the form of speech (isolated, connected, or continuous); EC summarizes the effects of extreme conditions (for example, in airborne uses, they would be vibration, stress, fatigue, acceleration); and AR is the accuracy requirement of the intended task.

Of course, the exact nature of this functional relationship between total difficulty of recognition tasks and the individual dimensions has not been determined. Difficulty increases monotonically with each dimension, but it has not yet been established which dimensions have the most effect of difficulty. (For example, how much change in noise level is equivalent to a doubling of vocabulary confusability, in terms of effects on overall task difficulty?)

What can be done currently is to focus consideration on which dimensions are controllably variable in the practical situation. For example, in aircraft, the speaker population is given by the crew selections, and is not variable or subject to change from task to task, so that it may be neglected in comparatively assessing tasks. The bandwidth will presumably be constant, and not relevant to assessing tasks within a specific mission. Speech recognition will initially be avoided in extreme conditions, and accuracy requirements are thus kept fairly uniform by not considering critical tasks within lists of candidate tasks. Amplitude variability and rate of speaking will affect recognition decisions, but are largely speaker-dictated, arbitrary, and uncontrollable, and hence not relevant to task selections. The remaining factors of interest in task selections are then: the confusability of the vocabulary, the noise conditions, and the form of speech. These dimensions have been given specific attention in assessing feasibilities of tasks (7).

Within the next few years, the only forms of speech that commercial recognizers will be able to handle will be isolated words and connected word sequences (like digit strings and strictly formatted word sequences), and the possibility of an initial form of fairly unrestricted continuous speech, such as the Harpy system handled (13). This author has developed a preliminary algebraic form of a feasibility metric, based on subjective scaling for six key issues first suggested by Honeywell (3). Feasibility was assessed for recognizers of isolated words, connected speech, structurally-restricted sentences, and unrestricted continuous speech. For each system type, a numeric rating was estimated for each of the following questions:

R1. Is the capability or device adequately developed?
R2. How much research is going on in this area?
R3. Is this an important capability to have?

R4. How long before it will be usable?
R5. Can you estimate end-user costs?
R6. What are the current miniaturization plans for this type of system, to allow cockpit implementation?

To arrive at one feasibility rating for each type of system, each scale was weighted by a factor assessing its relative significance to feasibilities for practical (airborne) applications. The formula used for combining these ratings into a single feasibility score was tentatively selected to be:

$$F = 3R_1 + 1R_2 + 1R_3 + 3R_4 + 2R_5 + 2R_6$$

The integers multiplying each R are intuitively-assigned weighting factors, based on the following reasoning. Feasibility is very closely tied to what has been "adequately developed", so R1 deserves prominence. The same is true of R4. User costs (R5) and microminiaturization for cockpits (R6) are of moderately high importance for the chosen airborne tasks, but not as crucial as adequacy and availability. The quality and amount of research (R2) is somewhat suggestive of what advances may be expected, but the two are not necessarily highly correlated, so that R2 is assigned a low weighting. Similarly, for now, it is difficult to assign an accurate judgment to the "importance" of each capability, or to argue to make such importance be a major factor in feasibility assessment, so R3 was assigned a low weighting.

A careful development of a statistical model would probably provide rigorously motivated assignments to R1 through R6, and to the proper weightings of each, thus improving the above formula. Perhaps later work will yield such refinements.

Once the relative feasibilities of different types of systems are assessed, we can assert that the feasibility of the type of system needed for each task is the basic feasibility of the task itself. (See Figure 3.) Since either isolated words or connected word sequences serve current tasks, feasibility of each option can be determined. Feasibility will be predominantly determined by the system rating, and the confusability of the vocabulary. Ultimately, it should be possible to assess the confusability of each vocabulary by determining the "distances" between pronunciations of each pair of words, and defining the confusability of the vocabulary as a function of such distances. Such a measure should give most attention to words that are closest in sound structure, so that the most confusable words add the most to the assessment of vocabulary confusability. Such a comprehensive assessment of vocabulary confusabilities is beyond current capabilities, so an initial scaling methodology has been based on estimating whether the vocabulary is "simple", "moderate", or "difficult" due to similar sound structures. This author thus subjectively assigned a number on a scale from one to seven for each of the six issues R1 to R6 for each system type and each of the three gross categories of vocabulary difficulty. Then, the formula was used to combine the weighted sum of ratings into a feasibility score for each system type and vocabulary type, yielding the basic scores shown in the NO NOISE column of Table I.

Table I also shows a general estimation of the effect of added noise on the feasibility of accurate recognition with various types of speech recognizers. Under GOOD environmental conditions, the signal-to-noise ratio might be +30dB or more. Such low-noise conditions assure that isolated word recognition with simple or moderately difficult vocabularies is highly feasible now; connected speech recognition is also highly feasible with simple vocabularies like the digits. Difficult vocabularies in isolated word recognition, and moderately difficult vocabularies in connected speech recognition, are feasible within the next few years (that is, of "medium feasibility" indicated by the lowered scores in for GOOD conditions and HARD vocabularies). Even simple vocabularies in task-restricted speech understanding and continuous speech recognition are of "medium" feasibility. When moderate amounts of noise are added (assuming primarily broadband noise like white noise), so that a signal-to-noise ratio of only about +10dB results, then the feasibilities of accurate recognition are reduced somewhat, as suggested by the estimated feasibility scores in the MODERATE noise column of Table I. Only simple isolated word recognition tasks are then of very high feasibility. This condition of moderate noise is expected to be representative of most phases of aircraft missions, or most assembly-line conditions. POOR signal-to-noise conditions may occur at some times in practical applications, and will probably reduce feasibilities to the point where only simple isolated word recognition tasks can be reliably accomplished. Under EXTREME environmental conditions, wherein the signal is buried in the noise and/or other environmental factors (stress, fatigue, vibrations, acceleration, etc.) markedly affect the stability of voice patterns, even the easiest speech recognition tasks will be of limited feasibility.

As shown in Figure 3, the final stage of task evaluation is to combine the analysis of utilities and feasibilities of voice into a final list of priorities for introducing voice into the most useful and feasible situations.

2.4 Inadequacies in Current Abilities to Assess Tasks

The methodology outlined for assessing tasks is admittedly in a preliminary stage of development. There are gaps in our ability to accurately assess feasibilities of various recognition capabilities, and gaps in the formal application of such assessments to the selection of recognition tasks. To date, we have:

TABLE I.
ESTIMATED FEASIBILITIES OF SPEECH RECOGNITION TASKS

FORM OF SPEECH	VOCABULARY	FEASIBILTIES FOR EACH ENVIRONMENT			
		GOOD	MODERATE	POOR	EXTREME
ISOLATED	EASY	72	60	48	31
ISOLATED	MEDIUM	60	47	33	19
ISOLATED	HARD	48	34	29	18
CONNECTED	EASY	50	38	28	22
CONNECTED	MEDIUM	32	28	23	18
CONNECTED	HARD	22	19	18	18
TASK	EASY	33	24	19	15
SENTENCE	MEDIUM	23	23	18	15
(SUS)	HARD	15	15	15	15
CONTINUOUS	EASY	23	20	15	12
CONTINUOUS	MEDIUM	19	18	15	12
CONTINUOUS	HARD	12	12	12	12

- Nothing like an adequate theory of the difficulties of various types of recognition tasks.
- Minimal knowledge of how to measure the confusabilities of various vocabularies.
- Only the crudest notion of how noise will affect task difficulty.
- Primitive abilities to estimate the time before usable versions of each type of speech recognizer will appear.
- Insufficient research being conducted, and no guarantee that much of that work will advance the feasibility of powerful speech recognizers.
- Only a vague notion of the ultimate importance of various types of recognizers.
- Decreasing costs for recognizers, but still relatively high projected costs for the more versatile recognizers of connected or continuous speech.
- Only the beginnings of a trend towards microminiaturization and ruggedization of recognizers, to match airborne needs or other challenging practical conditions.
- Little ability to systematically assess and compare the accuracies of commercial recognizers, and essentially no

tests in typical practical applications; also, we have little knowledge about needed accuracies for real tasks.

- No systematic methodology for integrating the effects of extreme conditions (vibration, acceleration, stress, preoccupation with other tasks, etc.).
- No representation in the current feasibility measures of the effects of talker variability, amplitude variations, changes in speech rate, and other dimensions of speech recognition task difficulty.
- No procedures for urging potential users to carefully analyze the detailed tasks they must undertake in the task situation, or for forcing the assessment of task utility and feasibilities for speech input.

As our abilities to assess utility and feasibility of voice input improve, there will be a corresponding increase in concern with accurately determining which devices will adequately meet actual neeeds. We next briefly consider how to select the best device for each practical task.

3. SELECTING THE BEST RECOGNIZER FOR YOUR APPLICATION

To select the best device for each task, one needs to know what is currently available, what devices are expected soon (i.e., in the time before a recognizer must be operational in the application), the detailed specifications of each device, and, perhaps most importantly, the appropriate selection criteria for rating the alternatives. Until recently, there was no public document which compiled together all the specifications of all the devices. However, now the book SELECTING, DESIGNING, AND USING SPEECH RECOGNIZERS provides listings of all important specifications of 1981 commercial speech recognizers, with guidelines regarding the primary selection criteria. No longer do potential users or purchasers need to devote extensive duplicative efforts to collecting brochures, technical articles, and application notes. Step-by-step procedures are provided for identifying relevant specifications and selecting the most suitable value for each such feature. Then, although no device will probably ever match every desired specification, the closest match can be determined, with weightings attached to the relative importances of the various features.

Here we will briefly outline the knowledge used and the method for aiding selections. In Section 3.1, the currently available devices are shown to distribute along a broad spectrum of costs, with highlights of devices very briefly outlined. The devices expected in the next few years are briefly speculated about in Section 3.2. Section 3.3 summarizes a few example specifications, and Section 3.4 suggests one priority list for guiding the selection criteria suitable for most applications.

3.1 An Expanding Spectrum of Recognition Devices

Figure 4 shows a primary feature forming a broad spectrum along which the commercial speech recognizers are distributed; namely the cost of the device. Speech recognizers range from $10 to $200 chips or chip sets, through $200 to $1000 hobbyist or consumer-oriented devices, through $1000 to $5000 single-board recognizers and "voice terminals", through $5000 to $50,000 fully-independent recognition systems, up to $30,000 to $100,000 special purpose systems. It is a challenge for the buyer to know what each device offers, and how much capability is really needed for each application. Almost all devices are designed for speaker dependent (i.e., trained) recognition of isolated words. Verbex has provided the only commercial speaker-independent systems, and they have also reported ability to accurately recognize connected word sequences (specifically, digit strings), in speaker-dependent mode. Nippon provided the first connected word sequence recognizer, which reportedly can accurately recognize sequences of up to five digits, or other restricted word sequences, without pauses between words. As used in such commercial terminology, connected word recognition is a limited capability short of the full ability to accept and recognize continuously spoken sentences. Trends are towards lower costs for higher performances, speaker independence, and connected speech recognition, but currently the more the user demands in terms of such flexibilities as high accuracy, speaker independence, or connected speech, the more he or she must pay for a suitable device.

A full description of the various devices is provided elsewhere (8). Here we shall only briefly highlight some of the contrasting features of the devices. At the low end of the spectrum, we find some of the most recent entries, such as the integrated circuit chips offered by Interstate Electronics and Weitek. Such chips are obviously suitable for use in low-cost consumer devices, like appliances and toys, and their claim of speaker independence may be presumed to be possible only at substantial sacrifice of recognition accuracy, or highly restricted vocabularies. Hobbyist devices, such as Heuristics introduced in the late 1970's, have opened up a wide market of hobbyists and consumer-oriented uses in personal computers, low-cost voice-actuated machines, etc. Some such devices, like the earliest Heuristics circuit boards and the Voicetek and Mimic boards, provide only the acoustical feature extraction capabilities, which provide patterns for an external computer to match with stored patterns, for recognition. Others include on the boards programmable read only memories which contain software for small-vocabulary word recognition. Accuracy may be restricted by the limited hardware and feature extractors, and may be subject to substantial variations due to environmental conditions. Field users have reported accuracies typically from 80 to 90%. While well over 10,000 such hobbyist devices have

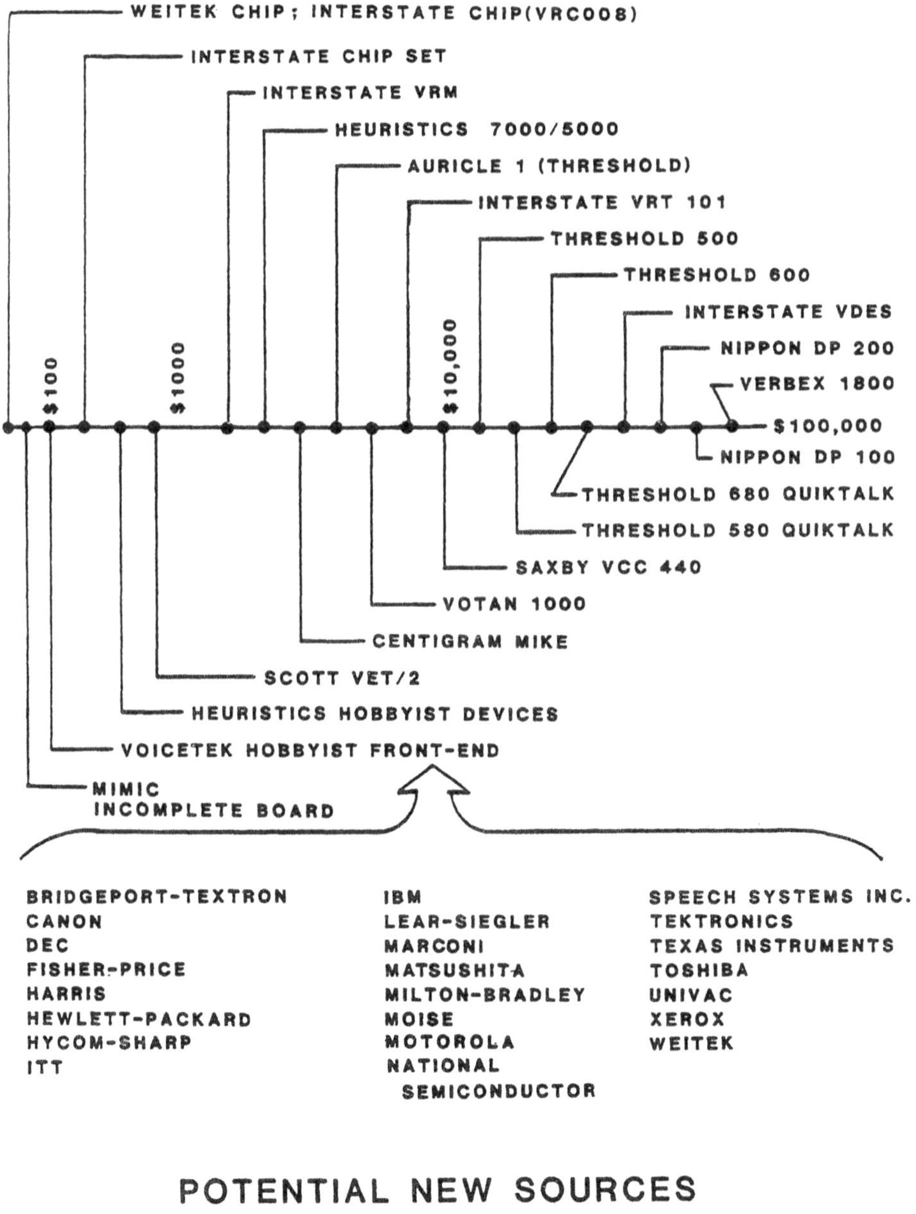

Figure 4. The rapidly-expanding spectrum of speech recognition products, and sources of new devices (cf. 8).

been sold, and this was the first aspect of the industry to ever make money, it is unknown how many of these devices are actually used for practical voice control of machines.

Low-cost recognizers from Scott Instruments permit 40-word voice control of microcomputers, and provide a graphic display of scores on all 40 words, so that a researcher can investigate what words are considered most similar, what score occurred on the correct word when an error is made, how to improve performance with supplementary algorithms, and how to improve the vocabulary by removal of confusable word pairs. Another interesting feature of these low-cost recognizers is that they compare the syllable count for each word with that for the input, to reduce the set of plausible matches and speed decisions. These features demonstrate that low-cost devices can include novel recognition procedures and good user facilities. Accuracies, however, have been under 90%, in previous tests (2).

The Interstate Electronics Voice Recognition Module (VRM) was the first of several high-performance single-board speech recognizers that now constitute the central interest of most new entries into this field. With the same algorithms as an earlier $20,000 device, the VRM reportedly attained the highest accuracy of several independently-tested boards (2). Other single-board recognizers include the Centigram MIKE, the Heuristics 5000 and 7000 devices, the Auricle I (from a division of Threshold Technology), and the Votan Models V1000, V2000, and V3000. Features to look for in these devices include a host-adjustable and/or hardware-adjustable reject threshold for rejecting pronunciations that appear ambiguous, a pre-amplifier with bypass (for testing with tape-recorder inputs), automatic rejection of bad training samples, hardware self-tests, head-mounted microphone, automatic gain controls, optional implementation in a self-powered subsystem form or incorporation within a computer terminal, reasonable (40-100 word) vocabularies, some noise and amplitude insensitivity, good interfacing capabilities for use with microcomputers or minicomputers or terminals, some indications of scores on alternative words (or identifications of "runners-up", etc.), and, of course, high accuracy in representative tests.

The earliest speech recognizers incorporated minicomputers within the device, at costs in the tens of thousands of dollars. From 1972 to 1978, Threshold Technology was the leading source of speech recognizers, and they probably still have more systems in daily use for practical voice entry of data and controls than any other manufacturer. Independent tests (2) show them to be quite accurate, and unmatched by lower-cost devices. The TTI 500 is a board that depends upon host computers to store and handle templates and recognition decisions, while the 600 is a CRT compatible voice data entry terminal incorporating a small computer. Models 580 and 680 permit very brief pauses (about 20ms) between words, and are called QUIKTALK, suggesting speed of data entry that approaches that of connected speech, but at some cost in naturalness. Real field experience is one of the highlights to look for in this field, for which Threshold is well known. The Threshold devices are also the only devices that explicitly use some phonetic feature detection procedures in their algorithms.

The first connected speech recognizer was the DP-100 offered by Nippon in 1978, at about $80,000. Prices have dropped considerably, and the new model DP-200 makes connected speech input posssible for around $28,000. The Nippon devices also were the first to introduce dynamic programming into commercial recognizers, at considerable gain in time normalization ability, and high reported accuracy. Verbex (formerly called Dialog Systems) also offers connected speech capabilities, and for years had offered the only speaker-independent isolated word recognizer. The Verbex recognizers have also been distinctive in their specific design for telephone use, although other recognizers have been tested over the telephone. also.

This brief overview of such a wide spectrum of devices is probably insufficient for guiding users, but a recent book presents over 100 specifications of almost all these devices, and offers guidelines for how to select among them (8).

3.2 Forthcoming Devices

The general trends towards lower costs and advanced capabilities like speaker independence or continuous speech are suggestive of a maturing and rapidly expanding technology. The user might expect many new devices to appear in the next few years, and any manufacturer must be ready for increased competition. In addition to the spectrum of commercial devices, militarized, ruggedized recognizers are being developed (e.g., at Lear-Siegler), and advanced algorithms include linear predictive analysis and dynamic programming. Noise handling and robustness in the presence of variations in amplitude, rate of speaking, speaker quality, etc., are all being addressed.

Other companies known to be developing speech recognition products, or rumored to be exploring such developments, include: Toshiba, Sharp, Matsushita, and Canon, in Japan; Bridgeport-Textron, Hewlett-Packard, IBM, ITT, Texas Instruments, and others in the United States; Marconi Avionics and Logica in England; and the developers of "MOISE" in France. All of those might be expected to announce products by 1983, with connected speech products from many by 1985. Many other groups are doing advanced development and research, that should also lead to other products within the decade before 1990. This author is convinced that some of the forthcoming devices will use techniques far advanced from the traditional filter banks, linear time normalization, strict acoustic template matching, and strict signal processing without speech knowledge that has dominated past work. The user may thus expect that whatever recognizers are purchased now may necessarily be replaced by more advanced, accurate devices within a few years.

3.3 Specifications of Available Devices

With such an impressive array of available (and forthcoming) devices, the user is challenged to select the best for his or her use. With the tables of over 100 specifications for devices, and the explicit listings of alternative values for each specification, as provided in my recent book (8), a user can make side-by-side comparisons, select desired features, and even define the characteristics for a new and better recognizer. It is beyond the scope of this chapter to tabulate all specifications and describe the procedures for making every decision in an informed manner. Users must consider physical characteristics such as whether the device is stand-alone or must be tied to a separate computer, how bulky and heavy it is, and how it interfaces to other devices. Task considerations must include specifying how much pause is needed (i.e., allowed rate of speaking), and what facilities are offered for applications programming. Cost is of course an important factor. Ability to handle speaker independence and connected speech, and details about training methods must be understood. Vocabulary sizes, word lengths, and syntax facilities must be characterized. The allowable microphone or input transducers must be identified, as must the noise levels and other channel conditions that can be accurately handled. Users must understand details about alternative recognition techniques, such as amplitude and time normalization, frequency filtering, feature extraction, and template matching procedures. The size and nature of reference patterns will affect accuracy. Since all manufacturers claim high accuracy, that crucial factor cannot be used for distinguishing devices until independent testing is done. Response processes and interfacing to other parts of a total system also need to be considered. While most users will want single-board recognizers that they can interface to some available computer, others may need the computation ability and the accuracy associated with complete systems or the distinctive abilities of special-purpose systems.

Table II lists recommended criteria for selecting a recognizer, organized into three groups of decreasing importance, with criteria within each group also ordered roughly in order of decreasing significance. Any chosen recognizer will probably have to agree with the user's requirements on all the criteria in the topmost group. Those features listed in the second group seem particularly important to assuring robustness and likelihood of accuracy under field conditions. Experience shows that the size

RECOMMENDED CRITERIA FOR SELECTING A SPEECH RECOGNIZER

- Form of Speech (IWR, CWR, SUS, CSR)
- Speaker Dependent or Independent
- Type of Device (Chip, Chip set,
  Incomplete Front-end Board,
  Single-Board Recognizer,
  Voice Terminal,
  Full Recognition System,
  Specialty Recognition System)
- Reported Accuracy
  (especially in independent tests)
- Vocabulary Size
- Cost

- Bits/word in Reference Pattern
- Features Extracted
- Minimum Total Interword Pause (IWR)
- Method for Adjusting Rejection Threshold
- Time Normalization Procedure
- Normal Bandwidth
- Possibility for Telephone Use
- Minimum Number of Training Passes

-Usual Training Sequence
-Manufacturer's Field Experience (# sold, etc.)
-Normal and Optional Response Methods
-Number of Frequency Filters
-Acceptable Noise Level
-Error Correction Procedures
-Standard and Optional Interfaces
-Standard Microphone
-Size and Weight

Figure 5. Recommended criteria for selecting a speech recognizer suitable for a defined application. Primary criteria are in largest letters, secondary criteria in medium size letters, and less-critical criteria are in the smallest lettering. Priorities, shown in decreasing order for an average application, may vary with each new application and its special requirements.

of the reference pattern and the features extracted markedly affect performance. Applications will vary in the demands they put upon rate of speech entry, ease of adjusting rejection thresholds, variability in timing, and bandwidth and transducer requirements. The other features listed in the bottom group of Table II will be non-critical but valuable criteria for choosing among available recognizers. A user may weight some of them as more important than others, and select, from those devices that meet the first group's conditions, those that best match the additional features of the final group in Table II. For some specific applications, some of these lower-ranked criteria should be specifically considered very significant. For example, airborne applications may make the size and weight, the acceptable noise level, and the microphone become unusually important.

An illustrative example of a device choice might be useful. If you know that the chosen recognizer has to handle speaker-dependent isolated word recognition, all the available devices might be eligible. If you determine that the device should be a board that interfaces to your available computer, only the Heuristics hobbist boards, the Scott devices, the IEC VRM, MIKE, Heuristics 7000, Auricle I, Threshold 500, and Votan products would appear to be candidates (and not all of them can interface to any arbitrary computer without special-option interfaces). If accuracy, as reported in the Doddington-Shalk independent test (2), must be above 97%, that reduces the candidates to the VRM and the Threshold 500. If vocabulary size must be 20 words, and costs under $5000, the only remaining candidate would seem to be the VRM, unless you are willing to wait until tests of the accuracy of the newer Auricle I and Votan products have determined their adequacy or inadequacy.

These device selection procedures, with full tabulations of specifications, and with weightings attached to the various factors, could be automated, for rapid, objective selections for a variety of applications.

4. EVALUATING THE PERFORMANCES OF RECOGNIZERS

A critical specification that is missing from previous work is the demonstrated recognition accuracy for each recognizer. In this section, we shall briefly summarize what is being done to reliably establish performance characteristics of available commercial speech recognizers, and to establish what factors influence recognition accuracy. We consider desired goals in performance evaluation (Section 4.1), some proposed conditions for testing (Section 4.2), and steps being taken to define and conduct standard tests (Section 4.3). This area of performance evaluation is considered one of the highest priority topics in speech recognition technology (5,8).

4.1 Desired Goals in Performance Evaluations

For the first ten years following the advent of commercial speech recognizers, we had no publicly available independent comparisons of the performances of available devices. Recently, two limited tests have been conducted (2,14), each under one set of experimental conditions. Interestingly, both studies found the same relative ordering of error rates of the various devices. However, recognizers ultimately need to be tested in a variety of representative conditions, and there are several critical questions to answer, including these:

1. What do we mean by "performances"?
2. What factors can influence performance?
3. How do those factors vary in real applications?
4. How does performance correlate with each factor?
5. How do the factors combine and interact, to determine performance?
6. How can we extrapolate limited test results to predict performance under untested conditions?
7. How can a theory of performance evaluation predict performances?

Although other performance measures, like data throughput rate, user satisfaction, etc., might be defined, perhaps the primary performance measure currently being investigated is the accuracy of recognition, or, conversely, the error rate in recognition. Ultimately, we would want to distinguish substitutionary errors from false rejections of acceptable utterances and false acceptances of sounds or speech forms that were not included in the set of expected utterances. In many recognition tasks, percentages of "correct detections" (or "hits" in correctly finding words or other units) should be determined, as should "false alarms" like incorrectly declaring a word (or other unit) is present when it actually is not present. Initially, most tests of recognizers may be expected to group all these types of errors together, and to discuss overall error rate (2,13,15).

Over 80 factors that may influence recognition accuracy have been identified and recommended for monitoring during performance evaluations (8,13,16). The question of "What causes speech recognizers to make mistakes?" may be ultimately more important than the relative assessments of specific transitory devices. A few limited tests have indicated that recognizers make fewer errors when extensively trained to the individual voice, confined to isolated words separated by clear pauses, and used with phonetically dissimilar words in small vocabularies, with nearly ideal environments, slow speech, low noise, experienced talkers, stable head-mounted microphone, and large stored bit patterns for templates of expected pronunciations. Generally, adult male voices are better handled than females or children, and some languages with simple syllabic structures, like Japanese, appear easier than languages with more variabilities in sound structures, like English. These and other generalities need extensive verification, and we must seek to characterize the relative effects of each factor that might influence accuracy. For example, which is more troublesome to recognizer accuracy: A doubling of the number of speakers or a 20 dB decrease in signal-to-noise ratio? Is it harder to extend from digit recognition to alphanumeric recognition, or to go from isolated digits to short connected strings of digits?

How these factors actually vary in real field applications also needs to be understood, since past experience shows that laboratory results do not usually extrapolate to performances in the field. Also of interest is how these various factors may interact, so that combined changes on several factors, in going from a laboratory test to a field use, might cause more errors than the individual factors might suggest. These issues are discussed elsewhere (8,13,15,16).

We still await any substantial evidence about how recognition accuracy correlates with most of the 80 or more variables. Two methods for determining such correlations are:

(A) Model the exact way in which factors may influence the statistics of decisions, and theoretically hypothesize performance relationships with each variable or set of variables; and

(B) Conduct controlled tests that experimentally determine the effects of each variable on recognition accuracy.

Ultimately, a comprehensive theory of performance relationships to all factors would be very desirable. Unfortunately, no such theory is yet developed, and the primary focus in performance evaluation studies has recently been on the design and running of controlled tests to determine performance characteristics.

Obviously, with over 80 factors influencing recognition error rates, and an expected need to test recognizers at several (at least three) values of each variable, a brute force testing of all combinations of conditions could require 3^{80} utterances to be processed! We have to find efficient ways of testing. In general, we desire any tests of recognizers to meet this set of desiderata: They must be honest, unbiased, independently-run tests, assuring objective assessments. To assure equal treatment of competing devices, the recognizers should all be tested on the same common tasks, preferrably including tasks of various difficulties. Not only do we want to know how many errors occur, but also why they occur, so we must focus on sources of errors in recognition. This requires monitoring of all the variables likely to influence performances. Standard tests should also be representative of field conditions, but this does not mean they should be conducted in the field with uncontrolled conditions; controlled tests can be done to carefully isolate the factors that cause each error. Then, our goal will be to generalize or extrapolate to untested sets of conditions, so that field tests will not yield surprisingly different results from those obtained in the laboratory, such as has often occurred in past work. We also need to assure the repeatability of tests of recognizers, as is basic in all scientific work; controlled conditions, archival speech databases, and reliable recording equipment will be needed. Extrapolation of past results to predictions of future performances is critically associated with assuring statistical adequacy. Large tests are usually suggested by the need to determine accuracies to within a small confidence interval; however, we can't wait for millions of utterances to be analyzed under each conceivable combination of conditions on each of the many variables.

Efficiency of testing is critical. One valuable way to reduce required data collections and analyses is to test recognizers under very difficult conditions, so errors quickly occur, showing relative performance levels with minimal data. Another aspect of that idea is statistical truncation, whereby analysis of a device's performance ends as soon as the device yields a substantial number of errors, such as 50 errors. The relative performance under typical field conditions can be tested first, and further analysis under the more extremal conditions only if performance is good enough to further study accuracy under various other conditions. We usually do not care whether a recognizer gets 90.1 or 90.2% accuracy, but that same 0.1% difference in performance can be very important between 99.9 and 99.8%. What we want is for the confidence interval around detected accuracy means to be a small fraction of the error rate. Another efficiency idea is to use a special form of multiple regression analysis, called "response surface methodology" (8,17), to tell how much data to collect at each position in the space of alternative conditions. Response surface methodology helps determine what values of each variable should be used in testing, such that results under the conditions on those variables can be reliably extrapolated to predict performances throughout the space.

4.2 Proposed Testing Conditions

Based on the goals and desiderata previously described, a four phase program for testing recognizers has been proposed (8,16). The earliest phase involves tests at the central conditions suggested by a center composite design of response surface methodology. Initial tests are recommended with the following conditions:

PRACTICAL TASKS: First test the low-cost boards and terminals, with isolated words, at several speech rates, beginning with an average rate of about 40 words/minute. Computer prompting can help determine the speaking rate.
HUMAN FACTORS: Test with 10 to 20 fully-described speakers, with random training sequences, but some study of the effects of other training sequences. Record 10 training samples per word, and numbers of test samples as dictated by a response surface methodology design. Include test samples recorded immediately after training, as well as at later times, like 4, 12, and 28 days later. Investigate performance with various numbers of training samples for each word.
VOCABULARIES: Include tests of the digits, alphanumerics (0-9 and A-Z), and other sub-vocabularies that efficiently test various sound structures. Include task-related vocabularies.
RECORDING CONDITIONS: Record 7½ ips analog tapes and/or digital tapes, in a quiet office, with 7000 Hz bandwidth, computer prompting and control, with a head-mounted Shure SM-10 microphone.
INPUT CONTROLS: Vary the amplitude of the input, perhaps from +6 dB to -6 dB or more, with a computer-controlled attenuator. To permit testing with various signal-to-noise levels, like 30, 15, and 0 dB, add noise electronically, using white, pink, and impulse noises in controlled conditions.
EFFICIENT TESTING: Use correlations from a statistical pre-test to focus on variables to assess in later tests, truncate testing of any device after a moderate number of errors has occurred, and apply a modified central composite design of response surface methodology to determine best testing conditions and amount of data to accumulate under each set of conditions.

Work is underway to conduct tests according to these conditions.

4.2 Towards Standard Tests of Recognizers

Several efforts are indicative of the eventual development of standard tests for assessing speech recognizers. In March, 1982, the U.S.A. National Bureau of Standards and the U.S.A. Naval Air Development Center sponsored a "Workshop on Standardization of Speech I/O Technology" (13), which was dedicated to advancing the cooperative efforts to define standard procedures for assessing speech recognizers and synthesizers. For several years, the IEEE Computer Society Subcommittee on Speech Recognition and Understanding has discussed the need for evaluation procedures. Initial tests under restricted conditions have been run (2,15). In the spring of 1982, a contract to conduct initial tests of recognizers under a variety of conditions was awarded to this author through Signal Technology Incorporated, in cooperation with the National Bureau of Standards. Several companies (Verbex, Hewlett-Packard, Texas Instruments) and leading researchers (Rex Dixon of IBM, Gerard Chollet of CIT/Alcatel, etc.) have active interest in this critical topic. Also, the RSG-10 Speech Recognition Subgroup of NATO has begun to comparatively test recognizers.

We may expect that such efforts will lead to standardized tests, endorsed by national or international standards groups, within a few years. Since speech recognition can only advance as much as our ability to distinguish successes and failures, to explain them, and to predict further performances, extensive high-priority work must be devoted to tests and theory that will help determine performance levels and the sources of errors in speech recognizers. Users eagerly await such theory, tests, and their resulting assessments, and the markets for speech recognition will probably never full open up without such reliable assessments being done. Also, designers of recognizers will need such knowledge of past problems if they are to develop even better, more advanced recognizers.

5. DESIGNING ADVANCED SPEECH RECOGNIZERS

We have considered how to select tasks that warrant voice input, some criteria for selecting the best of available recognizers, and procedures for evaluating speech recognizers. We also briefly considered what is wrong with speech recognition technology. This variety of knowledge is essential to deciding what is needed if new recognizers are going to be designed that improve upon current capabilities, increase accuracy, and truly serve the users. In this section, we shall briefly consider the value of designing new recognizers based on desired specifications (Section 5.1), how to make some straightforward improvements in current algorithms (Section 5.2), why it is crucial to introduce speech science into speech recognizer designs (Section 5.3), how prosodics-monitoring algorithms can aid recognition (Section 5.4), and the need for improved phonological and linguistic models in recognition (Section 5.5). As was noted in Section 1, most of speech recognition technology is WRONG today, and in need of these suggested advances. More detailed discussions are provided in a recent book (8).

5.1 Design According to Desired Specifications

It appears that most or all recognition devices have not been designed from complete sets of specifications based on user needs. If a new recognizer is to last as a highly-competitive device, it should: (A) be based on up-to-date, advanced state-of-the-art techniques; (B) provide attractive features not available in other devices; and (C) offer user facilities that assure improved integration into practical systems. Unfortunately, most recognizers have been based on the same ideas as previous devices, with little or more innovation, like "reinventing the wheel". Designers should endeavor to include up-to-date techniques like LPC, dynamic programming, focus on important phonetic features, advanced distance measures with phonetic weighting, etc. Users will be increasingly critical of outdated ten-year-old techniques like are used in many recognizers. Project leaders should ask why they expect to succeed and what unique capabilities they can offer.

Attractive features that will help assure competitiveness for new devices include: low ultimate costs, and cost effectiveness for the user; hardware self-test and fault isolation; rejection threshold that is precisely adjustable over a broad range, either by software command or by hardware control, permitting both stand-alone and host-dependent operation of rejections; continuous listening ability, to detect words even while previous ones are being analyzed; a control for rejecting or not rejecting bad training samples; easy selection of active sub-vocabularies and easy addition of new words; ASCII code for identifying the recognized word, rather than a mere vocabulary item number; a method for adjusting to or subtracting out the background noise; a sensitivity control; a training method that necessarily forces usual training sequences to be random, not consecutive repetition of the same word; an effective automatic gain control; easy transfer of templates to non-volatile memory; ability to use multiple-word phrases in the vocabulary; etc.

Recognizers should be designed more with the user in mind. The following are some facilities that would really aid the user in assessing and using speech recognizers: comprehensive performance evaluation tests run by the manufacturer and reported to the users, including information regarding performances under various signal to noise ratios; database tapes for sale as user options, for users to check equipment with standard conditions; providing of intermediate results (scores, runners-up, etc.); explicit advise on how a user can design vocabularies with minimal confusabilities among words; methods for assuring that the user is trained as the machine gets trained; applications programs and guidelines for field uses; explanations of limitations of the device, and guidelines for attempted extensions to more difficult tasks like speaker independence and connected speech; instructions on proper placement and use of microphones; etc.

Procedures for designing recognizers according to desired specifications, and for incorporating competitive features and facilities are discussed further in a recent book (8).

5.2 Straightforward Improvements in Current Algorithms

Without much risk, it is possible to make many valuable improvements in current algorithms. Traditional filter banks can be replaced by perceptually-weighted spectral analysis, such as critical-bandwidth filtering, mel-scale spacing, or linear prediction analysis; more focus on the high-amplitude and lower-frequency information would be useful, while de-emphasizing the spectral valleys; expensive spectral analyses need not be run everywhere in the speech; manner of articulation cues can be used to guide acoustic and phonetic analyses; gain control and automatic amplitude normalization need to be developed; instead of slicing the speech into equal size linear slices, analyses could focus on rapid transitions and spectral distances before spectral templates are matched; speaker normalization by simple non-linear frequency warping on calibration utterances could be attempted; noise cancelling algorithms can be incorporated; dynamic programming can be improved by focus on certain parts of the frequency spectrum and certain time regions, plus phonetic state detection before warping; island-driven analysis can avoid dependence upon word boundaries; etc. These improvements can in most cases be accomplished without major system re-designs (8).

5.3 The Need for Speech Science in Designs

Most speech recognizers are really just signal matchers, giving little or no consideration to the nature of speech and what is important to linguistic distinctions. Recognizers are limited in performances due to all the errors and inadequacies listed in Section 1 of this paper. Most of these limitations and errors can be eliminated by due consideration to what we do already know about speech. We know that all information is not equally important, that word boundaries are difficult to detect and poorly articulated, that speech is not a linear string of invariant bundles of easily-detected features, that phonetic and prosodic features can be detected from acoustic information, that regions like stressed syllables are islands of reliability, that certain parts of the speech are more affected by a change in rate of speaking than are others, that word confusions are predictable from similarities in phonetic structure, etc. Almost none of this knowledge is explicitly incorporated into current or previous algorithms. Nor has much consideration been given to how human perceive speech, as a model for a machine.

It is also striking that repeatedly during the three decades of developments of speech recognizers, there have been strict mathematical techniques that have led to the independent "discovery" of important features that turn out to be equivalent to those suggested earlier by linguistic and speech science research. For example, the "eigenvectors" for representing speech sounds turn out to closely resemble the acoustic correlates of such linguistic contrasts as voiced/unvoiced, manner of articulation, etc. Strict acoustic segmentation routines often divide utterances into the transitional and steady-state elements recognized as important in models of articulation and co-articulation of speech. It pays in analyzing the signal in any communication system to understand the nature of the source and the usual destination for the signal. Without such consideration of the nature of speech, recognizers will probably always be limited.

5.4 Integrating Prosodics into Recognition Procedures

The most neglected aspects of speech knowledge are the suprasegmental, or "prosodic", features, like intonation, stress patterns, and rhythm. Humans learn prosodics before they learn phonemes of a language, yet machines have been focused on localized segments. Human perceivers seem to make early decisions about large units like prosodically-marked phrases, and fill in phonetic details only where necessary to disambiguate sound structures. Machines have been designed oppositely, working from the "bottom up", to compose larger units from throughly analyzed local details. Humans exhibit hierarchical patterning of errors in perception, suggesting reliability in voicing decisions, manner of articulation categorizing, and syllable size judgments. Yet, machines giving equal attention to all data have shown much more random errors, less amenable to easy detection and correction. Humans use prosodic cues to sentence types, structural divisions, subordination, and structural modeling; no machines have attempted to effectively use such knowledge. It is time to change these unnecessary discrepancies.

Algorithms do exist for prosodic analysis (5,8,18). Acoustic prosodic analysis can begin with a simple filtering to obtain several useful energy contours. Speech can be separated into voiced and unvoiced regions from a simple threshold on a 0-450 Hz energy contour. Voiced peaks of sonorant (0-3000 Hz) energy, surrounded by substantial energy dips, can be identified as syllabic nuclei. Multisyllabic sonorant sequences can be further divided into syllables by other energy contours (like a 600-3000 Hz "intermediate sonorant energy" contour), and by detection of significant formant movements. Fundamental frequency (Fo) tracking can be accomplished by any of several "pitch trackers", such as by autocorrelation of the center-clipped signal, with energy thresholds used to provide Fo values only where the signal is clearly voiced. Syllables with high integrals of energy in the nucleus, plus local rises in Fo, can be identified as stressed, by a complex context-dependent algorithm (8,18). Phrase boundaries can be detected from fall-rise valleys in intonation and timing interruptions.

Prosodically-guided speech recognition can then proceed as follows. Detect phrase boundaries, find stressed syllables in each phrase, do phonetic analyses in the stressed "islands of reliability", align possible words around the reliable stressed vowel, analyze other reliably detected regions like sibilants, do further narrowing of sets of hypothesized words with such detected sound structures and with prosodic constraints like position-in-the-phrase, and, where necessary, analyze other detailed sound structures to eliminate remaining ambiguities.

Prosodics can also provide other guidelines to recognition

procedures. Interstress intervals can provide measures of speech rate, to help determine applicable phonological rules and to guide alignment processes and syntactic hypothesizing.

Several groups in Europe and Japan (cf. review in 8) have endorsed the idea of expanded work on prosodic aids to speech recognition, and some American work is beginning to incorporate prosodic aids.

The Need for Improved Phonological and Linguistic Models

Speech recognition technology has been plagued by compromises and expediencies. We have seen such compromises in the acknowledged but unaccomodated deficiencies in acoustic processing, time normalization, dependencies upon word boundaries, lack of prosodics information, etc. Other critical deficiencies are concerned with the phonological and linguistic modelling assumed in almost all recognizers. We must dispense with the erroneous and diversionary assumptions of linearity and invariance. Recognition systems should be based on a phonological theory that acknowledges, and has rules that explain, context dependencies. Such a theory should account for acceptable syllable structures, stress effects, rate of speaking, syntactic structural divisions, non-uniform phonetic clarity, phonological deletions and insersions, etc. There already are several theories that should handle such phenomena far better than the autonomous phonemic theories implicit in current recognition algorithms. A need in recognition technology is to attract the attention of good phonologists to the problems and needs of speech recognition work.

Another major deficiency in current recognition algorithms concerns the use of finite state grammars, and the passive use of grammar (both syntax and semantics) only to select acceptable word sequences from large numbers of combinations of alternative hypothesized words. Context-free grammars or some form of context-dependent phrasal grammar (like an ATN grammar or a layered finite state grammar) should be used, to assign phrase structures which can be compared with prosodic cues to structures. Measures of structural complexity are also in need of improvement. Semantic grammars (for which semantics is embedded within the syntax) may be one way to incorporate semantic constraints, but they have been found to severely limit the ability to readily adapt a recognizer to a new task with different semantics. Human perception models would suggest a more active role of syntax and semantics in early stages of recognition, rather than a mere "weeding out" of alternative words sequences after full phonetic analyses. Linguists and perception theorists should be able to develop better recognition strategies based on active, early use of syntax, semantics, prosodics, and phonological regularities.

6. TOWARDS A 'SCIENCE' OF RECOGNIZER DESIGN AND USE

In summary, we have seen that speech recognition technology is currently unnecessarily burdened with: errors in theory and practice; antiquated algorithms; ignoring of the variations in reliability and importance of the acoustic data; neglect or inability to use prosodics, active syntax, and semantics; insufficient public knowledge of performances of the various commercial devices; very little systematic analysis of what tasks really need or can profit from voice input; and a general disregard for user needs in the designs of devices. These are strong indictments of a field that wishes to expand its marketability and productivity. However, we have also considered a variety of solutions to these problems. They amount to saying that we need to turn the tinkerer's "art" of speech recognition to more of a "science", guided by user needs, systematic and comprehensive evaluations of devices, careful design of devices according to full specification of desirable features, and algorithms based on human perception processes and all that can be brought to bear on the problem from a variety of speech-related disciplines. It is doubtful that the marketeers' promises of a bright future for speech recognition technology will substantially materialize without a major change in approach by researchers, designers, manufacturers, evaluators, and users. Yet, the necessary tools to make major advances in speech recognition technology are beginning to be developed, and there is a definite potential to eliminate or alleviate previous limitations, and provide valuable voice interactions with the many machines that will affect our future lives.

7. REFERENCES

1. Chapanis, A. (1975), "Interactive Human Communication", Scientific American, 232, No. 3, 36-42.

2. Doddington, G. R., and T. B. Schalk (1981),"Speech Recognition: Turning Theory to Practice", IEEE Spectrum, 18, Sept., 1981,26.

3. Honeywell (1980), "Application of Advanced Speech Technology in Manned Penetration Bombers", Proposal from Honeywell Systems and Research Center to Wright-Patterson AFB. Minneapolis.

4. Lea, W. A. (1980). Trends in Speech Recognition. Englewood Cliffs, NJ: Prentice-Hall. Especially Chs. 1 and 8.

5. Lea, W. A. (1980b), Issues in the Selection and Practical Use of Speech Recognizers", Proceedings of the Symposium on Voice Interactive Systems: Applications and Payoffs, Dallas, TX, pp. 357-372.

6. Lea, W. A. (1981), Computer Recognition of Speech (book). Santa Barbara, CA: Speech Science Publications.

7. Lea, W. A. (1981), "Assessing the Feasibilities of Airborne Speech Recognition Tasks", prepared by Speech Science Publications, Santa Barbara, CA, for Honeywell Systems and Research Center, Minneapolis, MN, May 19, 1981.

8. Lea, W. A. (1982). Selecting, Designing, and Using Speech Recognizers (book). Santa Barbara, CA: Speech Science Publications.

9. Lea, W. A. (1981),"Speech Recognition, in Electronic Technology ... the Next 25 Years", Silver Anniversary Issue, EDN, October 14, 1981, pp. 168-172.

10. Lea, W. A. (1981), "A Research Program to Advance Airborne Uses of Voice I/O", prepared by Speech Science Publications for Honeywell Systems and Research Center, Minneapolis, MN.

11. Martin, T. B. and J. R. Welch (1980), "Practical Speech Recognizers and Some Performance Effectiveness Parameters", in Trends in Speech Recognition (W. A. Lea, Ed.). Englewood Cliffs, NJ: Prentice-Hall, 24-38.

12. North, R. A. (1981), "Application of Advanced Speech Technology in Manned Penetration Bombers", Final Report from Honeywell Systems and Research Center, Minneapolis, MN, on Air Force Flight Dynamics Laboratory (WPAFB) contract number F33515-80-C-3606, Wright-Patterson AFB, Ohio.

13. Pallett, D. (1982), Editor, "Proceedings of the Workshop on Standardization of Speech I/O Technology", sponsored by the National Bureau of Standards and Naval Air Development Center, March, 1982.

14. Reddy, D. R. (1976), Speech Understanding Systems: Summary of Results of the Five Year Research Effort at Carnegie-Mellon University, Pittsburgh. Also, cf. B. Lowerre and D. R. Reddy (1980), "The Harpy Speech Understanding System", in Trends in Speech Recognition (W. A. Lea, Ed.). Englewood Cliffs, NJ: Prentice-Hall, pp. 340-360.

15. Baker, J. M. (1981), "Issues and Answers in Evaluation of Automatic Speech Recognizers", Journal of the Acoustical Society of America, Supplement 1, Vol. 70, Fall, 1981.

16. Lea, W. A. (1982), "What Causes Speech Recognizers to Make Mistakes?", Proceedings of the 1982 International Conference on Acoustics, Speech, and Signal Processing (ICASSP), Paris, France, 2030-2033.

17. Lea, W. A. (1982),"Design of Efficient Standard Tests for Evaluating Speech Recognizers," Final Report on NADC Contract to Speech Communications Research Laboratory, Los Angeles, CA.

18. Lea, W. A., and F. Clermont (TO APPEAR),"Algorithms for Acoustic Prosodic Analysis."

INDEX

acoustic analyzer 207
   parameters 71
   phonetic analysis 139
      decoding 69
      transcription 245
   processor 232
   representations 71
   tube
adaptation to the speaker 129
adaptative channel vocoder 123
   PCM vocoder 31
   predictive coders 25
all pole model 2
alternative segmentation 162
analysis conditions 6
anatomical and physiological constraints 141
APC coder 31
area function 15, 141
arial II 269
articulatory model 139
asymetrical form 212
ATN grammar 274
augmented transition network (ATN) grammar 248
autocorrelation function 54
autocorrelation method 4

baseband coders 29
Bayes classifier 293
beam width factor 200

case grammar 262
CCD filters 223
centisecond labelling 80
cepstral coefficients 125
   analysis 208
classification 148
clustering 179
   vocoder 39
coarticulation 184, 199
communication theory 231
complexity of algorithms 309
computational complexity 314
connected word recognition 191, 205
consonant clusters 158
   recognition 89
contour analysis 292
conversation system 239
covariance method 4

database oriented system 240
decision units 154, 158
deep structure 263
demisyllables 75
   recognition 154
   segmentation 158
dialog procedure 265
dictionary knowledge 246
diphone 39, 75
   recognition 42
   synthesis 40
   templates 42
   vocoder 39
direction independent 248
dynamic interpolation 156
   programming 43, 149, 179, 191
   segmentation 146
   time warping 211

effectively computable functions 311
end points detection 86, 200
entropy 236
estimation of formant frequencies 7
experimentation frame 244

filter bank 224
   coefficients 2
final consonant clusters 158
   demisyllable 158
finite state parser 184
formant variability 16, 17
frequency band limitation 139
fundamental frequency 11, 50
   harmonic detection 58

generation of lexical hypotheses 166

high frequency regeneration 32
hypothesis search 235

identification task 289
initial demisyllable 158
island parser 248
isodata technique 181
isolated word recognition 139, 179, 192
Itakura distance 181

knowledge modules 239

large vocabulary 139
laser patent subtext 236
lexical correction 160
decoding 274
knowledge 166
tree 273
representation 258
level building procedure 213
linear prediction 1, 24, 121, 141
model 2
predictive coders 35
linguistic decoder 232
information 147
local constraints 212
loudness functions 156
low-pass filtering 59

markov source 233
medial consonant clusters 158
mel frequency cepstrum analysis 208
scale 72
microsegmentation 145
mixed-source model 24
models for speech recognition 217
of speech processes 233
mos technology 224
multi channel approaches 62
myrtille II 256

network model for lexical access 172
new raleigh grammar 233
noise shaping 28
noisy channels 29, 38
nonlinear time normalization 156
NP problems 317

parallel computer simulation 243
processing 239
parcor coefficients 6
pattern recognition problems 309
peak picking method 9
performance evaluation 356
perplexity 236
phonetic knowledge
representation 109
transcription 160
vocoder 40
physiological constraints of articulation 140
pitch determination 49
polynomial problems 323
pragmatics 262
pragmatic knowledge 249
preclassification 145
preemphasis 6
primitive syntax 197
pseudo-syllabic segments 168

quantization of LPC parameters 36
question answering application 249

recognition by listening 278
devices 351
reflection coefficients 6
relative redundancy 180
representation of knowledge 165
residual signal 7, 11

segmentation 17, 78, 145, 154
accuracy 106
errors 159
selecting a recognition system 349
semantics 249, 257
sentence recognition 259
short term analysis 53
slope condition 211

source of variability in the voice 286
speaker independence 139
  independent connected word recognition 179
  independent systems 147
  recognition 277
  trained isolated word recognition 179
  variability 181
special purpose chips 225
spectral analysis 122
  folding 33
  sensitivity 36
  time sample 233
spectrogram reading 101
spectrum derivative 75
speech coding 21
  data compression 22
  production model 2
  quality optimization 29
  synthesis 15
  transmission 21
  understanding 253
static segmentation 146
statistical clustering techniques 180
  modeling 231
strategy 248
structural analysis for pitch extraction 57
switched capacitor filter 223
syllabic segmentation 156
syllable 75
  description 271
  nuclei 156
  structure 154
symetrical form 212
syntax constraints 201
  directed 179
    level building 185
  knowledge 248

task complexity 235
  for voice I/O 340
  analysis procedure 342
technology for waveform based speech recognition 221
template matching procedures 80
time registration path 192
  warping adjustement 210
  window 6
toeplitz matrix 5
topologically sorted 185
training and tuning 91
  process 126

undecidable problems 316
units of recognition 75
utility of voice input 343

variable frame rate transmission 37
  order linear prediction 36
variation in speaking rates 184
verification 149
  task 289
very large scale integrated circuit technology 217
viterbi algorithm 246
vocal tract 141
    length 147
voiced un voiced discrimination 51
  excited type writer 231
vowel normalization 17
  recognition 88

waveform coders 23
whole word template matching 191
window shift 6

zero crossing densities 75

GPSR Compliance

The European Union's (EU) General Product Safety Regulation (GPSR) is a set of rules that requires consumer products to be safe and our obligations to ensure this.

If you have any concerns about our products, you can contact us on ProductSafety@springernature.com

In case Publisher is established outside the EU, the EU authorized representative is:

Springer Nature Customer Service Center GmbH
Europaplatz 3
69115 Heidelberg, Germany

Batch number: 10761697

Printed by Printforce, the Netherlands